LIBRARY
NETWORK

E. Donth

Relaxation and Thermodynamics in Polymers

Glass Transition

Ernst-Joachim Donth

Relaxation and Thermodynamics in Polymers

Glass Transition

Akademie Verlag

Author:
Dr. Ernst-Joachim Donth
Technische Hochschule Merseburg

With 76 pictures and 8 tables

1st edition

Editorial Director: Heike Höpcke
Production Manager: Sabine Gerhardt

Deutsche Bibliothek Cataloguing-in-Publication Data:

Donth, Ernst-Joachim:
Relaxation and thermodynamics in polymers: glass transition;
with 8 tables/Ernst-Joachim Donth. – 1. ed. – Berlin: Akad. – Verl., 1992
ISBN 3-05-501392-1

Akademie Verlag is a member of the VCH Publishing Group.

Printed on non-acid paper

Printing: The Alden Press
Bookbinding: The Alden Press

Preface

This book is intended to be a modern, but simple review for newcomers who wish to cooperate with polymer experts in the field of dynamics, relaxation, and thermodynamics: students, physicists, chemists, and material engineers. As this field is rapidly growing the author tries to maintain a certain distance from the matter. I thought about a cartographer making a coarsening in the scale 1 : 200 000 from a topographic map 1 : 25 000 with large white and grey spots. I had to decide: which are the main points (phenomena in a spatial scale 2 . . . 20 nm), which are the main connections (relations in space and time), and which are the details that may be dropped (many experimental details and configurational details along the polymer chains). However many basic results from related small-molecule systems are included to sharpen the view to the particularities of polymers.

The necessary repetitions – I chose the inductive method – are used to offer different views to the basic facts. Coarsening inquires a strong code. The author stakes on the intuitive power of the reader to understand a coarse situation best in space and time. It is a new paradigm to discuss relaxation in terms of both time and space; till now this was reserved for scattering. I often tried to make a rough, but consistent sketch even if the elements are not always cleared up. Although some important results from proud theoretical physics are included, the main part is at a descriptive level – there are still too many white spots in polymer science.

To keep a map readable only a small number of important aspects can be included. The subject is therefore restricted to widespread, common materials such as polyethylene (PE), polystyrene (PS) and similar polymers with flexible chains. Their study is necessary for the business with functional and special polymers, liquid crystalline polymers and the many other topics in modern polymer science.

Some general aspects of relaxation and thermodynamics are also the subject of the book. There are still basic problems e.g. with broad spectral distributions, and with the subsystem concept in thermodynamics, especially when the latter will be applied to small scales of order the correlation and structure lengths. The central subject of the book is the glass transition. One can turn the tables: Polymers represent not only a complication compared to the situation in simple liquids, but they mark, by their structure, some new length scales in the 2 . . . 20 nm range that can be used to study spatial aspects of thermodynamics and relaxation. The spatial scales are "spread" and differentiated in polymers so that they can separately be felt by different kinds of experiments. Additionally, the time scale can conveniently be controlled by relatively small changes in temperature.

For all the molecular and experimental details the reader is referred to standard books (e.g. Refs. 1–9). A descriptive introduction to the whole field of polymer science is for instance Ref. 10. There are excellent encyclopedic books or series, e.g. Refs. 11–14, and many journals that document the progress, e.g. "Macromolecules", "Polymer", or "Journal of Polymer Science". In this book about 400 references to original papers and books are included that can help the reader to find more information, also from the references therein, and so on.

I wish to express my thanks to colleagues that supported this book by reading the preliminary drafts of some chapters: Dr. M. Schulz and Dr. J.-U. Sommer in Merseburg, by edition and typing the manuscript: Dr. K. Schröter and Mrs. K. Herfurt, and by drawing the figures: Mrs. R. Dohnert; and to Mrs. H. Höpcke, reader in the Akademie Verlag, for the good collaboration.

Merseburg, March 29, 1992

E. Donth

Table of Contents

Introduction

In small-molecule liquids and ordinary glass-forming substances the characteristic lengths of interest to structure and dynamics are of order one nanometer: next neighbors, network elements, and so on. Only in critical states and in crystals are the interesting length scales much larger than 1 nm: critical correlations, grain size, decay of disclination fields etc. Flexible polymers with stable, covalently bonded chains, e.g. vinyl polymers, have, besides the monomer scale (bond length $l_0 \approx 0.15\,\mathrm{m}$, van der Waals chain diameter $\sigma \approx 0.5\,\mathrm{nm}$ with a number density of order 10 monomeric units per nm^3), further length scales: coil radius $R_0 \approx 20\,\mathrm{nm}$ monitored by chain length, Kuhn step length $l_K \approx 1.5\,\mathrm{nm}$, entanglement spacing $d_E \approx 7\,\mathrm{nm}$, fold length in crystal lamellas $l \approx 10\,\mathrm{nm}$ controlled by undercooling, and, in networks, a length describing the mean distance between the crosslinks.

In so-called inorganic polymers (glasses), as a rule, the networks are in dynamic equilibrium with bond breaking, and lengths 2 nm can only survive under special conditions.

Dynamics in polymers is influenced by the new length scales. The different scales of molecular movement are put in order by a general scaling principle: the larger the typical length, the larger the typical relaxation time.

Dynamic scattering (X-ray, neutron, light) can resolve both length and time scales, at least in principle.

Ordinary elastic scattering only resolves length scales, and relaxation only resolves time or frequency scales. It is difficult to complete the information without detailed models.

Equilibrium thermodynamic variables are integrals over these length scales or time (frequency) scales, e.g.

$$kT(\partial\bar{n}/\partial p)_T = 1 - 4\pi \int c(r) r^2\, dr, \tag{i}$$

or

$$C_p = k\overline{\Delta S^2} = k \int S^2(\omega)\, d\omega \tag{ii}$$

where the symbols will be defined later, see Eqs. 1.30, 2.54, and 7.1. As the general scaling principle is hidden by the integrations, it is difficult to see which length scale is responsible for thermodynamic variables.

The problem of what are the "particles" for polymer thermodynamics is also rather complex. In statistical considerations the "species", e.g. the units of placement on lattice sites, are often monomeric units or chain segments. But the presumed carrier of identity that defines the Gibbs factorial $\nu!$ in the configuration integral is the whole chain. Moreover, the element for thermodynamic analysis is the subsystem being, for the present, an intellectual construction that can have different, but not too small size. The analysis becomes rather difficult when this size is of order of the length scales listed above.

In the First Part of this book, tools for a dynamic and a thermodynamic treatment are described rather independently. Starting at Chap. 6, situations are described where both branches are more and more interweaved, such as in thermal glass transition, spinodal phase decomposition, and folded chain crystallization.

I Fundamentals

To familiarize the reader with macromolecular chains we shall consider four exercises about temporal and spatial aspects of Brownian motion.

(1) Fig. 1a shows a sketch of Perrin's original observation for a Brownian particle. The positions after constant time steps (Δt) are connected by straight lines. A simplified analysis (Ref. 15) is based on Fig. 1b. The integers 0, 1, 2, . . . , N are the time counts of a homogeneous time. [Of course, all symmetry properties of time and space refer to the average statistical situation in the materials described e.g. by distribution functions.]

Because of the space isotropy (i.e. no gradients) it is sufficient to consider one dimension with coordinate x. Then $x(t_i)$ is the position of the particle at time $t = t_i$, $i = 0, 1, 2, \ldots, N$. Put $t_0 = 0$ and $x(0) = 0$. Isotropy implies $\overline{x(t)} = 0$, or $\Sigma\, \overline{\Delta x_i} = 0$, with $\Delta x_i = x(t_i) - x(t_{i-1})$. More interesting is the mean square of $x(t)$,

$$\overline{x^2}(t) = \sum \overline{(\Delta x_i)^2} = \sum_{ij} \overline{\Delta x_i \Delta x_j}. \tag{1.1}$$

If Δt is large enough, then all succeeding x intervals are statistically independent, in other words, there is no correlation between them,

$$\overline{\Delta x_i \Delta x_j} = 0, \; i \neq j, \tag{1.2}$$

and we have

$$\overline{x^2}(t) = \sum \overline{(\Delta x_i)^2} = N\overline{\Delta x^2} \tag{1.3}$$

if all $\overline{(\Delta x_i)^2}$ are equivalent, e.g. equal in homogeneous space and time. There is no stochastic length larger than the the particle diameter.

Defining a diffusion coefficient

$$D = \tfrac{1}{2}\overline{\Delta x^2}/\Delta t \tag{1.4}$$

we obtain $\overline{x^2}\,(t) = 2Dt$, $t = N\Delta t$; for 3 dimensions we have

$$\overline{r^2}(t) = 6Dt. \tag{1.5}$$

The square of mean Brownian shift is proportional to time. Consider a

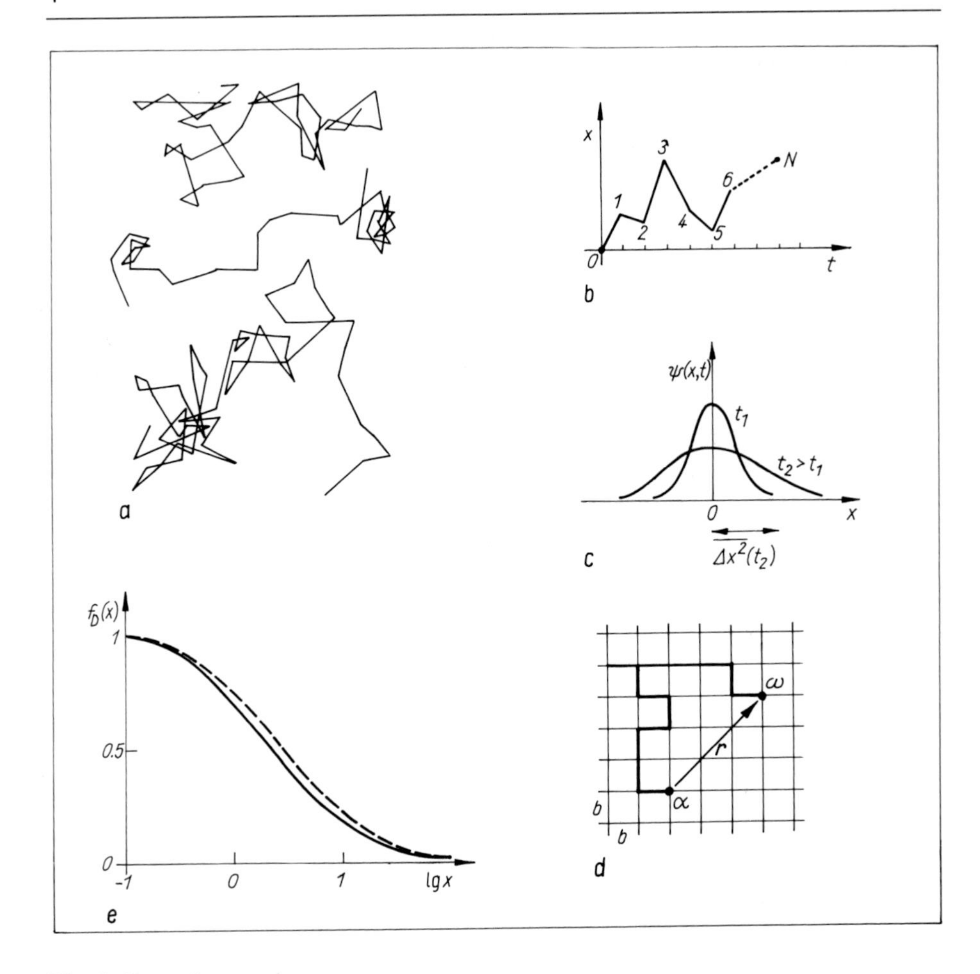

Fig. 1. Brownian motion.
a. Sketch of Perrin's observation. b. Brownian motion $x(t)$ in one dimension (for Eqs. 1.1 to 1.4). c. Diffusion of Brownian particles; $\psi(x, t)$ density. d. Random walk from α to ω on a square lattice with step length b. e. Debye structure factor, approximation Eq. 1.15b dotted; $x = q^2 R_{\text{gyr}}^2$, q scattering vector.

larger number of independent Brownian particles starting from $x = 0$ at time $t = 0$. Then their concentration $\psi(\boldsymbol{r}, t)$ is controlled by the diffusion equation

$$D\Delta\psi - \partial\psi/\partial t = 0 \tag{1.6}$$

with the Laplacian $\Delta = \nabla^2$. This equation implies a constant particle number, and the situation of Fig. 1c defines a current proportional to the ψ gradient $(\nabla\psi)$.

(2) Fig. 1a also stands for a snapshot of a flexible chain when the length b of segments is chosen to be so large that their orientations are statistically independent. To avoid mathematical difficulties connected with the time-space continuum (Wiener measure) the chain is considered to be a random walk (Ref. 16) on a primitive cubic lattice with step length b, see Fig. 1d. We follow the presentation of Ref. 4. Similar to the temporal case (1) we have

$$\boldsymbol{r} = \boldsymbol{b}_1 + \boldsymbol{b}_2 + \cdots + \boldsymbol{b}_N = \sum \boldsymbol{b}_i . \tag{1.7}$$

Hence the mean square *end-to-end distance* of the chain, R_0, is obtained analogously to Eq. 1.5,

$$R_0^2 = \overline{\boldsymbol{r}^2} = \sum_{ij} \overline{\boldsymbol{b}_i \boldsymbol{b}_j} = \sum \overline{\boldsymbol{b}_i^2} = N\boldsymbol{b}^2 . \tag{1.8}$$

For large N the whole statistics is determined by many small independent steps, which means by a Gaussian distribution. Consider many random walks starting at $\boldsymbol{r} = 0$. Be $p(x, y, z)$ – compare it to $\psi(\boldsymbol{r})$ in (1) – the corresponding probability density to find the chain end (N) at $\boldsymbol{r} = (x, y, z)$. Gaussian means

$$p(x, y, z) = p(x) \cdot p(y) \cdot p(z) - N^{-3/2} \exp\,(-3r^2/2Nb^2) \tag{1.9}$$

or $p(x) \sim N^{-1/2} \exp\,(-x^2/2\overline{x^2})$ with $\overline{x^2} = \overline{r^2}/3$, see Eq. 1.8, and with the measure dx.

A chain where Eq. 1.9 is also true for end-point distances of any (also small) parts is called a *Gauss chain*. In that case one can calculate the *radius of gyration* from Eq. 1.9, the result being $R_{gyr} = R_0/\sqrt{6}$.

Comparison of Eqs. 1.3 and 1.8 gives an equivalence between the temporal and spatial picture,

$$b^2 \Leftrightarrow 6D\Delta t \quad \text{or} \quad R_{gyr}^2 \Leftrightarrow Dt. \tag{1.10}$$

(3) Spatial correlations between the segments of a polymer chain in a solvent can be obtained by elastic scattering experiments. For Gauss chains one can calculate the correlations from Eq. 1.9. Expressed by a distribution $\phi(\boldsymbol{r}_i - \boldsymbol{r}_j)$, we have from Eq. 1.9

$$\phi(\boldsymbol{r}_i - \boldsymbol{r}_j) = \left(\frac{3}{2\pi b^2 |i-j|}\right)^{3/2} \exp\left(-\frac{3}{2}\frac{(\boldsymbol{r}_i - \boldsymbol{r}_j)^2}{|i-j|b^2}\right) \tag{1.11}$$

with a mean value proportional to the half of the exponent denominator,

$$\langle(\boldsymbol{r}_i - \boldsymbol{r}_j)^2\rangle = |i - j|b^2, \tag{1.12}$$

see also Eq. 1.8.

The scattering intensity of a volume with spatially isolated Gauss chain coils is proportional to the *structure factor* generally defined by

$$S(\boldsymbol{q}) = \langle\exp(i\boldsymbol{q}(\boldsymbol{r}_i - \boldsymbol{r}_j))\rangle. \tag{1.13}$$

In this situation the spatial correlation is in a way measured by a "standing" wave with scattering vector $\boldsymbol{q}$ of magnitude

$$q = (4\pi/\lambda)\sin(\vartheta/2), \tag{1.14}$$

where λ is the wave length of the beam and ϑ is the scattering angle corresponding to a momentum exchange $\hbar q$. The ensemble average $\langle...\rangle$ of Eq. 1.13 can be calculated for Gauss chains using the distribution ϕ of Eq. 1.11. We obtain the Debye structure factor

$$S_D(q) = \frac{1}{N}\int_0^N dn \int_0^N dm \exp(-(b^2q^2/6)|n - m|) = Nf_D(x) \tag{1.15a}$$

with

$$f_D(x) = \frac{2}{x^2}(e^{-x} - 1 + x) \approx \frac{1}{1 + x/3}, \quad x = q^2R_{gyr}^2. \tag{1.15b}$$

The exact and the approximate Debye function are compared in Fig. 1e. The decrease is "logarithmically broad". This term will generally be used when the interesting part of a function $f(y)$ extends to one logarithmic decade in y (i.e. a factor 1 : 10) or more. A nearly linear diagram is obtained for S^{-1} vs. q^2. When parameterized by polymer concentration it is called Zimm plot and can be used to determine R_{gyr}, for instance by extrapolation.

(4) The entropy ΔS linked with the possibilities of Fig. 1d is a thermodynamic measure for the spatial correlation of a chain. One must count the number $\psi(N)$ of possible N-step ways from α and ω in Fig. 1d. Then

$$\Delta S = k \ln \psi(N). \tag{1.16}$$

It depends on the end-to-end distance $\boldsymbol{r}$. Counting Gauss chains we obtain

$$\Delta S(\boldsymbol{r}) = \Delta S(r) = \Delta S(0) - k\frac{3r^2}{2R_0^2}. \tag{1.17}$$

Hence, the contribution to the free energy is

$$\Delta F(\mathbf{r}) = \Delta F(0) - 3kTr^2/2R_0^2. \tag{1.18}$$

"Small-scale" deformations, inside the coil interval $b^2 \lesssim r^2 \lesssim R_0^2$, though possibly concerning all N elements, give only a contribution of order one kT characteristic for one degree of freedom. This means that the chain coil is an extremely sensitive thing. Small forces are sufficient to add small displacements to relatively large coil deformations.

Usually, this ΔF of order 1 kT is only one contribution, among others, to the thermodynamic potentials being in total of order NkT per chain, $N \gg 1$. Then small deviations from the Gauss coil play a minor role. On the other hand, extension of the chain $r \to Nb$ or sharp folding that consumes considerable amounts of energy can give essential contributions from chain conformations. It is a certain independence of the large intermolecular segment-scale contributions and the tiny correlations in the coil scale that brings a good deal of interest to polymer science.

1. The 10-nm scale: structure elements of 2 . . . 20 nm size

The beautiful single chain picture of the preceding pages must be modified by additional structures of different size.

1.1 The 1-nm scale: ϱ level

Van der Waals diameters σ of monomeric units and short-range order form the scanning elements for the chain pictures of Fig. 1. The structure of simple small-molecule liquids, being a reference for short-range order, is mainly determined by the steep repulsion potentials between the particles. The functions describing this order oscillate and are damped in the range of nearest neighbors (distance $r_0 > \sigma$, but comparable). They are not logarithmically broad, and the last significant details can be observed at 1.5 . . . 2 nm, see Figs. 2a and b.

The short-range order can be described by correlations based on the particle number density $n(\boldsymbol{r}, t)$ that is determined by the pointlike positions of all particles,

$$n(\boldsymbol{r}, t) = \sum_j \delta(\boldsymbol{r} - \boldsymbol{r}_j(t)) \tag{1.19}$$

where δ is the Dirac delta. The van Hove correlation function can then be defined by (see e.g. G.H.A. Cole in Ref. 17, pp. 1–40)

$$\bar{n}G(\boldsymbol{r}, t) = \overline{n(0, 0) \cdot n(\boldsymbol{r}, t)} \tag{1.20}$$

where $\bar{n}$ is the mean number density, the number of particles per volume. Similar to all correlation functions, $G(\boldsymbol{r}, t)$ describes the (spatial ($\boldsymbol{r}$) and temporal (t)) *decay* of relations between the particles. $G(\boldsymbol{r}, t)$ is the probability density of finding a particle at position $\boldsymbol{r}$ at time t given that a particle (the black one in Fig. 2a) has the position $\boldsymbol{r} = 0$ at time $t = 0$. This function has two parts,

$$G(\boldsymbol{r}, t) = G_s(\boldsymbol{r}, t) + G_d(\boldsymbol{r}, t). \tag{1.21}$$

The self part G_s is the black-particle reference, in many cases not so important for structure correlations, and G_d is the distinct part. The latter, at $t = 0$, defines the static *radial distribution function* in the isotropic liquid

$$g(r) = G_d(r, 0)/\bar{n} \tag{1.22}$$

with the properties

$$\begin{aligned} g(r) &\to 1 \quad \text{for} \quad r \gg r_0, \\ g(r) &\to 0 \quad \text{for} \quad r \ll \sigma. \end{aligned} \tag{1.23}$$

The first peak describes the correlation with the next neighbors, the second peak corresponds to the second shell.

The range of $g(r)$ is larger than the range of intermolecular potentials $\varphi(r)$ because the particle hard cores must mutually arrange in condensed matter. The potential $\varphi(r)$ is often assumed to be acting additively between all point pairs $\boldsymbol{r}_i$ and $\boldsymbol{r}_j$. A *direct correlation function* $c(r)$ with short range, comparable to the $\varphi(r)$ range, can be defined by convolution with this arrangement (Ornstein Zernicke equation)

$$h(1, 2) = c(1, 2) + \bar{n} \int h(1, 3)c(2, 3)\, d\boldsymbol{r}_3 \tag{1.24}$$

where 1 means $\boldsymbol{r}_1$ etc., and $h(\boldsymbol{r}) \equiv g(\boldsymbol{r}) - 1$ is the total correlation function.

The structure factor $S(q)$ (Eqs. 1.13 and 1.14) is directly connected with the scattering intensity $I(q)$,

$$S(q) = (I(q) - I_0)/I_0 \tag{1.25}$$

with I_0 the intensity with no scatterer.

X-ray scattering e.g. involves momentum $\hbar q$ transfer between photons and atoms. For median scattering angles ($q\lambda \approx 1$, WAXS = wide angle X-ray scattering), $S(q)$ contains information about structures in the range of the X-ray wavelength (usually $\lambda \lesssim \sigma, r_0$). Small angle scattering SAXS ($q\lambda \ll 1$) gives structure information for larger-than-λ ranges. Structure information for ranges smaller than λ ($q\lambda \gg 1$ as e.g. from light scattering experiments, $\lambda = 500\,\text{nm}$) is rather global and contains no details for $r \ll \lambda$ (Tyndall effect). Examples are the calculation of R_{gyr} from a Zimm plot by light scattering, or of the correlation length from critical light scattering.

Structure function $S(q)$ and spatial correlations ($g(r)$, $c(r)$) are connected by spatial Fourier transformations. Therefore, $S(q)$ cannot be considered to be a true correlation function damped with increasing q; $S(q)$ is rather similar to a spectral density over q. In isotropic materials one finds

$$g(r) = 1 + (2\pi^2\bar{n}r)^{-1} \int_0^\infty q(S(q) - 1) \sin(qr)\, dq, \tag{1.26}$$

$$S(q) = 1 + 4\pi\bar{n} \int_0^\infty r^2(g(r) - 1) \frac{\sin qr}{qr}\, dr, \tag{1.27}$$

$$c(r) = (2\pi^2\bar{n}r)^{-1} \int_0^\infty q(1 - S^{-1}(q)) \sin(qr)\, dq \tag{1.28}$$

with $S^{-1}(q) = 1/S(q)$. Eq. 1.28 uses the fact that the Fourier (or Laplace) transform of a convolution equals to the product of transforms. The Fourier transform of the Ornstein Zernicke equation, Eq. 1.24,

$$h(q) = c(q) + nc(q) \cdot h(q) \tag{1.29}$$

looks therefore simpler than Eq. 1.24.

The typical structure factor $S(q)$ of small-molecule liquids is sketched in Fig. 2c. In contrast to the Debye function for Gauss chains (Fig. 1e, with about $g_D(r) - 1 \sim r^{-1}$ for $r < R_0$) we see that now $S(q)$ is not logarithmically broad, like $g(r)$ of Fig. 2b. The oscillation reflects the short-range particle arrangement in a few orders of $\sin(qr)/(qr)$.

The situation for flexible polymers is depicted in Fig. 2d. An additional element is the covalent bond length $l_0 \approx 0.15\,\text{nm} = 1.5\,\text{Å}$ along the chains. This length dominates the second peak of $S(q)$ for neutron scattering. The WAXS structures in amorphous PVAC and slightly ordered PVC (Fig. 2e) show some order in the Angstrom scale. "Subtracting" this intramolecular structure and correcting for differences of scattering strength (form factors) for different atoms we would, however, obtain $S(q)$ and $g(r)$ like as for argon, Figs. 2b and 2c. [There is a world of indices in polymer structure discriminating between different pairs of atoms, isotopes, components etc. which can be used to obtain many details of the structure by applying different scattering (e.g. neutron, X-ray) or probes (e.g. ^{1}H, ^{13}C . . . NMR and so on).]

The static correlation functions uniquely determine the thermodynamic equilibrium variables for simple liquids. From $g(r; p, T)$ or $c(r; p, T)$ one can calculate the number density $\bar{n}(p, T)$ using the compressibility equation quoted in the introduction,

$$\begin{aligned} kT(\partial\bar{n}/\partial p)_T &= 1 + 4\pi\bar{n} \int_0^\infty h(r; p, T) r^2\, dr \\ &= 1 - 4\pi\bar{n} \int_0^\infty c(r; p, T)\, r^2\, dr. \end{aligned} \tag{1.30}$$

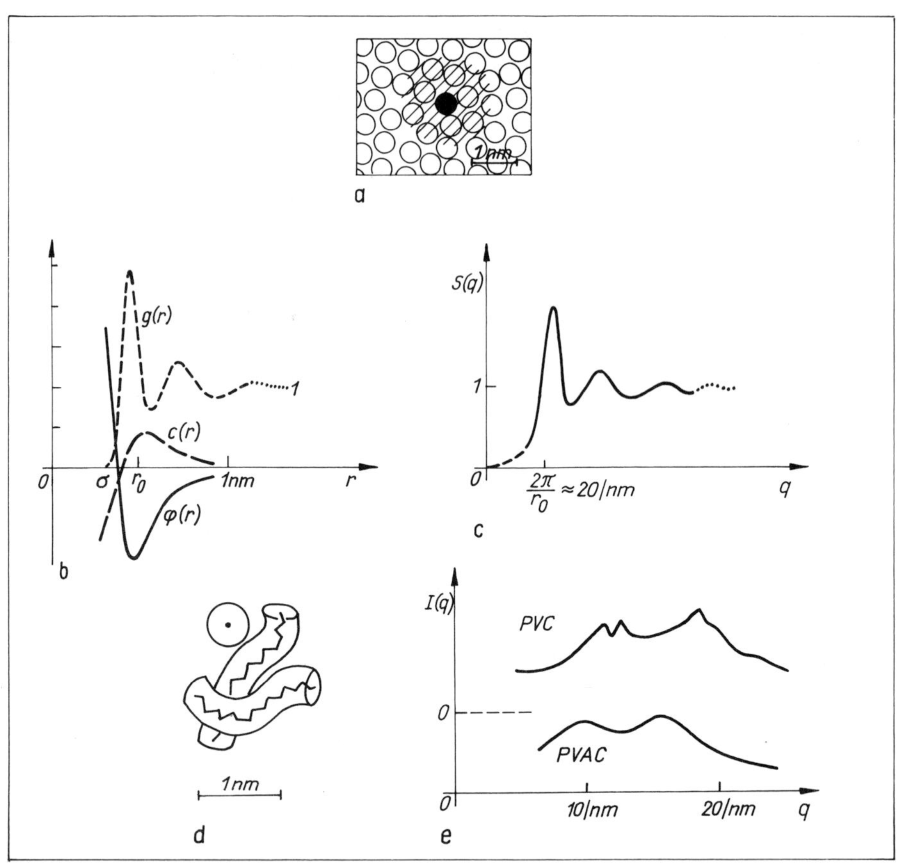

Fig. 2. The ρ level structure.
a. Short range order (hatched) in a liquid around an arbitrarily selected particle (black). b. Radial distribution function $g(r)$, direct correlation function $c(r)$, and intermolecular potential $\varphi(r)$ for small-molecule liquids (such as Argon). c. Structure factor $S(q)$ corresponding to the situation of parts a and b, see Eqs. 1.26 to 1.28. d. ρ level structure in amorphous polymers. e. Corresponding WAXS scattering intensity in the scale of $q \approx 10/\text{nm}$ for amorphous polyvinyl acetate PVAC and weakly ordered polyvinyl chloride PVC.

Using a caloric equation, e.g. for configurational internal energy,

$$\Delta u(p, T) \approx \frac{3\bar{n}(p, T)\, kT}{2} + \frac{4\pi\bar{n}^2(p, T)}{2} \int_0^\infty g(r; p, T)\varphi(r) r^2 \, dr \tag{1.31}$$

which is exact for pairwise additive intermolecular potentials $\varphi(r)$, all thermodynamic variables can be calculated. In contrast with the polymer coil function, Eq. 1.18, the thermodynamic properties from the 1-nm scale are very sensitive to small changes in $g(r; T, p)$ or $c(r; T, p)$.

A dynamic scattering experiment explores the correlation function $G(r, t)$ with a "current" wave $\exp(i\boldsymbol{qr} - i\omega t)$ involving both momentum $\hbar\boldsymbol{q}$ and energy $\hbar\omega$ transfer. The time, for instance, is also accessible in time-of-flight measurements of neutrons undergoing scattering. At present, unfortunately, correlations resolved simultaneously in time and space can be managed in the 1 or 10 nm scale only for times $t \leqslant 1\,\mu s$, corresponding to frequencies $\nu \geqslant$ MHz. [If we are content with only temporal information, for instance from photon correlation spectroscopy, typical relaxation times or even relaxation spectra of molecular correlations in the nanometer scale can also be obtained for $t \geqslant 1\,\mu s$ from dynamic light scattering, with scattering vectors larger than 100 nm].

The *structure function* or the intermediate scattering law

$$S(q, t) \; = \; (2\pi)^{-1} \int (G(\boldsymbol{r}, t) - \bar{n})\, e^{i\boldsymbol{qr}}\, d\boldsymbol{r}, \tag{1.32a}$$

with $S(q, t = 0) = S(q)$, describes the temporal decay of density correlation in the q space. The *scattering function*

$$S(q, \omega) \; = \; \int S(q, t)\, e^{-i\omega t}\, dt \tag{1.32b}$$

does not explicitly describe the decay of any correlation. The explicit decay of r and t correlations in $G(r, t)$, or r correlations in $g(r)$, or t correlations in $S(q, t)$, is restrained in favor of q and ω structures ("periodicity") in $S(q)$ and $S(q, \omega)$.

The pure time aspect of correlation decay is the subject of linear response (Chap. 2). Briefly, in a formula such as

$$B(t) \; = \; \Delta B \cdot \overline{\Delta n(t)\Delta n(0)}/\overline{\Delta n^2} \tag{1.33}$$

the decay of density correlation for a whole subsystem is reflected.

One may be surprised to find so many pages on small-molecule liquids in the beginning of a polymer book. But, referring to the parable of the cartographer in the Preface, the 1-nm scale objects are the scanning points for a map of 10-nm scale objects. The 1-nm scale is the scale of monomeric units that is the field for secondary relaxation, physical aging and ordinary glass transition, and from here originate important contributions to thermodynamic variables, also for polymers. The term ϱ *level* (ϱ from density) will be used as a short hand to characterize such contributions from the 1-nm scale. In contrast, all phenomena that can be described by order parameters $\psi(r, t)$ and their gradients belong to the 10-nm scale, or larger, and will be characterized by the term ψ *level*. The mutual relations between ϱ and ψ level make, as mentioned above, a good deal of interest in polymer science, even if not all 10-nm phenomena of interest can be described by order parameters, e.g. entanglements and networks.

The statistical mechanics and dynamics in terms of chain conformations is described e.g. in Ref. 18, 19 and 20, respectively.

1.2 Classification of length scales in polymers

Consider the monomer units as the particles and count the ϱ level up to about 2 nm. Then we have about 100 particles in a cube of size 2 nm. This is large enough to define a subsystem, i.e. we can use thermodynamics at the upper limit of the ϱ level.

The 10-nm scale reaches up to the coil diameter R_0 (or a few R_0). According to Eq. 1.8, R_0 is monitored by the chain length. Typical values for vinyl polymers are $R_0 \approx 7$ nm for a degree of polymerization of about $Z = 100$ ($M \approx 10^4$ g/mol), and $R_0 \approx 22$ m for $Z = 1000$ ($M \approx 10^5$ g/mol). As the chains usually form interpenetrating coils, we have many chains in a sphere of diameter R_0. Their number increases with $Z^{1/2}$ and is called Flory number n_F; a typical value is $n_F \approx 50$.

Structures $\gtrsim R_0$ are often declared as morphology, although there is no sharp borderline. This scale is important for technical applications of polymers, e.g. the domain size (μm) in blends with high impact strength, or the size of crystalline spherulites (100 nm up to mm).

This book is mainly devoted to the 10-nm scale, i.e. from 2 nm up to (some) R_0.

1.3 Macromolecular coils

Consider the coils of Figs. 1a and 1d. A definition of the step length is given by the *Kuhn step* l (or l_k) which can uniquely be calculated from the mean

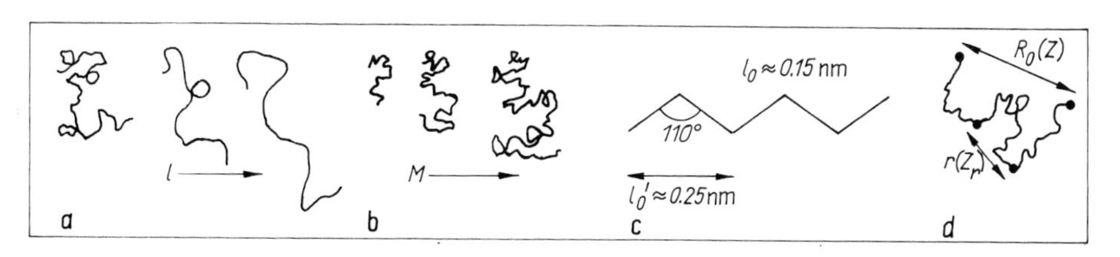

Fig. 3. Coils of flexible chains.
a. Dependence on Kuhn step length l for given chain length (M, N, or Z). b. Dependence on chain length for given Kuhn step. c. Parameters of covalent bonds in the backbone of vinyl polymers. d. Mean chain end-to-end distance (R_0 chain radius, coil radius or coil diameter) and mean geometric distance r between two segments separated by Z_r monomeric units, for Gauss chains.

square chain end-to-end distance R_0 and a chain contour length L using the Gaussian formula

$$R_0^2 = l^2 N, \quad \text{and } L = lN, \tag{1.34}$$

where, more precisely, L is the length of the extended chain straightened without distorting the bonding angle being about 110° for vinyl polymers. The number of Kuhn steps N can also be calculated from R_0 and L by Eq. 1.34.

For a given substance, the length of an unbranched chain can equivalently be defined by the molecular mass or weight M, or the degree of polymerization Z, the number of monomer units,

$$N \sim M \sim Z \tag{1.35}$$

which symbols will often be used as synonyms.

Eq. 1.34 is based on the statistical independence of orientation for neighbored segments. The Kuhn step is therefore a measure of the persistence length ($\approx l/2$): how many monomeric units along a chain are needed to forget the original orientation. The smaller l the larger the chain flexibility (Figs. 3a and 3b).

The Kuhn step for vinyl polymers (such as polystyrene) is about 1.5 nm corresponding to about ten covalent C–C bond lengths ($l_0 \approx 0.15\,\text{nm} = 1.5\,\text{Å}$) or 6 monomer units ($l_0' \approx 0.25\,\text{nm} < 2l_0$, see Fig. 3c). This relationship is usually expressed by the characteristic ratio

$$C_\infty = R_0^2 / N_0 l_0^2 \tag{1.36}$$

with N_0 the number of main-chain skeletal bonds, $N_0 = 2Z$ for vinyl polymers. For more complicated polymers the definitions include particularities of chemical configuration, see Ref. 18. The value of C_∞ is a measure of chain stiffness scaled by l_0. For Gauss chains, C_∞ is the ratio of Kuhn step to skeletal bond length, or, if R_0 is reduced by Z and l_0 instead of N_0 and l_0, C_∞ is the number of monomeric units per Kuhn step.

Thinking in monomer units instead of Kuhn steps, Ferry's *structure length* a is more practical for Gauss chains,

$$r = a\sqrt{Z_r} \tag{1.37}$$

where r is the mean geometric distance between any two monomeric units of a Gauss chain separated by Z_r monomeric units. From Figs. 3c and 3d, and from Eq. 1.36 we see that $a = (ll_0')^{1/2} \approx 0.6 \ldots 0.7\,\text{nm}$. Of course, Eq. 1.37 holds only for larger $Z_r \gtrsim 25$ because even for freely rotable bonds, conserving the bond angle, the persistence length would be larger than l_0'. From Fig. 3d we see that $r = R_0$ for $Z_r = Z$, of course.

The Kuhn step, and similarly a or C_∞, is usually related to states where the real chain is approximately a Gauss chain as for the so-called θ state in dilute solution or for concentrated systems. Their dependence on solvent, or temperature and pressure, is often rather weak, of order ten per cent for "moderate" changes. For instance, in melts the general trend is decreasing of l or a with increasing temperatures.

In good solvents the chain coil is considerably expanded in a non-Gaussian manner. For vinyl polymers, the mean thermal fluctuation of bond length and bond angle is only a few per cent.

1.4 Polymer concentration

A polymer solution is called dilute when the concentration c or the volume fraction ϕ of polymers is so small that the coils do not mutually overlap. Figs. 4a–c (Ref. 6) show the situation in good solvents. Consider increasing polymer concentration. The coil-overlap onset-concentration c^* is the smaller the larger the chain length. The corresponding polymer density can be estimated from $\varrho^* \approx M/((4/3)\pi R_{\text{gyr}}^3 n_\text{A}) \sim M^{-1/2}$ with n_A the Avogadro number. For $c > c^*$ the polymer coils interpenetrate one another. Then the term semi (semidilute, semiconcentrated) is used as long as the local variations of polymer concentration or density $\varrho(x)$ remain large in the $x < R_0$ scale. The density profile shows regions with high $\varrho(x)$ alternating with

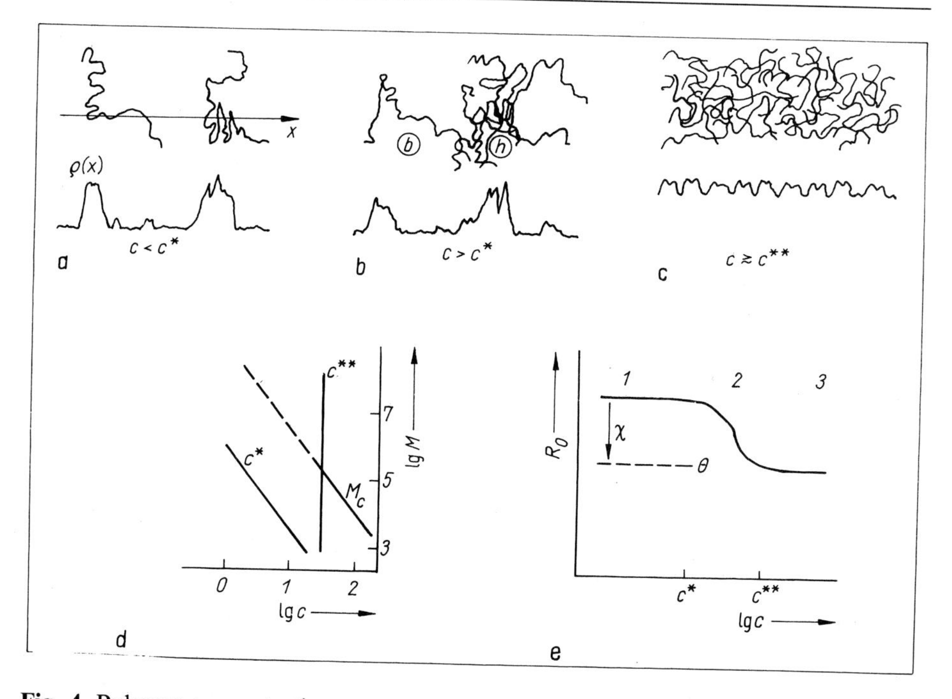

Fig. 4. Polymer concentration.
a. Dilute solution. $\rho(x)$ ρ-level polymer density profile. b. Semi regime above coil-overlap onset concentration c^*; *h* heap, *b* blob. c. Concentrated system. d. Regions in the log M – log c plane for a system like PS-toluene; c in g/dl, M in g/mol. e. Concentration dependence of coil radius R_0 in good solvents, χ Flory Huggins parameter. 1 dilute, 2 semi, . . . , 3 concentrated regime.

regions without or with only a few polymer units within (*nanoheterogeneity*). [Similar effects are also expected for semidilute polymer–polymer blends of compatible components.] A polymer system is called concentrated, $c > c^{**}$, when the concentration profile is only granulated in the ϱ level, Fig. 4c. Being mainly a concentration effect, c^{**} does not (or little) depend on chain length, see Fig. 4d.

The structure in the semi-regimes is rather complicated and changes with concentration. The ϱ level structure in the concentrated regions and in the space in between is modified with increasing concentration of the other component. Furthermore, the local coil structure within the chain coils is also different: more Gaussian in the higher concentrated regions (heaps), as in the

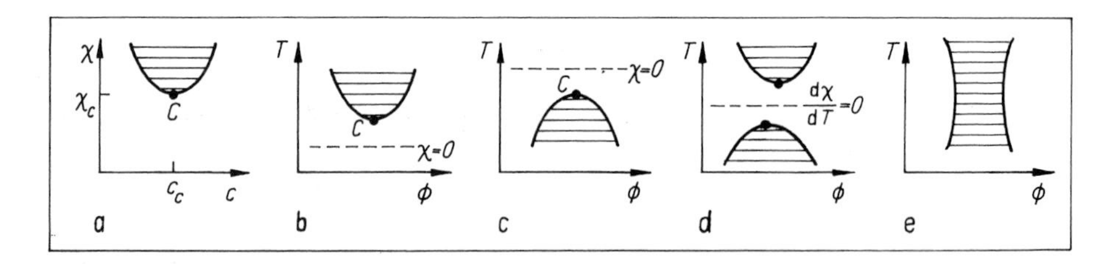

Fig. 5. Phase decomposition in polymer systems.
a. Theorist's diagram: Flory Huggins parameter χ vs. polymer concentration, C critical point. Good solvents have $\chi < 0$. b. LCST behavior, C lower critical solution temperature. c. UCST behavior, C upper critical solution temperature. d. State diagram with UCST and LCST, between them is a temperature with a χ minimum. e. No complete miscibility for all temperatures (hourglass).

melt, and more extended in the interspaces (de Gennes blobs). The size of heaps increases and the blob size decreases with polymer concentration. Many people believe that for long chains entanglements can also occur in the semi-regimes (Ref. 21), the higher the polymer concentration the lower the critical molecular mass $M_c(c)$ needed for entanglement (Fig. 4d). Then we have at least two additional lengths (besides glass transition correlation ξ_a and coil radius R_0): blob/heap size and entanglement spacing, both changing with polymer concentration. These lengths prevent uniform scaling of the regimes between c^* and c^{**}. Both relaxation and thermodynamics are varied by many details in this concentration range.

The *quality of a solvent*, related to a particular polymer, is characterized by the Flory Huggins parameter χ describing, roughly speaking, the (free) energy cost relative to kT for dissolution of one segment or monomeric unit (see Secs. 3.3–3.4 for details). A good solvent is characterized by $\chi < 0$, gain of free energy. The dilute state with $\chi = 0$ is called θ *state*, its temperature for zero pressure is the θ temperature.

Phase decomposition (mixing gap) can be observed in bad enough solvents, see Fig. 5. In terms of χ, the criterion for phase decomposition in a polymer system is $\chi > \chi_c$, with $\chi_c \approx 1/2$ for a small-molecule solvent. The critical solution point is $\{\chi_c, c_c\}$ with a rather low critical polymer concentration $c_c \sim 1/\sqrt{N}$. On the left hand side of the mixing gap, $c \ll c_c$, the polymer coils are rather collapsed, see below.

The ordinary state diagram (c vs. T, shortly cT, or ϕT diagram) is obtained when the temperature dependence of $\chi(T)$ is inverted to $T(\chi)$. Close by χ_c, a lower critical solution temperature is obtained for $d\chi/dT > 0$ (LCST type in Fig. 5b), an upper critical solution temperature for $d\chi/dT < 0$ (UCST type in Fig. 5c). The occurrence of both types in one system (Fig. 5d) is connected with an extremum (minimum) of $\chi(T)$ between, corresponding to a rather small temperature dependence orf χ. Phase decomposition for all temperatures (Fig. 5e) is generated by positive $\chi > \chi_c$ at all temperatures. The parameter χ can get values of order 1 or larger. Similar state diagrams can be observed in polymer-polymer systems.

The dependence of the coil radius on polymer concentration ($R_0(c)$) is sketched in Fig. 4e, above. As already mentioned, in the θ state of dilute solution we find nearly Gauss coils. $R_0(c)$ increases with increasing quality of the solvent (then χ becomes negative with larger amounts). The expansion can reach several ten per cent of R_0. It is not caused by an enlargement of the Kuhn step but is generated by a subtle repulsion between chain segments of one chain. Though one segment seldom meets with a segment separated from it by a longer distance along the chain, the interaction is rather effective because the coil structure is rather sensitive (see the comments on Eq. 1.18). Since two monomer units cannot occupy the same place we have repulsion in good solvents (excluded volume effect). On a lattice we can consider a random self-avoiding walk which gives

$$R_0 \sim M^{0.6} \tag{1.39}$$

instead of exponent 0.5 for Gauss chains. We observe a chain expansion for the same persistence length as before.

In bad solvents (large χ beyond the θ state) the coil radius becomes smaller than in the θ state. The free energy cost of dissolution is decreased for higher local polymer concentration, where the probability for polymer-solvent contacts is smaller. Effectively, this is an attraction between the monomer units that can lead to coil collapse.

In concentrated systems there is no difference between chain and "solvent": the latter consists of the same monomer units as the test chain itself. Seen from the ψ level, they meet only with "themselves". There is no gain or loss of free energy, no interaction, and if one tried to introduce a χ parameter one would find $\chi \to 0$ for $\phi_p \to 1$. This means finding a coil radius and coil structure similar to the θ state, $R_0 \sim M^{0.5}$. In concentrated systems, chain expansion or contraction with temperature or pressure is mainly caused by a larger or smaller Kuhn step length, $l(T)$ or $l(p)$.

The crossover between dilute and concentrated polymer systems, Fig. 4e, is rather complicated because the additional lengths vary with concentration and permit different semi-regimes. Studying increasing concentration the theorists discuss the crossover in terms of a gradual "screening" of the subtle long-range segment interaction between the chain segments in the dilute solution.

1.5 Entanglement

Beyond a *critical molecular mass* M_c the melt viscosity η sharply increases with M,

$$\eta \sim M^{3.4} \text{ for } M > M_c. \tag{1.40a}$$

This phenomenon (Fig. 6a) is connected with the term chain entanglement

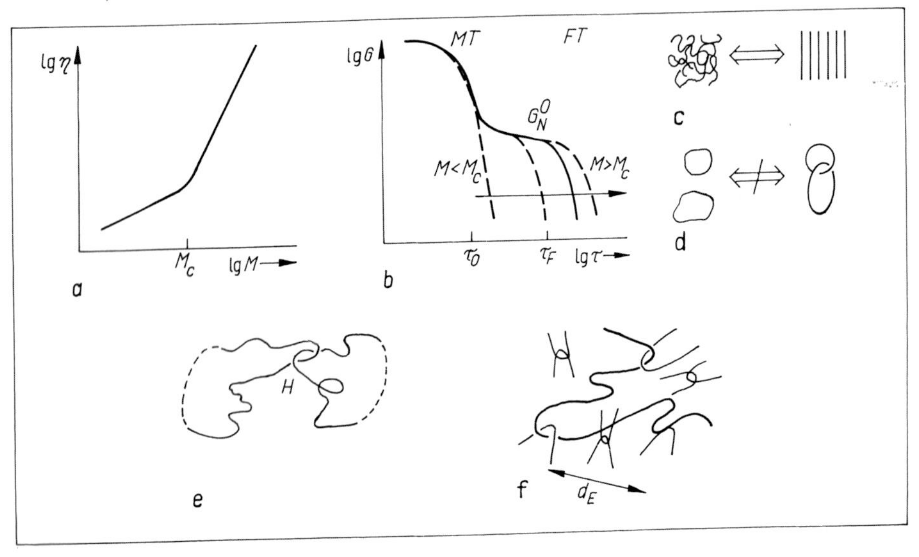

Fig. 6. Entanglements.
a. Critical molecular weight M_c (entanglement onset) for viscosity η. b. Shear relaxation modulus $G(\tau)$. Plateau zone ($G = G_N^0 \approx$ const, G_N^0 plateau zone modulus) between main transition MT and flow transition FT. c. Interpenetrating coils and lined-up chains are topologically equivalent. d. Knots can topologically be defined only for closed chains (ring polymers). e. Transient network hooks with transient topology. f. Density of entanglement points and entanglement spacing d_E.

(for classical reviews see Refs. 22, 23). It can also be observed by considerable changes of the shear curves for $M > M_c$, see Fig. 6b. A *plateau modulus* of order $G_N^0 \approx 1\,\mathrm{MPa}$ develops from the general decrease of shear relaxation modulus as function of relaxation time, $G(\tau)$. The transition from glass like behavior ($G \approx 1\,\mathrm{GPa}$) to flow behavior ($G \approx \exp(-\tau/\tau_0)$) is then divided into two parts: the main (or glass) transition MT where the material is softened to a rubber-like behavior, and the flow (or terminal) transition FT describing the crossover to fluidity. The width of the plateau zone increases with the same exponent as in Eq. 1.40a,

$$\tau_F/\tau_0 \sim M^{3.4}, \tag{1.40b}$$

where τ_0 and τ_F are indicated in Fig. 6b.

A further typical molecular mass, the entanglement m.m., can be calculated from G_N^0,

$$M_e = \rho RT/G_N^0 \tag{1.41}$$

connecting the entanglement with the idea of a temporary network (cf. with Eq. 3.2b and Fig. 7b, below). Both M_e and $M_c \gtrsim M_e$ are of order 10^4 g/mol, see Table I. Using Eq. 1.37 for Gauss coils a new geometric length can be defined, the *entanglement spacing*

$$d_E = aZ_c^{1/2}, \tag{1.42}$$

with Z_c the degree of polymerization corresponding to M_c. A typical value is $d_E \approx 7\,\mathrm{nm}$ for $Z_c = 100\,\mathrm{nm}$ and a structure length $a \approx 0.7\,\mathrm{nm}$.

To characterize the entanglement in terms of mutually interpenetrating coils, concepts like hooks, knots, or loops are used; generally one speaks of topological restrictions. The last term needs an explanation: Mutually interpenetrating linear chains are topologically – in the mathematical sense – completely equivalent to a stack of extended chains, Fig. 6c, since the space can continuously be deformed from the coils to the stack without changing any mathematical neighborhood. Knots can be topologically defined only for closed loops, Fig. 6d. Only for time intervals where the chain ends cannot diffuse to the hook or knot, the chain ends can be mentally connected to a ring without changing the dynamic situation in the vicinity of the knot (dotted lines in Fig. 6e). A topology can therefore be defined only for small time intervals: we find a transient or temporary knot topology. Therefore the entanglement is a structure with temporary topological elements.

If the entanglement could uniquely be contracted to a definite number of

Table I. Parameters of some conventional polymers

Name	M_0	l_k	C_∞	a	T_g	M_c	G_N°
App. 2	g/mol	nm	–	nm	K	kg/mol	MPa
PDMS	74	1.0	6.5	0.65	146	25	0.3
PE	28	1.2	7	0.6	250	5	2.1
PP	42	1.1	6	0.55	244	7	0.5
PIB	56	1.05	5	0.6	205	16	0.25
cis PI	68	0.9	5	0.7	200	7.5	0.45
PS	104	1.7	10	0.75	373	35	0.20
PVAC	86	1.6	9.5	0.7	305	25	0.35
PVC	62.5	1.2	6	0.6	347	6.5	1.0
PTFE	100	3.9	24	1.1	400	13	3.0
PET	192	1.1	4.2	1.3	342	3.3	0.10
PMA	86	2.1	8	0.65	280	24	0.25
PMMA	100	1.7	7	0.6	378	32	0.6
PEO	44	0.8	4	0.5	220–250	4.5	0.065
PPO	58	0.9	5	0.6	213	8	0.7
PC	242	3.0	2.5	1.1	418	5	1.5
Se	79	0.5	1.0	–	303	35	0.6

The data are selected from Ref. 38a, b, c. The chemical configuration is explained in App. 2. PC polycarbonate from bisphenol A, Se selenium, M_0 molecular mass of monomeric unit, l_K Kuhn step, C_∞ characteristic ratio, a structure length, T_g glass temperature ($\beta(U)$ for PE and PTFE, for PP in the amorphous phase of a sample with 55% crystallinity, PEO cf. Fig. 51d), M_c critical molecular mass for entanglement, G_N° plateau zone modulus.

simple tight knots of definite topology (*entanglement points*), then the entanglement spacing Eq. 1.42 is the average geometric distance between such points neighbored on one chain. Since many chains are interpenetrating (Fig. 4c) the mean distance between arbitrary entanglement points (also for different chains) is smaller, a typical value is about 3 nm, see Fig. 6f.

An ab initio calculation of M_c is still missing. Consider increasing chain length. The problem is to find a larger transient structure with lifetime τ_F that can, for $M > M_c$, survive shorter lifetimes $\tau_E < \tau_F$ for its temporary topological elements. In other words we must find temporary structures obeying the general scaling principle: the larger the slower. Obviously, this is not possible for too small M, since a small-number structure cannot survive a breakdown of one element.

Correlations for M_c refer to two possible reasons: Topological elements, e.g. a hook (see Ref. 24 and preceding work by Privalko, Graessley and others) and packing effects (Ref. 25).

(i) A hook model between two chains respectively, discussed on the background of a broad set of data, leads to

$$Z_{\text{eff}} \approx 3C_\infty^2 \sim (l_K/l_0)^4, \tag{1.43}$$

where l_K is the Kuhn step and Z_{eff} is the number of effective skeletal bonds in an entanglement strand. Accordingly, for a given bond length l_0, the critical length is the larger the greater the chain stiffness. The high power in Eq. 1.43 could be the consequence of two things: the greater l_K/l_0 the rarer the eye and the longer the chain part forming the hook.

(ii) The addition of a solvent causes the entanglement spacing to increase and the chain packing to decrease. Correlations give formulas like $M_c \phi^{2b} \approx$ const, or $d_E^2 \phi^{b'} \approx$ const, with ϕ the polymer volume fraction and b, b' positive exponents of order 1. Models stressing the importance of packing effects lead, in a way, to the inverse dependence on C_∞,

$$Z_{\text{eff}} \sim \rho^{-2} C_\infty^{-3}, \tag{1.44}$$

because higher flexibility favors the packing.

The temperature dependence of M_c is weak and is of order 10% for large temperature differences.

1.6 Polymer networks

When the flexible chains are crosslinked at the ends or at other points, then we have a network with a fixed nontrivial topology, Fig. 7a. An additional characteristic length is defined by the average distance of crosslinks, for a fixed topology given by an average chain length (or molecular mass M_c) between two neighboring crosslinks. At the end crosslinking prevents the material flow, cf. Fig. 7b, the limit value of the shear modulus can be estimated from a fomula like Eq. 1.41,

$$G_\infty = \rho RT/M_c, \tag{1.45}$$

decreasing with higher M_c. For weak crosslinking, M_c (network) > M_c

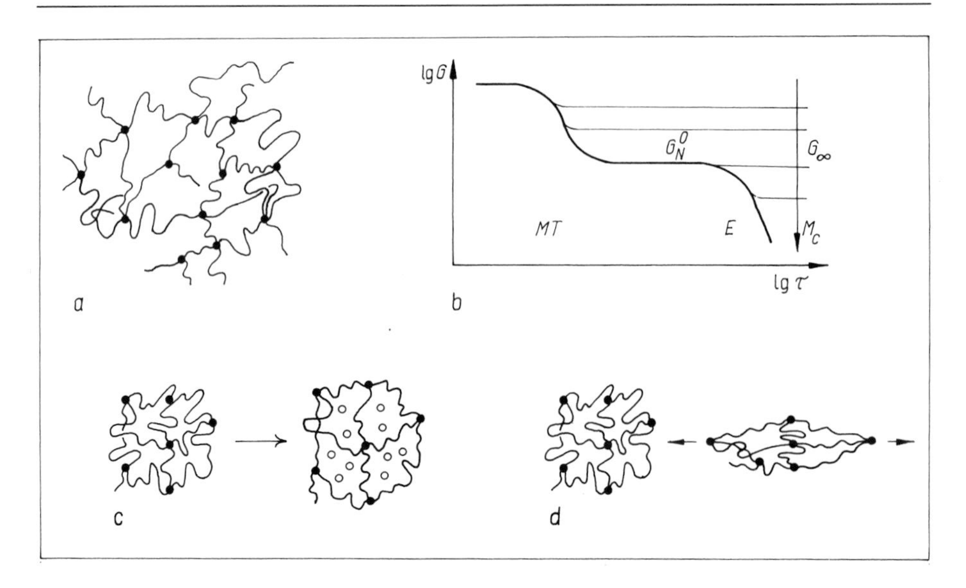

Fig. 7. Networks.
a. Topology (hypothetically). b. Shear relaxation modulus for increasing average molecular weight of chain parts between crosslinks, M_c. The signals from flow transition (E) depend on the entanglement onset ($= M_c$ in the sense of Figs. 6a, b. Do not confuse this physical M_c with the chemical M_c from crosslinks). The shear modulus in networks remains finite, $G \to G_\infty > 0$ for $\tau \to \infty$. c. Swelling. d. Stretching.

(entanglement), we have two transitions, MT and FT, similar to the situation in entangled polymer melts (Fig. 6b), before the network modulus (G_∞) is reached at large relaxation times. In highly crosslinked materials, the so called duromers, with $G_\infty > G_N^0$ and $M_c(n.w.) < M_c(E)$, the flow transition is completely suppressed.

Dynamics and thermodynamics of networks are modified by several particularities. The functionality of crosslinks, i.e. the average number of chain parts emerging from the crosslink, is a topological property. Free dangling ends also belong to topology. Swelling means the storage of solvent molecules, oligomers or polymers that are not chemically connected with the network (Fig. 7c). Weak networks (so-called elastomers) can easily be stretched to a large amount, up to several hundreds per cent, with no essential change of volume, of course (Fig. 7d). This rubber elasticity as well as swelling does not alter the network topology.

1.7 Chain folded crystallites

Polymers with sufficient regularity of chemical configuration along the chain (tacticity) can crystallize (Refs. 8, 26). The typical, though not the only, form of crystallites developed from rather different situations in solutions as well as in melts are layer crystals (Fischer, Keller) that are often idealized as chain folded lamellas, Fig. 8a and b, see Ref. 27–29. This is a unique feature occurring only for polymers: the molecular chain structure and the lamellar habit are intrinsically linked.

The real coil structure of long chains in lamellar stacks, Fig. 8c and d, where the lamellas are separated by more or less amorphous layers, is governed by two facts. (a) Surprisingly, the mean chain end-to-end distance in melts and stacks is comparable, and (b) although there is no regular adjacent back folding like Fig. 8a, there remains some clustering of stems. Both facts are observed by neutron scattering in partially deuterated materials, see Ref. 30 and 31, respectively.

Several new lengths are introduced by lamellas or lamellar stacks. The lamellar thickness (l) is approximately equal to the stem length, if the stems are not too oblique. A typical order is $l \approx 10\,\mathrm{nm}$, l depends on the conditions of crystallization. It is larger for higher crystallization temperature (T_c), i.e. for smaller undercooling, $\Delta T = T_m^\circ - T_c$, with T_m° the ideal melting temperature for $l \to \infty$. Furthermore, l can increase by subsequent growth of layer thickness moderated by complicated molecular motion in the crystallite.

A further length (l_{if} of order 1–2 nm, Fig. 8d) is the interface thickness between lamellas and the amorphous phase. The problem is that the flux of chains emanating from the covering face must be diminished substantially for attainment of random disorder in the amorphous phase. This demands a length at least of order the Kuhn step. The structure is perhaps to be modified by entanglements displaced by the lamellar crystallization and by chain ends diffusing out from the core of lamellas.

A length of 2 nm is still sufficient to define thermodynamic subsystems in the ϱ level, i.e. with monomeric units as particles. The interface becomes an interphase with rather independent dynamic and thermodynamic properties. Neglecting the gradients we then have Mandelkern's three phase model, Fig. 8d, with an additional length: the amorphous layer thickness (l_a), or the long period $L = l_a + l_c + 2l_{if}$ defining the repeat unit of a stack. The tendency

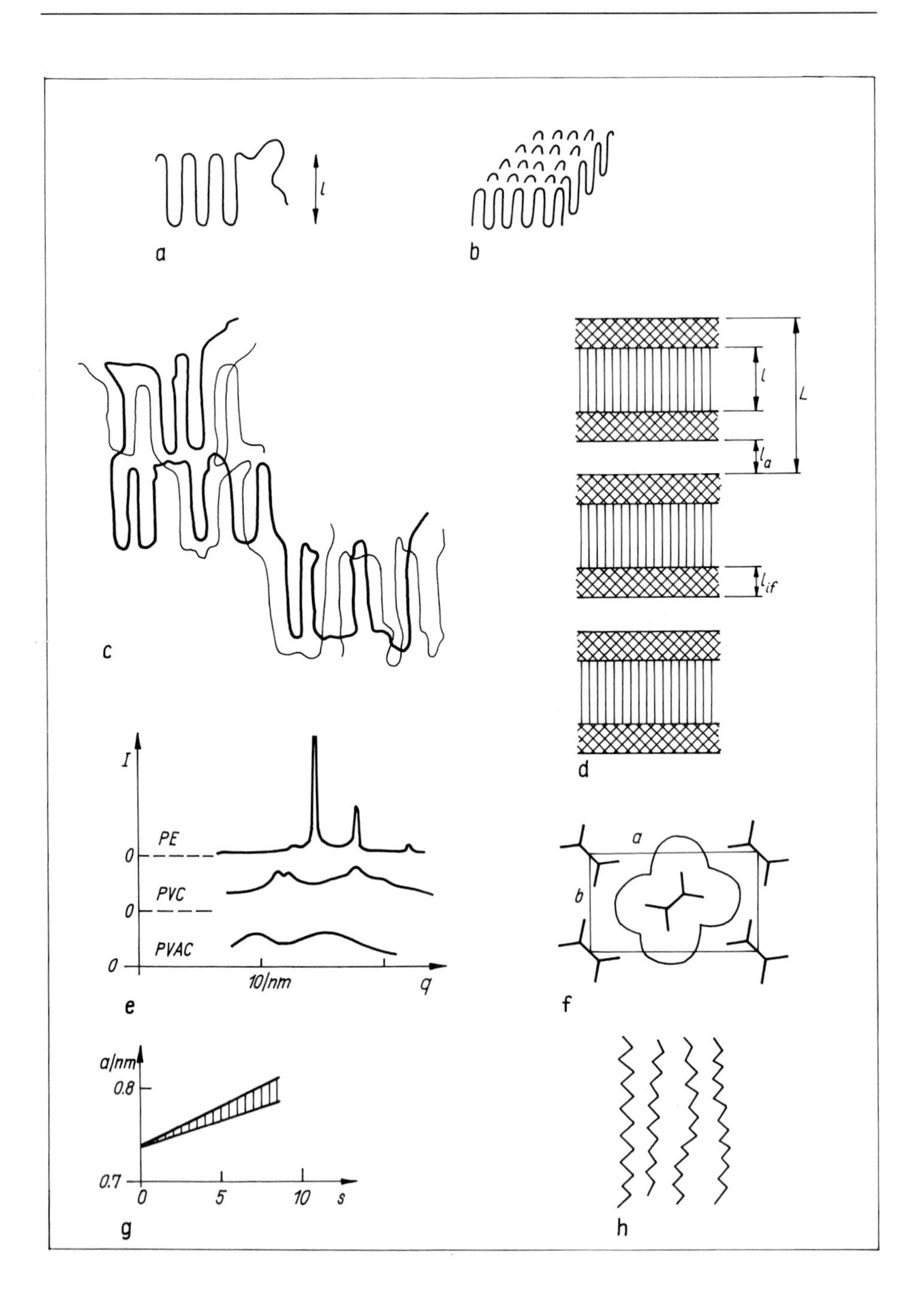

l
a
b
c
l
L
l_a
l_{if}
d
I
PE
0
PVC
0
PVAC
0
10/nm
q
e
a
b
f
a/nm
0.8
0.7
0
5
10
s
g
h

for equilibrating is thickening of lamellas and decreasing of interphase thickness.

Similar to small-molecule materials the unit cell of polymer crystallites can by determined from wide angle X-ray scattering (WAXS, Fig. 8e). Fig. 8f shows the result for polyethylene PE with the following parameters: structure Pnam, orthorhombic symmetry, $1 * 2/1$ helix, 2 monomeric units per cell, $a = 0.74\,\mathrm{nm}$, $b = 0.493\,\mathrm{nm}$, $c = 0.254\,\mathrm{nm}$ ($\approx l_0'$), c-axis parallel to the stem orientation.

In general, the structure of the lamella is not so compact as expected from trying to arrange chains with hard van der Waals cores in a condensed state. At least in one direction, for PE parallel to the a axis, the crystal can easily be extended up to several per cent. This is enough for inclusion of some small side groups in the crystal lamella (Fig. 8g, see Ref. 32). The a parameter also correlates with lamellar thickness l. The a value is larger for smaller l, see Ref. 33.

The ease of unit-cell widening enables some motional elements to occur, e.g. an 180° rotational jump accompanied by translation by one methylene group ($l_0'/2$) along the c-axis that leaves the chain in crystallographic register (Refs. 34 and 35). The cooperativity of such elements permits possibilities for some disorder in the crystallites, see Fig. 8h (dynamic *con*formational *dis*order – CONDIS (Ref. 36) – combined with long range positional and oriental order in the distribution functions). One can make a list of increasing disorder: crystal (no disorder), condis crystal, plastic crystal (orientational disorder), liquid crystal (position disorder), and melt. Each disorder is large enough to allow a glass transition in the phase.

Fig. 8. Chain folded crystals (layer crystallites).
a. Ideal adjacent back folding. b. Ideal lamella in three-dimensional space. c. Test chain in a more real situation (two-dimensional picture, from Ref. 28). d. Model for a lamellar stack accessible for thermodynamic analysis, with crystalline, interface (if any, or interphase), and amorphous (a) layers. e. SAXS Intensities for semicrystalline PE, weakly ordered PVC, and amorphous PVAC, see also Fig. 2e. f. Crystal unit for PE. g. Enlargement of the a axis by incorporation of side groups in the PE chains, e.g. methyl groups; s number of side groups per 100 C atoms of the PE backbone. h. Configurational disorder CONDIS of lamellar stems, schematically.

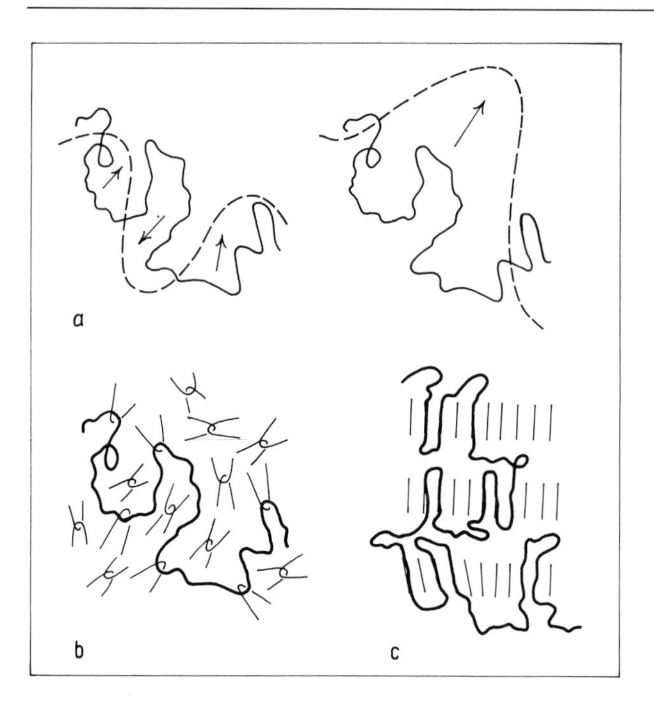

Fig. 9. Comparison of three states.
a. Dilute solution. The arrows indicate normal (Rouse) modes of different wave lengths. b. Entangled melt (see Fig. 6f). c. Semicrystalline state (see Fig. 8f).

1.8 Structure and dynamics – an outline

Consider the three selected states of Fig. 9a–c. In the dilute solution we observe 3 modes of chain dynamics: (i) The local mode in the $\lesssim 1$ nm range, i.e. a conformation change with respect to one bond (1-bond motion accompanied by the compliance of the next chain parts – spreading out of bond length and angle distortion in the per cent range – and by participation of the next solvent molecules, see Ref. 37), (ii) normal (in the sense of perpendicular) modes of different mode lengths up to the coil diameter, as indicated in Fig. 9a, and (iii) the diffusion of the chain as a whole. Additionally we find the motion of solvent molecules modified in the vicinity of polymer chains and being cooperative at low temperature (dynamic glass transition).

In the amorphous entangled melt (Fig. 9b) the molecular motion of chains is modified as follows. (i) The local modes have a higher activation energy, including perhaps 2-bond motions. (ii) The normal modes are heavily modified, especially at low temperatures. The transition from local modes to chain diffusion is separated into two parts, see Fig. 6b; Ref. 38. (A) *Main transition* MT. This is, for shorter times τ, a cooperative motion of about 100 monomeric units, the dynamic glass transition, followed by the modified normal modes including some flow confined in the entanglement spacing, hindered at larger τ by the topological restrictions of the entanglement. (B) *Flow transition* FT, separated from MT by the plateau zone. The FT is a slow fluctuative tangling and loosening of entangelement points, probably accompanied by a certain cooperative movement in the spatial scale of the coil radius, or even larger. At high temperatures the cooperativity of glass and flow transition breaks down, and we find, besides the local modes, normal modes similar to those in dilute solution, but additionally, in the FT, the entanglement fluctuation. The cooperativity is changed in the direction of collectivity (e.g. tube models), but there is also a weak mode coupling remaining. (iii) The self diffusion of the whole chain is dramatically lowered by the entanglement.

In semicrystalline materials (Fig. 9c) the time or temperature range of the cooperative motion largely depends on the phase in which they occur. The higher the order, as in the succession amorphous, interface, and crystalline, the larger the time scale (many orders) and the higher the typical temperatures (glass transitions) for a given time scale (several ten Kelvins). At higher temperatures the chains can also be passed or dragged through the lamellas.

Let us conclude this section (and chapter) with two general remarks.

(i) Precondition for a slow, large-scale motion is the working of a quick, small-scale motion. This condition is usually fulfilled because of the general scaling principle (see Introduction and Sec. 2.8, below) accordingly larger modes correspond to larger times or lower frequency. In simple cases the small-scale motion can give the time beats (attempt rate) for the large-scale motion (success rate).

(ii) Nanoheterogeneity and structure heterogeneities in the 10-nm range result in heterogeneities of mobilities and mode parameters and/or types. So, for instance, the one glass transition zone in small-molecule liquids is split into several components for polymers, sometimes with a fine structure. This phenomenon is called *multiplicity* of glass transition.

2. Linear response

Temporal aspects of thermodynamics form the subject of this chapter. Consider a thermodynamic subsystem. The real, primary observables of a thermodynamic experiment are the susceptibilities. The thermodynamic variables that are constituents of the energetic fundamental form can be calculated from the susceptibilities. These observables are, after spreading over the time (or frequency) axis, represented by moduli and compliances.

Consider, for instance, a shear experiment. The thermodynamic variables are the extensive deformation, expressed by the shear angle γ, and the intensive shear stress σ, see Fig. 10a. The energetic contribution to the linear form of the First Law is $\sigma d\gamma$ or $\gamma d\sigma$, multiplied by the volume of the subsystem. For a shear stress relaxation experiment, $\sigma(t)$, the observable is the shear modulus G, for a shear creep experiment, $\gamma(t)$, the observable is the shear compliance J. For the sake of simplicity, the term observable will sometimes be applied also to the variables.

For small enough perturbation the response of the subsystem is linear and causal although the molecular motion inside the subsystem is (highly) non-linear and necessarily chaotic. Linear means that the responses of two arbitrary perturbation programs are linear superpositions of the individual responses (Boltzmann's superposition principle), and causal means that the current response (the observable at actual time t) is only influenced by the program in the past, $t' \leqslant t$.

The situation for step programs is shown in Figs. 10b and c. The *relaxation* experiment gives the intensive variable $\sigma(t)$ after a step of the extensive variable $\gamma(t)$ with γ_0 step height. The ratio

$$G(t) \equiv \sigma(t)/\gamma_0 \tag{2.1}$$

is called (shear) relaxation *modulus*. Inversely, the creep or *retardation* experiment gives the extensive variable $\gamma(t)$ after a step of the intensive variable $\sigma(t)$ with σ_0 step height. The ratio

$$J(t) \equiv \gamma(t)/\sigma_0 \tag{2.2}$$

is called (shear) retardation *compliance*. Typical time scales are called relaxa-

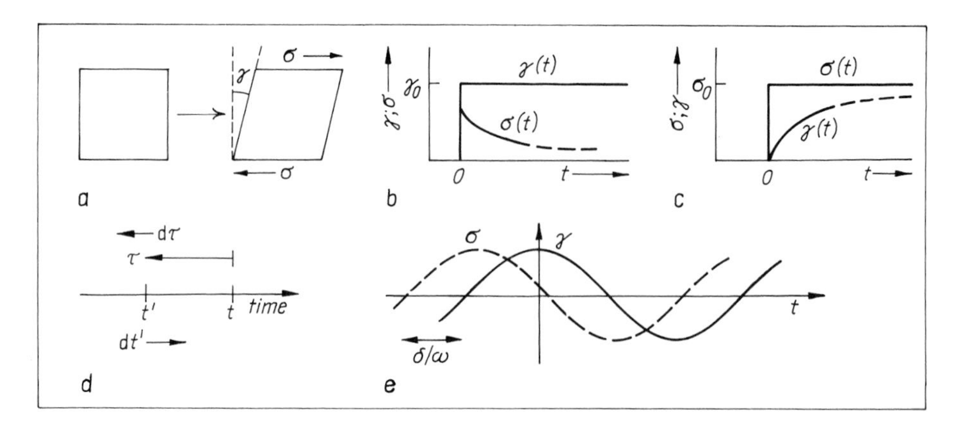

Fig. 10. Linear response for shear.
a. Shear experiment, γ shear angle, σ shear stress. b. Shear stress relaxation. c. Shear deformation retardation (creep experiment). d. The times: t actual time, t' time in the past, $\tau = t - t'$ backward look. e. Stationary periodic experiment, δ (shear) loss angle.

tion and retardation times (τ_R, $\tau_{R'}$), respectively. The linearity and causality properties of the thermodynamic experiments for arbitrary programs (in the linear region, of course) are uniquely expressed by convolutions,

$$\sigma(t) = \int_{-\infty}^{t} G(t - t')\dot{\gamma}(t')dt' \equiv G * \dot{\gamma}, \tag{2.3}$$

$$\gamma(t) = \int_{-\infty}^{t} J(t - t')\dot{\sigma}(t')dt' \equiv J * \dot{\sigma}. \tag{2.4}$$

The linearity is expressed by the additive properties of the integrals. The causality can be seen from the integration limits: t' covers the whole history up to the actual time t. The "memory" functions $G(\tau)$ and $J(\tau)$, $\tau = t - t'$, evaluate the shear history given by the rates $\dot{\sigma}(t') = d\sigma/dt(t')$ and $\dot{\gamma}(t') = d\gamma/dt(t')$, respectively. The time situation of a convolution is indicated in Fig. 10d. It is a useful exercise, using $\gamma(-\infty) = 0$ and the relation

$$\int_{-\infty}^{t} dt' \Leftrightarrow \int_{0}^{\infty} d\tau, \tag{2.5}$$

to prove (Ref. 39) the following equations:

$$G * \dot{\gamma} = \int_{0}^{\infty} G(\tau)\dot{\gamma}(t - \tau)d\tau = \frac{d}{dt}\int_{-\infty}^{t} G(t - t')\gamma(t')dt'$$

$$= G_g\gamma(t) + \int_{-\infty}^{t} \dot{G}(t - t')\gamma(t')dt', \tag{2.6}$$

where the "glass modulus" is the short time limit of $G(\tau)$, $G_g = G(\tau \to 0)$. The long time limit is called equilibrium modulus, $G_e = G(\tau \to \infty)$.

The susceptibilities, i.e. moduli and compliances, are usually represented as a function of logarithmic time, $\lg\tau$, or, after Fourier transformation, of the logarithm of frequency, $\lg\omega$. Regions in $\lg\tau$ or $\lg\omega$ where G and J change considerably, are called relaxation or retardation zones, or regions, generally *dispersion zones*. This term will be restricted to the time/frequency picture, additional aspects related to the scattering vector q will be discussed in Sec. 6.10.1 (transport zone).

Occasionally the formula

$$\Delta\sigma(t) = G_g\Delta\gamma(t) + \int_{-\infty}^{t-\varepsilon} G(t - t')\dot{\gamma}(t')dt' \tag{2.7}$$

is used, see e.g. Ref. 40. This is equivalent to Eqs. 2.3 and 2.6 if we agree upon that $\Delta\sigma$ and $\Delta\gamma$ are small increments where $\Delta\gamma(t)$ is separated from the integral by the exclusion of a small time element ε at t, $\varepsilon \leqslant \tau_R$,

$$\Delta\gamma = \int_{t-\varepsilon}^{0} \dot{\gamma}(t')dt', \tag{2.8}$$

and $\Delta\dot{\gamma}(t') = \dot{\gamma}(t')$ otherwise.

Using the simultaneous substitution $\sigma \leftrightarrow \gamma$ and $G \leftrightarrow J$, Eqs. 2.6 to 2.8 are also true for compliances.

Complex variables are used for a stationary periodic experiment (Fig. 10e). This is indicated by an asterisk (that does not mean the order to go to complex conjugate variables). Then

$$\gamma^* = \gamma_0 \exp(i\omega t), \quad \sigma^* = \sigma_0 \exp(i(\omega t + \delta)). \tag{2.9}$$

The corresponding susceptibilities,

$$G^*(\omega) = \sigma^*/\gamma^* = G'(\omega) + iG''(\omega), \tag{2.10}$$

$$J^*(\omega) = \gamma^*/\sigma^* = J'(\omega) - iJ''(\omega), \tag{2.11}$$

are called *dynamic* modulus and compliance, respectively. For a given thermodynamic state, G^* and J^* depend on the frequency (ω, or $\nu = \omega/2\pi$) in any dispersion zone.

The tangent of the loss angle, $\tan\delta$, is called the *loss factor*. From Eqs. 2.10–11 we see that

$$\tan\delta(\omega) = G''/G' = J''/J'. \tag{2.12}$$

Modulus and compliance of energetically conjugate intensive and extensive variables are connected by

$$G^*(\omega) \cdot J^*(\omega) = 1(\text{real}) \tag{2.13}$$

as can be seen from the definitions Eqs. 2.10–11.

Usually the real parts are called elastic or *storage* parts, the imaginary ones *loss* parts. The latter are related to the heat production connected with the periodic experiment (energy dissipation for shear). The Second Law implies

$$G''(\omega) \geqslant 0, \quad J^*(\omega) \geqslant 0 \tag{2.14}$$

which relation is sometimes called the dissipation theorem.

If resonances are excluded (i.e. we consider only responses that can additively be combined from positive exponential decays or switch-ons, cf. Eqs. 2.43–2.44 below), then $G' \geqslant 0$, $J' \geqslant 0$, and from Eqs. 2.12 and 2.14 we obtain $\tan\delta \geqslant 0$, the γ sine peak comes earlier than the σ peak in Fig. 10e. This does not reflect any directed causality between σ and γ – both Eqs. 2.3 and 2.4 are entirely equivalent relative to causality – but the equation $\tan\delta \geqslant 0$, in connection with the different signs in Eqs. 2.10 and 2.11 and the Second Law Eq. 2.14, reflects the different behavior of the extensive variable $\gamma(t)$ and the intensive variable $\sigma(t)$ in Figs. 10b and c. Excluding resonances means that $G(\tau)$ increases and $J(\tau)$ decreases with τ, for a given state.

From our daily experience we know that there are many interesting nonlinear phenomena with polymers. Examples are the large stretching ability of rubber, non-Newtonian flow (e.g. a reduction of viscosity with increasing shear rate (important for extrusion, milling, mixing, rolling, calandering, painting)), or the Weissenberg effects (normal stress from shear), see Ref. 41. If connected with some fluidity of the material they are subject of rheology (e.g. Ref. 42). The molecular mechanisms investigated by linear response are also important for nonlinear phenomena (e.g. by means of a time shift or reduction induced by a shear rate) but there are additional aspects of nonlinearity that are rather difficult to handle and are not included in this book.

2.1 Mechanical models for viscoelasticity

A shear (or bulk) response is called viscoelastic if loss and storage parts of modulus or compliance are of comparable order of magnitude, i.e. $\tan\delta$ not too small and not too large. There are a lot of mechanical models for viscoelasticity combined from two kinds of elements.

(1) A spring is purely elastic and describes the instantaneously acting Hooke law with modulus and compliance being constant and real ($\tan\delta = 0$)

$$\sigma(t) = G_0\gamma(t), \quad G(\tau) = G^* = J^{*-1} = J^{-1}(\tau) = G_0 = \text{const.} \tag{2.15a}$$

(2) A damper is purely viscous and describes the instantaneously acting Newton law with purely imaginary dynamic responses ($\tan\delta = \infty$)

$$\sigma(t) = \eta_0\dot{\gamma}(t),\ G^* = i\omega\eta_0 = J^{*-1}, \quad G(\tau) = \eta_0\delta(\tau),\ J(\tau) = \tau/\eta_0. \tag{2.15b}$$

Comparing these elements one can see why a situation with $\tan\delta > 1$ is called liquid-like and with $\tan\delta < 1$ solid-like.

Even simple connections of two or three elements can describe quite different viscoelastic properties, see Fig. 11a–c.

The *Maxwell liquid* (or element) is a serial connection of spring and damper. Then the compliances can be added similar to the deformations

$$J(\tau) = G_0^{-1} + \tau/\eta_0,\ J^*(\omega) = G_0^{-1} - i/\omega\eta_0. \tag{2.16}$$

From Eq. 2.13 we obtain $G^* = J^{*-1}$, and from a Fourier transformation (see Eq. 2.13, below) we obtain $G(\tau)$,

$$G^*(\omega) = G_0 i\omega\tau_0/(1 + i\omega\tau_0),\ G(\tau) = G_0\exp(-\tau/\tau_0) \tag{2.17}$$

with $\tau_0 = \eta_0/G_0$ as the characteristic relaxation time. The loss factor $\tan\delta = 1/\omega\tau_0$ is larger than 1 for low frequencies, $\omega\tau_0 < 1$, explaining the name liquid. The hydrodynamic regime for $\omega\tau_0 \ll 1$ is called *Newton liquid* with

$$G' \sim \omega^2,\ G'' \sim \omega,\ J' \equiv J_e^0 \sim \omega^0,\ J'' \sim \omega^{-1} \tag{2.18}$$

where $J_e^0 = 1/G_0$ (from Eqs. 2.16–2.17).

The *Voigt Kelvin body* (or element) is a parallel connection of spring and damper. Now we can add the moduli with the result

$$G(\tau) = \eta_0\delta(\tau) + G_0,\ G^*(\omega) = G_0 + i\omega\eta_0, \tag{2.19}$$

$$J(\tau) = G_0^{-1}(1 - \exp(-\tau/\tau_0)),\ J^* = 1/G^*(\omega). \tag{2.20}$$

The loss factor $\tan\delta = \omega\tau_0 < 1$ for $\omega\tau_0 < 1$, that is why the name body is used; $\tau_0 = \eta_0/G_0$ (as before) is better called a retardation time now.

The *standard model* is represented in Fig. 11c and describes a transition

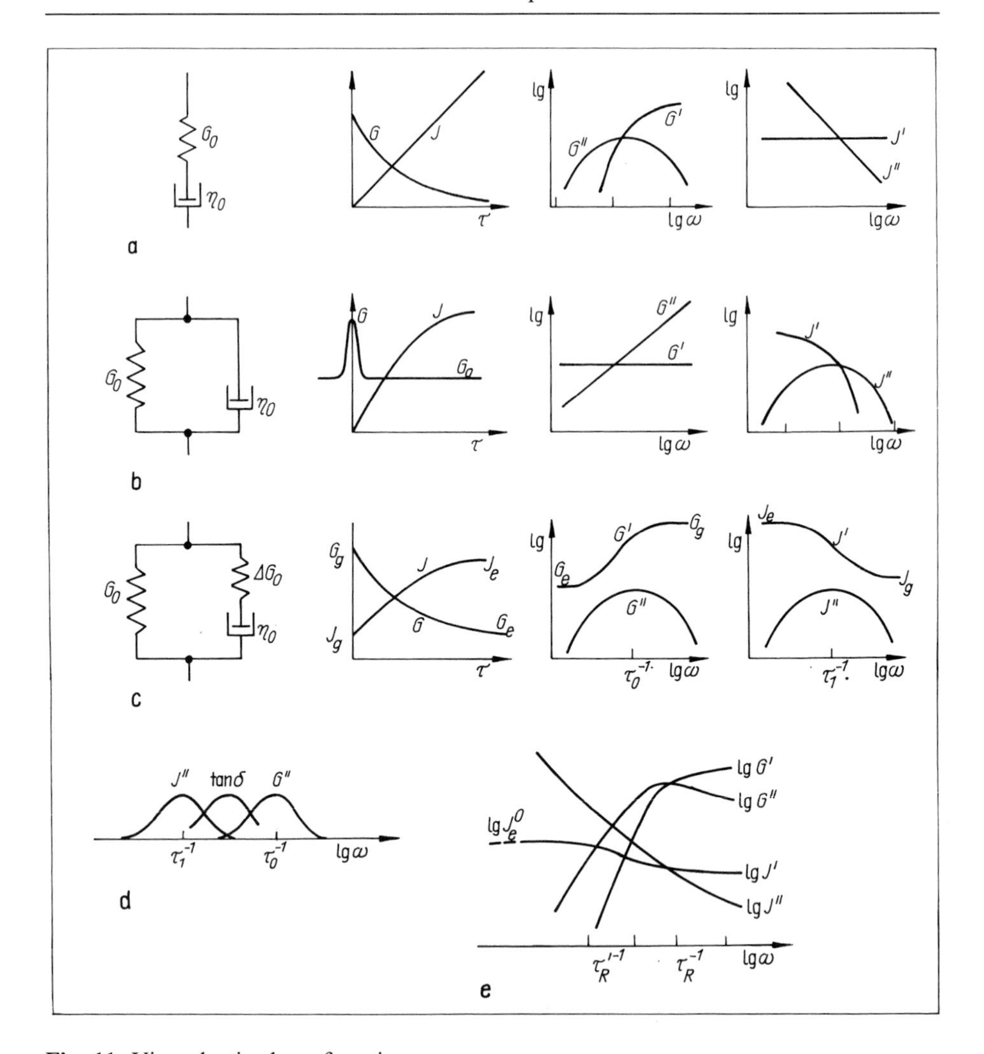

Fig. 11. Viscoelastic shear functions.
a. Maxwell liquid. b. Voigt Kelvin body. c. Standard model (with relatively small relaxation intensity, $\Delta G/2\sqrt{G_g G_e} < 1$). d. Relative positions of the (semilogarithmic) Lorentz curves for loss compliance J'', loss factor tan δ, and loss modulus G'', for a standard model with large relaxation intensity, $G_g/G_e = \tau_1/\tau_0 \approx 100$. e. Log-log plot of shear curves for a dynamic glass transition in a small-molecule glass former. The lg ω scale is in decades (factor 10). τ_R relaxation, τ_R' retardation time, J_e^0 steady state compliance.

with both finite glass ($G_g = J_g^{-1}$) and equilibrium responses ($G_e = J_e^{-1} > 0$). Their differences are generally called *relaxation* and *retardation intensities*,

$$\Delta G = G_g - G_e, \quad \Delta J = J_e - J_g = G_e^{-1} - G_g^{-1}. \tag{2.21}$$

The response can be calculated by adding the moduli of the spring G_0 and the Maxwell element $\{\Delta G_0 = \Delta G, \eta_0]$. One obtains *Debye formulas*:

$$G(\tau) = \Delta G e^{-\tau/\tau_0} + G_e, \; J(\tau) = J_e - \Delta J e^{-\tau/\tau_1}, \tag{2.22}$$

$$G^*(\omega) = G_e + \Delta G i\omega\tau_0/(1 + i\omega\tau_0), \tag{2.23}$$

$$J^*(\omega) = J_e - \Delta J i\omega\tau_1/(1 + i\omega\tau_1). \tag{2.24}$$

Observe the interesting fact that there are two characteristic times, one for the modulus

$$\tau_0 = \eta_0/\Delta G_0, \text{ relaxation time,} \tag{2.25a}$$

and another for the compliance

$$\tau_1 = \tau_0 \cdot (G_g/G_e) > \tau_0, \text{ retardation time,} \tag{2.25b}$$

where $G_e = G_0$ and $G_g = G_0 + \Delta G_0$, of course. In contrast with the Maxwell and Voigt Kelvin elements, for the standard model both loss parts and $\tan\delta$ have the same form

$$\text{const} \cdot \omega\tau_K/(1 + (\omega\tau_K)^2) \tag{2.26}$$

but different time constants τ_K. Plotted against $\lg\omega$ Eq. 2.26 gives a symmetric curve (Lorentz curve) with a maximum of const/2 at $\omega\tau_K = 1$ and a half-width of $\Delta \lg \omega = 1.144$. Changing τ_K results in form-invariant shifts along the $\lg \omega$ axis.

The Lorentz curve for $\tan\delta(\lg \omega)$ is exactly in the middle between the Lorentz curves of $G''(\lg\omega)$ and $J''(\lg\omega)$, see Fig. 11d. The maximum values are

$$G''_m = \Delta G/2, \quad \tan\delta_m = \Delta G/2\sqrt{G_g G_e}, \quad J''_m = \Delta J/2 \tag{2.27a}$$

at the frequencies

$$\omega\tau_0 = 1 \text{ (for } G''\text{)}, \quad \omega\tau_1 = 1 \text{ (for } J''\text{)} \quad \text{and} \quad \omega\sqrt{\tau_1\tau_0} = 1 \text{ (for } \tan\delta\text{)}. \tag{2.27b}$$

Fig. 11e shows the dynamic shear responses at a dynamic glass transition of a small-molecule glass former. None of the three mechanical models gives a qualitatively correct picture, in particular the width of the G'' peak for real glass transition is broader than for the Maxwell element, and we find a finite step for J' without a finite G_e.

2.2 Discussion of the mechanical models

The mechanical models can help to visualize the complications of viscoelastic behavior, especially for people which can grasp the combined effect of springs and dampers. Enlarging the number of elements, and kinds of connection, broad dispersion zones can also be modelled. The idea of spectra for viscoelastic behavior (Sec. 2.4, below) surely originates from such models.

The reverse is not true: A given curve $G(\tau)$, for instance, if different from the standard relaxation, does not uniquely determine a mechanical model. Additional conventions about structure and relations between the elements are needed, see again Sec. 2.4.

Electrical analogies are not useful because either dissipative (resistance) and nondissipative (capacitance, inductance) elements, or the structure, compared to the mechanical, must be changed.

A mechanical model must not be transferred to the ϱ level of molecular motion since this level is to describe by reversible fluctuations, and no viscosities. The particles move chaotically according to Newton's laws. Fourier transformations modelling the experimental situation (Wiener Khinchin theorem) result in spectral densities for observables where different frequency regions correspond to different molecular modes or mode parameters. Even if a dispersion zone could well be modelled by a Maxwell fluid a distribution of frequencies or times rather than a single relaxation time would be appropriate. The same is true for any exponential decay or Lorentz curve. The single viscosity parameter of mechanical models is an integral over a distribution of times (correlation function) in the ϱ level.

On the contrary, the ψ-level description of chain motion uses some kind of mechanical models. They are usually modified by spatial arrangements for the elements such as chains made from entropy springs and beads with Stokes friction, as for the Rouse model (see Sec. 5.3 below).

2.3 General linear response

The connection between relaxation and retardation, and between time and frequency representation is described in this section. Consider an extensive observable (x, "displacement") and an intensive observable (f, "force") of a subsystem, energetically conjugated by the contribution to the First-Law fundamental form (fdx or xdf). The compliance is defined by a convolution for x,

$$f(t')\ J(\tau)\ x(t), \quad x = J * \dot{f}, \tag{2.28}$$

and the general behavior at a dispersion zone is shown in Fig. 12a. The dynamic compliance is obtained, from Eq. 2.28, by a *Fourier Laplace transformation* of $\dot{J}(\tau) = dJ/d\tau(\tau)$ covering the whole dispersion zone,

$$J^*(\omega) = J_g + \int_0^\infty \dot{J}(\tau)\, e^{-i\omega\tau} d\tau = J'(\omega) - iJ''(\omega), \tag{2.29}$$

where the glass limit compliance J_g is extracted to assure the convergence. The "area" obtained by the integration is hatched in Fig. 12b.

Complementarily, the modulus is defined by a convolution for the intensive f (see Figs. 12c, d)

$$f(t)\ G(\tau)\ x(t'), \quad f = G * \dot{x}, \tag{2.30}$$

$$G^*(\omega) = G_g + \int_0^\infty \dot{G}(\tau)\, e^{-i\omega\tau} d\tau = G'(\omega) + iG''(\omega). \tag{2.31}$$

Observe (from Figs. 11d, e and 12a–d) that – inside the same dispersion zone – the extreme of the compliance are at longer times and lower frequencies than for the modulus. The differences on logarithmic scales ($\lg\tau$, $\lg\omega$) are the greater the larger the ratio of glass and equilibrium values ($\lg(G_g/G_e)$, $\lg(J_e/J_g)$). For shear, e.g., at both the main and flow transition of polymers this difference amounts to 3 or 4 orders in τ or ω corresponding to 3 or 4 logarithmic decades. Generally, the modulus underlines the transition to the solid, and the compliance underlines the flow onset which two phenomena are not the same in broad viscoelastic dispersion zones.

Analyzing shear for a true flow transition ($G_e = 0$) then one must take care for the convergence of the integrals, e.g. by two retractions,

$$J^*(\omega) - J_e^0 - i/\omega\eta = i\omega F[J(\tau) - J_e^0 - \tau/\eta] \tag{2.32}$$

where F stands for the Fourier Laplace transform. The symbol $J_e^0 = \lim J'(\omega)$ for $\omega \to 0$ (see also Eq. 2.18 and Figs. 11e and 12e) is called *steady state compliance* or recoverable compliance of a liquid, $J_e^0 > 0$ is typical for a flow transition (or a glass transition in small-molecule materials). It can

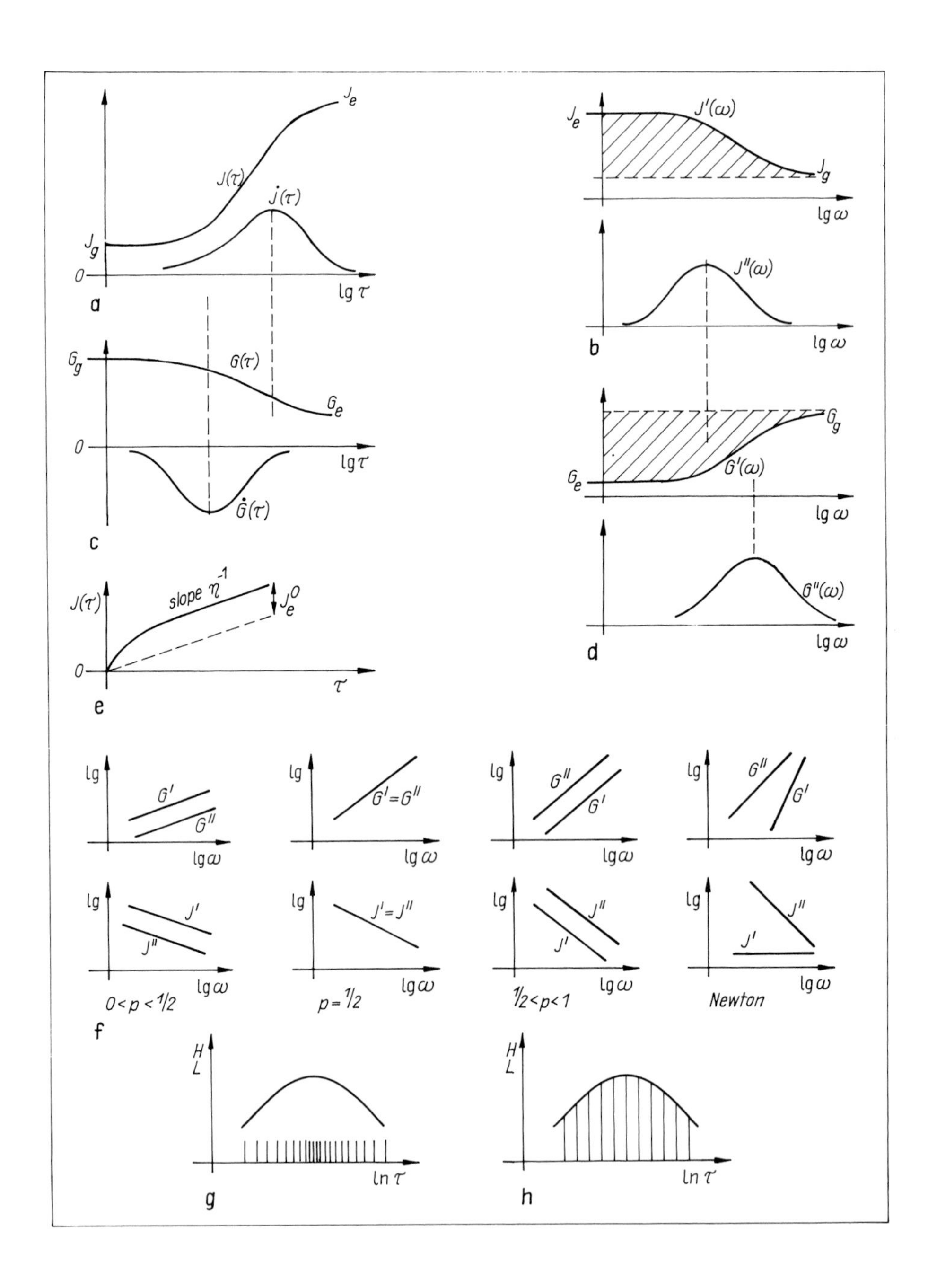

J_e
$J(\tau)$
$\dot{J}(\tau)$
J_g
0
lg τ
a
J_e
$J'(\omega)$
J_g
lg ω
$J''(\omega)$
lg ω
b
G_g
$G(\tau)$
G_e
0
lg τ
$\dot{G}(\tau)$
c
G_g
$G'(\omega)$
G_e
lg ω
$G''(\omega)$
lg ω
d
$J(\tau)$
slope η^{-1}
J_e^0
0
τ
e
lg
G′
G″
lg ω
G′=G″
0<p<1/2
p=1/2
1/2<p<1
Newton
J′
J″
J′=J″
f
H
L
ln τ
g
h

also be calculated from the shear modulus by an integration over the relevant dispersion zone,

$$J_e^0 = \eta^{-2} \int_0^\infty d\tau \, \tau G(\tau), \; (G_e = 0) \tag{2.33a}$$

where the viscosity is given by

$$\eta = \int_0^\infty d\tau \; G(\tau). \tag{2.34}$$

Another useful formula is

$$J_e^0 = \lim_{\omega \to 0} G'(\omega)/G''(\omega)^2. \tag{2.33b}$$

From Fig. 12e we see that J_e^0 is formally defined by

$$J_e^0 = \lim_{\tau \to \infty} \left(J(\tau) - \tau/\eta \right) \tag{2.33c}$$

but we should observe that J_e^0 is related to the whole transition.

Modulus and compliance are connected by Eq. 2.13, $J^*(\omega)G^*(\omega) = 1$. Transformation into the time picture gives

$$\int_{-\infty}^{t} G(t - t')\dot{J}(t')dt' = 1, \quad \int_0^t J(t - t')G(t')dt' = t. \tag{2.35}$$

These formulas can be grasped from Eqs. 2.28 and 2.30 by the slogan: effect as cause must give cause as effect. From Eq. 2.35 we see that always $0 \leqslant G(\tau)J(\tau) \leqslant 1$ if both variables are taken at the same time τ. This product has a minimum in the dispersion zone that reaches rather small values for large dispersion intensities.

Real and imaginary part of dynamic compliance – the same is true for the modulus – are connected by the *dispersion relations* of Kramers and Kronig being purely mathematical implications of the linear and causal material

---

Fig. 12. Linear response scheme.
a. Compliance for extensive observables, $J(\tau)$. b. Dynamic compliance $J^*(\omega) = J'(\omega) - iJ''(\omega)$. c. Modulus for intensive observables, $G(\tau)$. d. Dynamic modulus $G^*(\tau) = G'(\omega) + iG''(\omega)$. e. Illustration of steady state compliance J_e^0 for flow transition. f. Dispersion relation for power laws. $0 < p < 1/2$ solid-like behavior, $p = 1/2$ Rouse case. $1/2 < p < 1$ liquid-like behavior. For the Newton liquid all four exponents are different. g. Realization of a spectrum by a distribution of elements with constant intensities over the $\ln \tau$ axis. h. Realization by adjustment of element intensities at equidistant $\ln \tau$ values.

equations 2.28 and 2.30. Defining a Hilbert transform (see Ref. 43) by the principal value integral

$$H[f(\xi)](\omega) = \frac{1}{\pi} ⨍ \frac{f(\xi)d\xi}{\xi - \omega}, \quad H^{-1} = -H, \tag{2.36}$$

the dispersion relations are

$$J'(\omega) - J_g = H[J''(\xi)], \quad J''(\omega) = -H[J'(\xi) - J_g], \tag{2.37a}$$

$$G'(\omega) - G_g = -H[G''(\xi)], \quad G''(\omega) = H[G'(\xi) - G_g]. \tag{2.37b}$$

From the linear-response network of formulas it follows that from one function of the set

$$\{G(\tau), G'(\omega), G''(\omega), J(\tau), J'(\omega), J''(\omega)\}, \tag{2.38}$$

if known for about 5 logarithmic decades of τ or ω over a dispersion zone (or about 10 . . . 15 decades for main and flow transition in polymers), any other function of the set can be calculated, at least in principle. Some procedures to enhance the accuracy when the start function is known only in two or three decades are described e.g. in Refs. 2 and 44.

The dispersion relations Eqs. 2.37 are not as mysterious as they appear. Extensive tables can be found in Ref. 43, and the following two examples are to get some familiarity with them.

(1) If there is some power law over more than 1 logarithmic decade of ω, with exponent p, $0 < p < 1$, then we have there

$$J' \sim \omega^P, \; J'' \sim \omega^P, \; G' \sim \omega^{-P}, \; G'' \sim \omega^{-P} \tag{2.39a}$$

with $\tan\delta = G''/G' = J''/J' \approx \tan(p\pi/2)$. This means parallel straight lines for J' and J'', and for G' and G'', in a log-log plot, see Fig. 12f, the vertical distance between storage and loss parts is

$$\Delta\lg\tan\delta = \lg\tan(p\pi/2). \tag{2.39b}$$

Measuring both real and imaginary part the exponent p can be determined easier from this distance than from the slope. Care must be taken for $p > 0.9$ because Eq. 2.39a goes over to Eq. 2.18 as $p \to 1$.

(2) For logarithmic broad curves – hardly any details within one decade – the Hilbert transformation acts similar to a logarithmic differentiation: Staverman Schwarzl approximation. Written down for the moduli, e.g., one finds

$$G''(\omega) \approx (\pi/2)\,(dG'(\xi)/d\ln\xi) \quad \text{for } \xi = \omega, \tag{2.40}$$

$$G'(\omega) - G_g \approx -(\omega\pi/2)\,\{d[G''(\xi)/\xi]/d\ln\xi\} \quad \text{for } \xi = \omega, \tag{2.41}$$

$$\delta(\omega) \approx (\pi/2)\,\{d[\ln|G^*|(\xi)]/d\ln\xi\} \quad \text{for } \xi = \omega. \tag{2.42}$$

In case of emergency one has, of course, to look for computer programs with reasonable procedures for smoothing and extrapolating, for handling of flow, and for making accessible measurements at different temperatures, e.g. by WLF scaling, etc.

The general connection between linear response and linear thermodynamics of irreversible processes (TIP) is difficult. The former tries to describe broad time or frequency spectra directly, and the latter looks for a link between flux and force including rate, acceleration, and perhaps higher derivatives. Only for the restricted case of rate is the connection simply mediated by the standard model, see e.g. Ref. 45 chapter 12. If small, a gradient can be modelled by a set of subsystems with smoothly varying state parameters shifting the broad spectra (see Ref. 40). A general connection using a spectrum of internal parameters λ for linear TIP gives nonlinear, complicated formulas between the λ and τ spectra (Ref. 46).

2.4 Relaxation and retardation spectra

A viscoelastic spectrum is a distribution of exponential decays or of exponential switch-ons over the logarithmic time scale, $\ln\tau$, to model the relaxation modulus or the retardation compliance in broad dispersion zones. In other words, the spectrum is a distribution of Maxwell elements for the modulus, or of Voigt Kelvin elements for the compliance. To make the distribution definite, i.e. one-dimensional, two limiting cases are in use. Either we think of a distribution of elements with the same intensity (Fig. 12g), or we think of elements at fixed, equidistant $\ln\tau$ values and adjust their intensities (Fig. 12h, see Ref. 46a). The equilibrium value G_e for the modulus and the glass value J_g and the flow contribution τ/η for the compliance are extracted. The reduction to exponential decays excludes resonances (weakly damped vibrations) from the description, although their observables can also be included into the scheme of linear responses (inclusive the FDT). Thus, we have a *relaxation spectrum H* defined by

$$G(\ln\tau) = G_e + \sum_i G_0\, e^{-\tau/\tau_{0i}} \rightarrow G_e + \int_{-\infty}^{\infty} d\ln\tau_0\, H(\ln\tau_0)\, e^{-\tau/\tau_0}, \tag{2.43}$$

and a *retardation spectrum* L by

$$J(\ln \tau) = J_g + \tau/\eta + \int_{-\infty}^{\infty} d \ln \tau_1 \, L(\ln \tau_1)\,(1 - e^{-\tau/\tau_1}). \tag{2.44}$$

Both spectra can be incorporated into the set of Eq. 2.38.

Similarly to the differences between modulus and compliance, the relaxation spectrum underlines the transition to solid-like behavior at shorter times, and the retardation spectrum underlines the flow onset at larger times. To recall this the dummy variables in Eqs. 2.43 and 2.44 are chosen differently (τ_0, τ_1).

For dispersion activities with moderate intensities (ΔG, ΔJ) and non-zero equilibrium values (G_e, J_e) normalized spectra ($\tilde{H}$, $\tilde{L}$) can be used,

$$H = \Delta G \cdot \tilde{H}, \quad \int \tilde{H}\, d \ln \tau_0 = 1, \tag{2.45}$$

$$L = \Delta J \cdot \tilde{L}, \quad \int \tilde{L}\, d \ln \tau_1 = 1. \tag{2.46}$$

Using the dynamic responses for the standard model, Eqs. 2.29 and 2.24, and after some arithmetic the distribution can be expressed by

$$\begin{aligned}
G'(\omega) &= G_e + \Delta G \int \tilde{H}(\ln \tau_0)\omega^2\tau_0^2/(1 + \omega^2\tau_0^2) \cdot d \ln \tau_0, \\
G''(\omega) &= \Delta G \int \tilde{H}(\ln \tau_0)\omega\tau_0/(1 + \omega^2\tau_0^2) \cdot d \ln \tau_0, \\
J'(\omega) &= J_g + \Delta J \int \tilde{L}(\ln \tau_1)/(1 + \omega^2\tau_1^2) \cdot d \ln \tau_1, \\
J''(\omega) &= \Delta J \int \tilde{L}(\ln \tau_1)\omega\tau_1/(1 + \omega^2\tau_1^2) \cdot d \ln \tau_1.
\end{aligned} \tag{2.47}$$

Eq. 2.47 is also a convenient framework for conversion within the set of Eq. 2.38. The responses can be correlated by a suitable ansatz for $\tilde{H}$ (or $1 - \tilde{L}$), for instance by the Kohlrausch Williams Watts (KWW) formula (stretched exponential)

$$\tilde{H} = \exp\,[-(\tau/\tau_0)^{\beta}],\; 0 < \beta \leqslant 1 \tag{2.48}$$

becoming a Debye relaxation for $\beta = 1$. The spectrum is broadened for smaller β parameter, its spectral width $\Delta \lg\omega \sim \beta^{-1}$. The relation $\beta < 1$ is occasionally used as a synonym for "broader than Debye relaxation".

Since the conception of spectra originates from mechanical models the spectra do not have a direct molecular interpretation. The spectra accentuate the peculiarities of response because H and L approximately correspond to logarithmic differentiations of G and J, respectively. The experiment gives, by means of the FDT, spectral densities and correlation functions. The latter can also directly be correlated by functions like Eq. 2.48, and some care should be taken to find out the variable that is really used for adjustment in a given paper.

2.5 Correlation function and spectral density

We consider now variables that can directly be determined from a thermodynamic experiment (see Secs. 2.6 and 3.7, below). Consider the stationary fluctuation of e.g. an extensive observable $x(t)$ in a subsystem, with the additive normalization $\overline{x(t)} = 0$, i.e. the equilibrium average (x_e) subtracted. The *correlation function* $x^2(\tau) \equiv \varphi_x(\tau)$ is defined by a time average,

$$x^2(t) \equiv \varphi_x(\tau) = \lim_{T \gg \tau'_R} \frac{1}{T} \int_{(T)} dt\, x(t)\, x(t+\tau) \equiv \overline{x(t)x(t+\tau)}. \quad (2.49)$$

To cover all correlations of a dispersion zone with characteristic retardation time τ'_R the experiment time T must be larger than τ'_R. The correlation function $\varphi(\tau)$ tells us how the observable at times $t \pm \tau$ is correlated to the observable at t; $\varphi = 0$ ($\tau \neq 0$) means statistical independence of $x(t+\tau)$ and $x(t)$. For the stationary case ($\varphi(\tau)$ does not depend on the time where the interval T is chosen) the correlation function has the following properties that are merely mathematical implications of the definition Eq. 2.49 (see Ref. 47):

$$\varphi(\tau) = \varphi(-\tau), \overline{x^2} = \varphi(0), \quad |\varphi(\tau)| \leqslant \varphi(0). \quad (2.50)$$

Observe that the time conception of response ($J(t)$) and correlation function ($\varphi(\tau)$) is different. For $J(\tau)$ one can imagine a clock standing by the creep experiment, for $\varphi(\tau)$ the time τ is however composed from many intervals τ at $t_1, t_2, \ldots$, so to say from many experimental acts, correlated in a "vital" internal experiment. The correlation function of an intensive observable is defined analogously,

$$f^2(\tau) \equiv \varphi_f(\tau) = \overline{f(t)f(t+\tau)}. \quad (2.51)$$

Converting the time picture into the frequency picture we come to the *spectral density* defined by the Fourier components

$$x^2(\omega) = \frac{1}{2\pi} \int_{-\infty}^{\infty} x^2(\tau)\, e^{i\omega\tau} d\tau, \quad (2.52a)$$

$$f^2(\omega) = \frac{1}{2\pi} \int_{-\infty}^{\infty} f^2(\tau)\, e^{i\omega\tau} d\tau. \quad (2.52b)$$

The following properties are implications of this definition, again for the stationary case (Ref. 47),

$$\overline{x^2} = \varphi(0) = \int_{-\infty}^{\infty} x^2(\omega)\, d\omega, \quad x^2(\omega) = x^2(-\omega),\; x^2(\omega) > 0, \tag{2.53}$$

where negative ω are formal constructions resting on the symmetry $\varphi(\tau) = \varphi(-\tau)$ of Eq. 2.50. Negative ω can be avoided by use of the Fourier cosine transformation,

$$x^2(\tau) = \varphi(\tau) = 2\int_0^{\infty} d\omega \cos(\omega\tau)\, x^2(\omega),$$
$$x^2(\omega) = \pi^{-1}\int_0^{\infty} d\tau \cos(\omega\tau)\, \varphi(\tau). \tag{2.54}$$

The spectral density for intensive variable fluctuation is defined analogously. The interesting property $x^2(\omega) > 0$, $f^2(\omega) > 0$ comes from the concept of positive definiteness underlying the correlation function and has led to the name density. (Observe that $\varphi(\tau) < 0$ is allowed). Typical plots of correlation function and spectral density are shown in Figs. 13a, b. For polymers one has to pay attention to subtle flank effects. The multiplicity of glass transition implies steps in both functions that for shear can only be represented in log-log plots because of the different orders in magnitude for time/frequency as well as for the functions themselves, see Figs. 13c, d. Finally we should observe the essential differences between response and stationary fluctuation, see Figs. 13 d, e. The response is characterized by a time arrow ("after" the step), by thermodynamic causality in the sense of Eqs. 2.3 and 2.4, and by dissipation (or heat for the case x = entropy and f = temperature). In contrast, the thermodynamic fluctuation is reversible (although in vital time), correlative, and is not connected with any dissipation.

2.6 Fluctuation dissipation theorem (FDT)

The importance of an equation is the greater the more different the conceptions underlying the right and the left hand side. The *fluctuation dissipation theorem* (FDT, Nyquist 1928, Ref. 48) is, therefore, one of the great equations in physics since such different concepts as fluctuation and response are considered to be equivalent. For energy conjugate variables (xdf or fdx

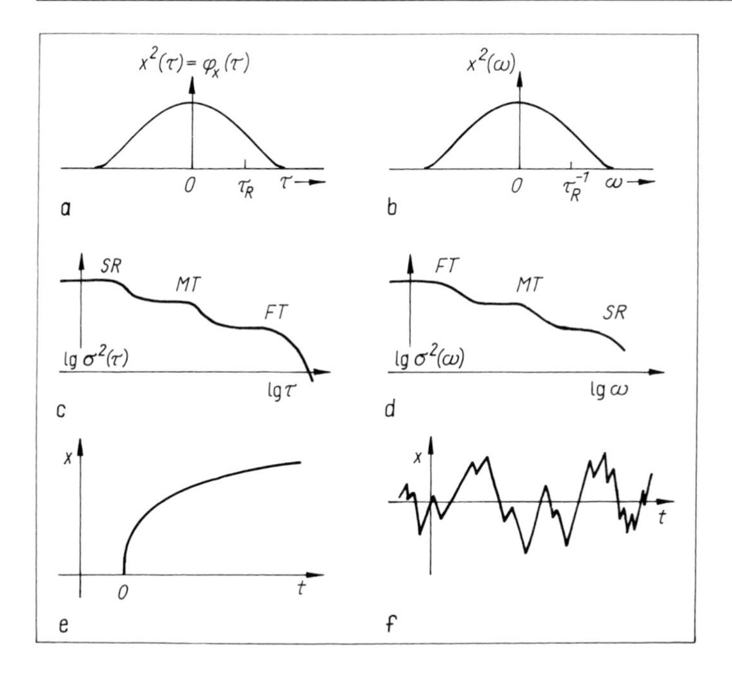

Fig. 13. Correlation function and spectral density.
a. Typical behavior of a correlation function $\varphi_x(\tau) = x^2(\tau)$ in a simple dispersion zone of a solid. b. Corresponding spectral density $x^2(\omega)$. c. Shear stress correlation function in an amorphous polymer. SR secondary relaxation, MT main transition, FT flow transition. d. Corresponding spectral density. e & f. Comparison of response and fluctuation.

in the First Law) in the classical (non-quantum mechanical) case the FDT states that

$$x^2(\tau) = -kT(J(\tau) - J_e), \tag{2.55}$$

$$f^2(\tau) = kT(G(\tau) - G_e), \tag{2.56}$$

$$x^2(\omega) = kT\,J''(\omega)/\pi\omega, \tag{2.57}$$

$$f^2(\omega) = kT\,G''(\omega)/\pi\omega. \tag{2.58}$$

In the linear region, the correlations of fluctuations on the l.h.s. directly determine the response on the r.h.s. The ratio is given, generally speaking, by the "molecular quantum" kT = correlation divided by response (Nyquist theorem). The term "dissipation" in the FDT comes from the loss parts of response in Eqs. 2.57 and 2.58.

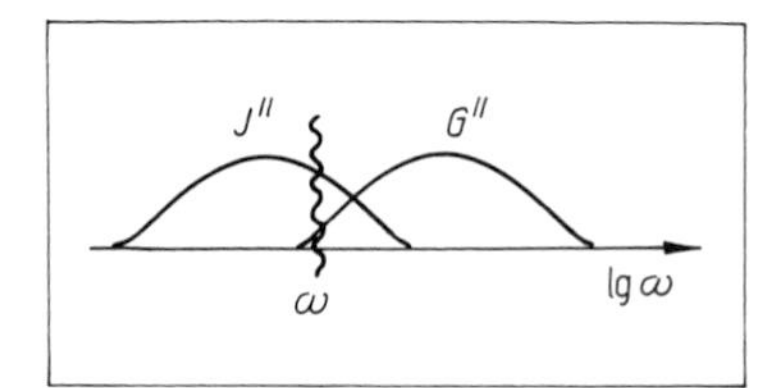

Fig. 14. ω identity of the FDT.

Integration over the frequency region of a dispersion zone gives the retardation intensity (see also Eq. i. of the Introduction)

$$\Delta J = J_g - J_e = \overline{x^2}/kT = \varphi_x(0)/kT$$

$$= (\partial x/\partial f)_{f\to 0} = \frac{2}{\pi}\int_{\omega=0}^{\infty} J''(\omega) d\ln\omega = \frac{2}{kT}\int_0^{\infty} x^2(\omega)\, d\omega. \quad (2.59)$$

An analogous equation is true for relaxations (ΔG), since the FDT is rather symmetric relative to the intensive–extensive complementarity.

In a sense, the FDT can be considered as an equation for a thermodynamic experiment: The result of the experiment, the observable, is a susceptibility represented by the responses J and G. The response realizes correlations that are not disturbed in the linear range.

This means that the following picture is misleading here: The external program would generate a new state that relaxes into equilibrium. The FDT instead shows that the external program makes observable only such correlations that also occur in the equilibrium, with no program; for subsystems any possible "external program" is already contained in the stochastic fluctuative "disturbances" from the heat bath (i.e. from the equilibrium fluctuations of environment or from inside). This situation will be called "internal experiment" (see also Sec. 3.7 below). Thus the conceptions of fluctuation, relaxation or retardation, dispersion, and molecular dynamics are synonymous in the framework of the (linear) FDT.

Nonlinearity means that the disturbance of subsystems is not contained in their natural disturbance from dynamic equilibrium.

There is a rather trivial but important aspect regarding the frequency relation between fluctuation and response, see Fig. 14. In the mechanical standard model (Fig. 11d) we observe two distinct peaks, one for the compliance J'' and the other for the modulus G''. One could think, again misleadingly, that this model is one uniform process with only one internal time where the two times (τ_0, τ_1) result only from the different disturbances of stress relaxation and creep experiments. The FDT, however, is much sharper. Any frequency ω of the response side of the FDT ($J''(\omega)/\omega$ or $G''(\omega)/\omega$) is exactly the same as in the fluctuation side ($x^2(\omega)$ or $f^2(\omega)$). This property is called *ω identity*.

It is not the point here that and how the compliance can be converted into the modulus, and inversely (see Eqs. 2.13 and 2.35), but the point is that the J'' and G'' peaks are really at different frequencies and, therefore, that compliance and modulus are differently sensitive to different frequencies, and that the different peak frequencies are differently sensitive to the parameter field of molecular motion or the length scales. Modulus and compliance from the same term of the linear form in the First Law are different activities (see Secs. 2.7 and 2.8, below).

Since the ω identity is also true in the quantum-mechanical amplification of the FDT (see e.g. Ref. 45) – but not in the amplification of the time picture as could be suggested by Eqs. 2.55–56 – it seems that the frequency plays a particular role in molecular situations (see also Secs. 2.8 and 3.7, below).

The extraordinary role of the frequency, as compared to the scattering vector q, is also demonstrated in the FDT for the scattering function (Refs. 49, 50),

$$S(q, \omega) \quad = \quad kT\chi''(q, \omega)/\pi\omega, \tag{2.60}$$

where $\chi''(q, \omega)$ is the loss part of a suitably defined susceptibility. The frequency is displayed by the extra factor ω^{-1} while the scattering vector behaves like an index when Eq. 2.60 is compared to Eq. 2.57: Eq. 2.60 is a q partition of Eq. 2.57.

Doing without positive, the "integral" FDT, Eq. 2.59, can be decomposed into r components. After Fourier transformations, $\omega \rightarrow t$, $q \rightarrow r$, and putting $t = 0$, we obtain from Eq. 2.60 the compressibility equation Eq. 1.30, a partition into r components. Thus, at least for case of density fluctuations, the compressibility can be discussed as a sum of contributions from Δr or Δq intervals.

Consider two situations for a logarithmic broad dispersion zone: (1) An ensemble of mutually isolated local mechanisms corresponding, each by itself, to independent standard models where the broad spectrum comes from a broad distribution of τ_0 parameters (activation energy distribution), and (2) a cooperative process including e.g. 100 particles where the broad spectrum comes from the mode parameter distribution inside the cooperativity range. The FDT, and therefore the thermodynamic experiment cannot distinguish between the two situations. Without q resolution from scattering we need indirect arguments to explore cooperativity from linear response.

Onsager's cross relations between cross susceptibilities (see e.g. Ref. 45) are time or frequency decompositions of the thermodynamic Maxwell relations, e.g. $(\partial S/\partial p)_T = -(\partial V/\partial T)_p$. They are based on properties of cross correlation functions $\langle xy(\tau)\rangle$ and on fluctuation reversibility. It should be underlined that x and y must come from the stock of observables that are contained in one fundamental form of the First Law.

2.7 Different activities

A solution of the cooperativity problem just mentioned is usually tried with the help of a molecular model permitting the prediction of different experimental results such as shear, bulk, dielectric . . . moduli and/or compliances. Partial adjustment of model parameters and/or model structure usually improves the prediction and, finally, the belief in the model's significance is the better the larger the time scale where the response can be reproduced and the higher the number of adjustable responses. This leads to the concept of *activities*. Referring to the FDT, a thermodynamic experiment does not determine a unique spectrum per se, but one single response. Different experiments determine different responses, also in the same state of the subsystem. We find different correlation functions or spectral densities from the corresponding responses, such as the fluctuation of dielectric polarization from the compliance $\varepsilon^*(\omega)$, or the fluctuation of a subsystem temperature from the temperature modulus $K_T{}^*(\omega)$, see Tab. II. [After modification, some NMR responses can also be included in this table. Magnetization is an extensive variable. The longitudinal NMR relaxation time T_1 can approximately (Ref. 51) be connected with a local field fluctuation. This means that $1/T_1(\omega)$ behaves like a field loss modulus.]

Denoting the activities by A, B, C, . . . ; such as $A^2(\omega)$ for the spectral density or $R_A(\tau)$, $R_A^*(\omega)$ for the response, then the main experience is that different activities are not necessarily at the same frequency, also within one single dispersion zone, at the same temperature, density, composition etc., see Refs. 52 to 54: They can have different shapes and different typical times or frequencies, see Figs. 15a and b. From this point of view, conjugate compliance and modulus are also different activities, see Fig. 14 above.

On the other hand, a dispersion zone does correspond to a certain basic

Table II. Examples for activities[1]

fluctuation of	program	response	activity symbols
entropy S	$T(t)$	entropy compliance	$J_s(\tau)$, $J_S^*(\omega) \equiv C_p^*(\omega)$
temperature T	$S(t)$	temperature modulus	$K_T(\tau)$, $K_T^*(\omega) = 1/J_S^*(\omega)$
volume V	$p(t)$	bulk compliance	$B(\tau)$, $B^*(\omega)$
pressure p	$V(t)$	compression modulus	$K(\tau)$, $K^*(\omega) = 1/B^*(\omega)$
shear angle γ	$\sigma(t)$	shear compliance	$J(\tau)$, $J^*(\omega)$
shear stress σ	$\gamma(t)$	shear modulus	$G(\tau)$, $G^*(\omega) = 1/J^*(\omega)$
dielectr. polariz. P	$E(t)$	dielectric permittivity	$\varepsilon(\tau)$, $\varepsilon^*(\omega)$
electric field E	$P(t)$	dielectric modulus	$M(\tau)$, $M^*(\omega) = 1/\varepsilon^*(\omega)$

[1] referring to the First-Law fundamental form

$dU = Tds - pdV - V\sigma d\gamma + VEdP$.

General extensive compliances J in the sense of this book are $J_S = C_p = V\varrho c_p$ with c_p specific heat capacity (J/K · kg), $V \cdot B$, $V \cdot J$, and $\alpha(\omega) = V\varepsilon_0(\varepsilon - 1)$.

molecular mechanism with typical modes in a typical time or frequency interval for a given state. This interval can be defined independently from the thermodynamic experiment e.g. by a computer simulation from molecular mechanics. This frequency interval will be called *primary spectrum.* The connection between this spectrum and the thermodynamic activities from experiments is given by the ω identity of the FDT. For classical conditions (no $\hbar$ in the equations) this concept can also be applied to times τ. Observe again that there is no ω or τ shift in this connection.

The distinction between the one primary spectrum and the many activities is of crucial importance for the analysis of the WLF equation (Secs. 4.2.3 and 6.3 below). Consider, for one dispersion zone, the shift of activities and of primary spectrum with the state e.g. with increasing the temperature. Then we perhaps observe that all the activities of the dispersion zone are guided by one universal scaling, lg $\omega(T)$. This can only be explained by the fact that it is the primary spectrum that is universally guided. Then from the ω identity

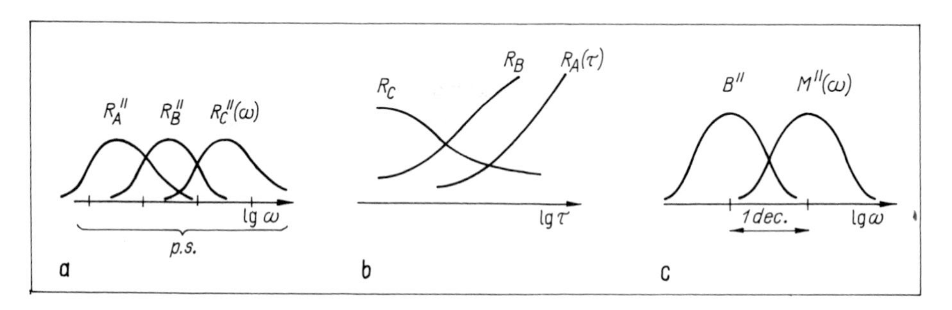

Fig. 15. Different activities.
a. Loss parts of different activities (A, B, C) in one dispersion zone with cooperativity of molecular motion, schematically. The horizontal brace indicates the primary spectrum p.s. of the zone. b. Corresponding responses. A, B compliances, C modulus. c. Comparison of loss bulk compliance $B''(\omega)$ from dynamic dilatometry and loss part of longitudinal wave modulus $M''(\omega)$ from dynamic light scattering (photon correlation spectroscopy) for PVAC at about $T = 50°C$, normalized to the same height.

it follows that not only the maxima of $R_A^*(\omega)$, $R_B''(\omega)$ etc. but also the whole curves are guided by the universal scaling law for lg $\omega(T)$.

How to handle activities which are composed from two simpler activities? Examples are the longitudinal wave modulus

$$M(\tau) = K(\tau) + (4/3)G(\tau) \tag{2.61a}$$

(interesting for ultrasonic experiments) or the linear creep compliance

$$D(\tau) = (1/3)J(\tau) + (1/9)B(\tau) \tag{2.61b}$$

where $D^* = 1/E^*$ with E^* the Young modulus. There are no problems as long as only such linear formulae are used. But, as shown in Fig. 15c (Ref. 55), one can try to compare e.g. the bulk compliance $B''(\omega)$ to the longitudinal modulus $M''(\omega)$ in the main transition. In general, the conversion is not linear, and the calculation would give quite different results for starting from frequency or the time picture (Ref. 44). For example

$$E^*(\omega) = 9K^*(\omega)G^*(\omega)/(3K^*(\omega) + G^*(\omega)) \tag{2.61c}$$

gives

$$\int_0^t E(t - u)\,[3K(u) + G(u)]\,du = 9\int_0^t K(t - u)G(u)\,du \tag{2.61d}$$

quite different from an equation that would follow from Eq. 2.61c when

$X^*(\omega)$ is simply substituted by $X(\tau)$, $X = E, K$, or G. The rule is (Ref. 56): always start from the frequency picture. This corresponds to the ω identity of the FDT.

2.8 General scaling principle (GSP). Mode lengths

This section is devoted to the following hypothesis: In general, independently from the activities, different regions of the primary spectrum (Δ lg ω at lg ω) can be connected with different typical lengths that characterize typical molecular modes. Then the larger lg ω the smaller the mode length (*General Scaling Principle*, GSP).

Introducing the Fourier scattering vector q instead of length, this principle means that there is a *dispersion law* $\omega(q)$ with $d\omega/dq > 0$. Furthermore, this principle could also mean that in case of several dispersion zones with typical pairs (ω_1, q_1), (ω_2, q_2) always $q_1 > q_2 > \ldots$ if $\omega_1 > \omega_2 > \ldots$.

Of course there is some uncertainty in ω and q because of the parameter distribution or the cooperativity of molecular motion in disordered materials (its so-called micro-Brownian character). For relaxations we can assume that the width of the Lorentz curve (Eq. 2.26) is a reasonable measure, i.e. $\Delta \lg \omega \approx 1$, the corresponding uncertainty of lg q can be estimated from the dispersion law, $\omega(q)$.

Always remembering this uncertainty we will define lgω as molecular *mobility*, similar to the concept as used for NMR relaxations. An interval Δ lg ω at lg ω will be called primary *mode*, and its typical length corresponding to lg ω by the dispersion law is called *mode length* λ, see Fig. 16g.

Then, as a consequence, we can expect the order $\lambda_A > \lambda_B > \lambda_C$ for the activities of Figs. 15a, b, and the order $\lambda_J > \lambda_G$ for compliance J and modulus G of Fig. 14 if the dispersion zone has cooperative character, or, more general, a dispersion law.

The main problem of the dispersion law relative to linear response is that the latter does not give any direct information about q regions, i.e. about mode lengths. In the common sense, the thermodynamic response experiment instead refers to the so-called thermodynamic limit $q \to 0$. The question, however, whether the response indirectly contains information about length scales or not is the question of selective sensitivity of different activities to different mode lengths.

Example 1. As will be shown in Sec. 5.4 the shear response of Rouse modes

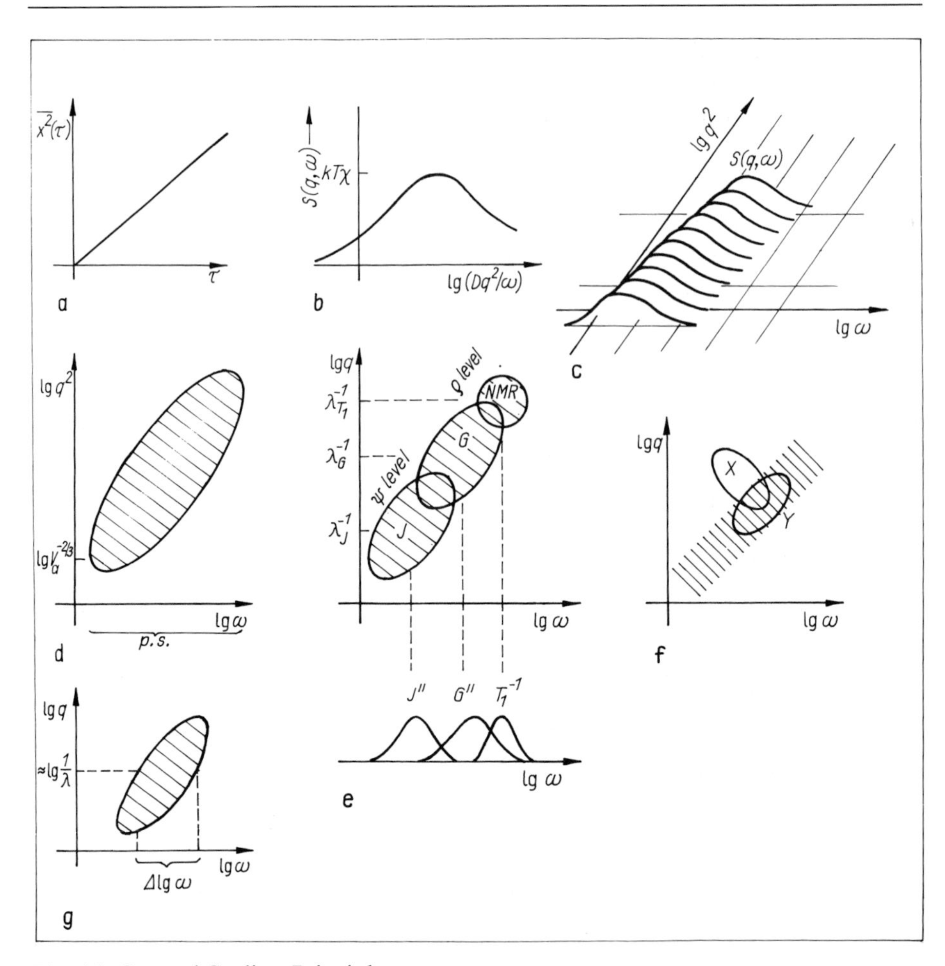

Fig. 16. General Scaling Principle.
a. Scaling of "compliance" $\overline{x^2}(\tau)$ for Brownian motion. b. Scaled scattering function for Brownian motion. D diffusion coefficient, χ susceptibility. c. Dispersion law for the transport zone of Brownian motion visualized by the scattering function. d. General dispersion law of a dispersion zone for relaxation and retardation. The brace indicates the primary spectrum p.s. The small-q limit is given by the sensitivity of thermodynamic experiments (see Sec. 3.7, $\xi_a \sim V_a^{1/3}$ cooperativity or characteristic length). e. Sensitivities of different activities to different regions of the dispersion law. The projection on the lg ω axis gives the responses and the projection on the lg q-axis the reciprocal mode length (regions). J and G refer to shear. f. Hypothetical counter example violating the general scaling principle. h. The concept of mode length λ: "Δ lg ω at ω = mode".

is rather broad, and increasing relaxation times can be linked with increasing mode lengths according to $\tau \sim Z_r^2 \sim \lambda^4$.

Example 2. Consider the Brownian motion of a particle. The mean shift according to Eq. 1.5, $r^2(t) = 6Dt$, is quasi-extensive and can be interpreted as a compliance, see Fig. 16a. Provided we have an activity A sensitive to λ_A, and another activity B sensitive to the length λ_B, then the typical times for the responses would be τ_A and τ_B with the scaling $\tau_A/\tau_B \sim (\lambda_A/\lambda_B)^2$. Inversely, knowing only τ_A and τ_B, the relation of length scales can be estimated from this proportionality, and the absolute λ could be estimated from the diffusivity D.

The latter example can be ramified. Investigate the Brownian motion with a current wave and consider the scattering function. For the evolution of concentration $\psi(\mathbf{r}, t)$ we have Eq. 1.6, i.e.

$$\partial\psi/\partial t - D\Delta\psi = 0. \tag{2.62a}$$

The correlation function for this order parameter,

$$S(r, t) = \overline{\psi(\mathbf{r}, t)\psi(0, 0)}, \tag{2.63}$$

with the time average over an approximately homogeneous subsystem at $\mathbf{r}$, obeys the same diffusion equation

$$(\partial/\partial t - D\Delta)S(r, t) = 0. \tag{2.62b}$$

The FDT defines the boundary conditions for this equation,

$$\lim_{q\to 0} S(q, t = 0) = kT\chi, \tag{2.64a}$$

with χ a susceptibility. [In the real space picture we have, after definition of a conservative current density

$$j = \nabla\psi, \nabla j(\mathbf{r}, t) + \partial_t\psi(\mathbf{r}, t) = 0, \tag{2.64b}$$

for χ the following (Kubo) integral

$$D \cdot \chi = \frac{kT}{6}\int dr \int_{-\infty}^{\infty} dt\, \overline{j(\mathbf{r}, t)j(0, 0)}. \tag{2.64c}$$

Other activities (A, B, . . .) need the definition and calculation of other current correlations with j_A, j_B, . . .]. Eq. 2.62b can be integrated in the (q, z) space with z the complex frequency (see Ref. 49 or App. 1), that is $\partial/\partial t \to z$, $\nabla \to q$,

$$S^{-1}(q, z) = (ikT\chi)^{-1}(z + iDq^2). \tag{2.65}$$

The return to the real frequency ω needs some care on the complex plane. Then

$$S(q, \omega) = 2kT\chi\, Dq^2/(\omega^2 + (Dq^2)^2). \tag{2.66}$$

Plotted against log x, $x = Dq^2/\omega$, we obtain a Lorentz curve, Fig. 16b. The maximum is at $x = 1$, i.e.

$$\omega = Dq^2 \tag{2.67}$$

corresponding to the scaling law of Example 2. In total, we find a Lorentzian wall over the lg Dq^2–lg ω plane with the dispersion law Eq. 2.67 and an uncertainty of about one logarithmic decade, see Fig. 16c. The vertical projection of this wall defines a stripe in the lg q^2–lg ω plane, its width indicates the uncertainty of the definitions for modes and mode lengths.

In general, the stripe for an arbitrary dispersion zone can have another dispersion law and a modified uncertainty. The stripe for the diffusion is only restricted at large q values corresponding to mode lengths of the ϱ level (e.g. 1 or 0.5 nanometers). For small q it can be continued to large λ, and this can be measured by scattering experiments. But in a thermodynamic (linear response) experiment, the activities will become very small for the flow zone (e.g. $G' \sim \omega^2$, $G'' \sim \omega$, see Eq. 2.18) so that the stripe is also restricted from below by the sensitivity of the thermodynamic experiment (mode length ξ_a corresponding to a volume V_a, see Fig. 16d). This point is further discussed in Secs. 3.6, 3.7 and 6.10.1.

The general scaling hypothesis gives information about mode length from linear response if the different activities are oriented to the stripe such as indicated in Fig. 16e. The different activities, e.g. the peaks of shear loss compliance $J''(\omega)$, shear loss modulus $G''(\omega)$, and longitudinal NMR relaxation time, $T_1^{-1}(\omega)$, are projections on the lg ω-axis mediated by the FDT. This is the meaning of the thermodynamic limit $q \to 0$. We can see from Fig. 16e how the different activities reflect different q regions, that is different mode lengths λ_J, λ_G and λ_{T1} in the dispersion zone.

A counter example is indicated in Fig. 16f. It does not represent a very probable situation for a response X.

As for the activities represented in Fig. 16e we can argue as follows. The mode lengths sensitive to NMR relaxation are short and correspond to large q and ω in the dispersion law, ultimately because of the short range of dipolar and quadrupolar interaction between the nuclear spins. The main NMR activity corresponds to the ϱ level, and therefore molecular-mechanical models are very helpful. [Of course there are also spurious NMR sensitivities to larger length scales corresponding to smaller q and ω of the dispersion zone. A review for the state of the art is Ref. 57. Special NMR methods can also deliver valuable information about networks, entanglements etc.] Shear response can be measured over many orders of intensity delivering information about larger q or ω regions. The modulus underlines shorter, and the compliance longer, length scales. This means that, especially for the latter, the models for shear are coarse and usually in the ψ level.

At present the concept of general scaling (dispersion law) in the 10-nm range can only be checked by inelastic scattering for times smaller than microseconds ($\nu > 1\,\mathrm{MHz}$), e.g. measuring $S(q, t)$ by neutron spin echo or time-of-flight methods. [Slow evolutions in structure (ψ level, e.g. spinodal phase decomposition or crystallization) can, of course, be followed in the minute scale, e.g. with X rays from a synchrotron.] Therefore, at present, the cooperative molecular motion in well shaped main and flow transitions below the splitting point ($\lesssim$ MHz) cannot be resolved in q or real space and we must rely on indirect length information from linear response with only time or frequency resolution.

Further details about the length scale of thermodynamic subsystems are discussed in Secs. 3.5 to 3.9 below.

3. Thermodynamics

Spatial aspects of thermodynamics form the subject of this chapter. The central concept is the subsystem and the relationship to its environment. At the end of the chapter we are led to temporal aspects again (functional subsystem).

In the first four sections we consider simple examples – phantom network and Flory Huggins mixture, each ideal in a sense – to draw the attention to some peculiarities of polymer thermodynamics.

3.1 Network thermodynamics

Consider a small (linear) elastic deformation of a weakly crosslinked network. A network of Gauss chains is called a phantom network because there is no obstacle against chain crossing. Consider one Gauss chain. Its contribution to the free energy of deformation is of order one kT (Eq. 1.18) which corresponds to one degree of freedom, and is of entropic nature (entropy elasticity). The equation of state for an ideal atomic gas is also of entropic nature,

$$pV = nkT. \tag{3.1}$$

Therefore, a similar equation is expected for entropy elasticity of networks although gas expansion and rubber stretching are different things. The problem is the carrier of identity, that is the unit that can substitute the place of one gas atom. If all chains are crosslinked then we have only one big molecule surely too large for one k. Choosing one monomeric unit or one Kuhn step as identity carrier, however, we would also be confronted with the one k in Eq. 1.18 from the other side because there were too many units. A reasonable compromise seems to be the choice of an average network chain between two crosslinks (molecular mass M_c) as identity carrier, see Ref. 58. Then, similar to Eq. 3.1, we have

$$f = \frac{\rho RT}{M_c} D \quad \text{or} \quad f = \frac{\nu}{V} kTD \tag{3.2a}$$

with f the elastic stress, D the (small) deformation, ν the number of network chains, and V the volume. The elasticity (shear or Young) modulus is then

$$G = \rho RT/M_c. \tag{3.2b}$$

The same order of magnitude is obtained from differentiation of Eq. 1.18 for the free energy of one single Gauss chain (of the phantom network). Putting $r^2 = x^2 + y^2 + z^2$ then

$$\partial F/\partial x = 3kTx/R_0^2 \approx kT\lambda/R_0, \quad \lambda = x/R_0. \tag{3.3}$$

In this treatment, instead of the identity question, an assumption is needed how the external deformation is distributed to small chain parts. Assuming that the ensemble of crosslinks is deformed affinely to the material sample (this means that the space defined by the crosslinks is deformed exactly as the sample itself, for $(x, y, z) \rightarrow (\lambda_1 x, \lambda_2 y, \lambda_3 z)$), then the formula Eq. 3.3 can be applied to the deformations of the mutually interpenetrating and crossing chain parts of the phantom network. Therefore, for all chains, i.e. for the crosslinked material, apart from a numerical factor, Eq. 3.2a or b is again obtained.

3.2 Discussion

It is well known that weakly crosslinked polymers of flexible chains can largely be stretched, up to several 100 per cent (rubber elasticity). The tensor of deformation ratios (with principal values λ_i, $i = 1, 2, 3$) is used for description, a ratio of lengths with and without stress (other name: strain parameter). The typical behavior for simple extension is shown in Fig. 17a. There are two main problems: (1) the exact relationship between the shear modulus near $\lambda = 1$, G, and structure parameters of network and chains, the so-called front-factor problem, and (2) the explanation of the curve form for large λ and for $\lambda < 1$. A more sophisticated estimation than Eq. 1.18 includes the Gauss distribution $\phi(\boldsymbol{r}_i - \boldsymbol{r}_j)$ of Eq. 1.11. Substituting $\boldsymbol{r}_i - \boldsymbol{r}_j$ by $\boldsymbol{R}$ for chain "end-to-end" distance (between the crosslinks of a phantom network) then, with the entropy density $S = \phi \ln \phi$, the free energy of rubber elasticity for affine deformation is

$$F = -kT \int d^3\boldsymbol{R}\, \phi(\boldsymbol{R}) \ln \phi(\lambda \boldsymbol{R}) \tag{3.4}$$

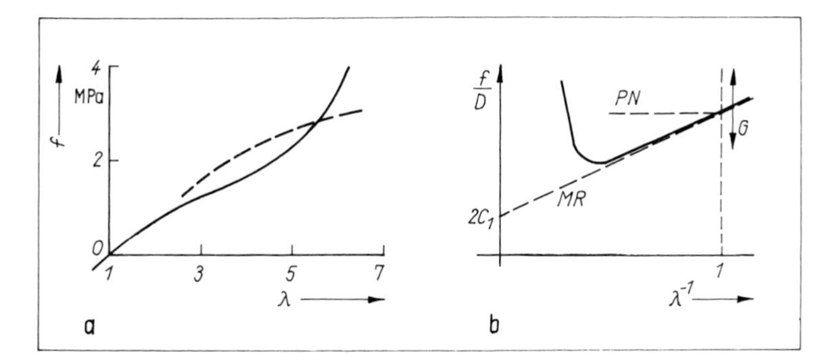

Fig. 17. Rubber elasticity of networks.
a. Stress f vs. strain ratio λ for a typical gun-vulcanized rubber. The slope of the curve at $\lambda = 1$ is the modulus G. The broken curve is $G(\lambda - 1/\lambda^2)$. b. Mooney plot. PN Gaussian phantom network, MR Mooney Rivlin equation, Eq. 3.11, ↕ G symbolically indicates the front factor problem.

where λ is the λ tensor

$$\lambda = \begin{pmatrix} \lambda_1 & 0 & 0 \\ 0 & \lambda_2 & 0 \\ 0 & 0 & \lambda_3 \end{pmatrix} \tag{3.5}$$

with the principal deformation ratios. The integral of Eq. 3.4 (see e.g. Ref. 58a) gives the Kuhn Flory formuia

$$F = (1/2)\nu kT\Sigma\lambda_i^2 \tag{3.6}$$

with a front factor that is one per definition. Large deformations are now described by the first invariant of the λ tensor. As for the invariants consider a deformation with principal axes not parallel to the x, y, z axes. Then the tensor contains off-diagonal elements. But all experimental results must be invariant against a virtual rotation of the coordinate system. Being a scalar, the free energy can only depend on scalar combinations of tensor components that are invariant against the virtual rotation. One finds three invariants. Expressed in terms of principal values they are

$$I_1 = \sum \lambda_i^2, \quad I_2 = \sum_{i<j} \lambda_i\lambda_j, \; I_3 = \lambda_1\lambda_2\lambda_3 = V(\lambda)/V_0, \tag{3.7}$$

with $V(\lambda)/V_0$ the relative volume expansion. For isotropic extension (such as for swelling, Sec. 9.5) we have $\lambda_1 = \lambda_2 = \lambda_3 = \lambda = l(\lambda)/l_0$, i.e. $I_1 = 3\lambda^2$, and, from the Kuhn Flory formula (Eq. 3.6) we obtain

$$\Delta F = \tfrac{3}{2}NkT(\lambda^2 - 1) \tag{3.8}$$

with $F(\lambda = 1)$ subtracted. For simple extension ($\lambda_1 = \lambda, \lambda_2 = \lambda_3 \neq 1$) with no compressibility assumed ($I_3 = 1$) we have $I_3 = \lambda\lambda_2^2 = 1$, $\lambda_2 = \lambda_3 = \lambda^{-1/2}$, $I_1 = \lambda^2 + 2/\lambda$, and from Eq. 3.6,

$$\Delta F = NkTw(\lambda) \quad \text{with} \quad w(\lambda) = \frac{\lambda^2}{2} + \frac{1}{\lambda} - \frac{3}{2}, \tag{3.9}$$

normalized to $w(1) = 0$.

Thus, for general reasons, ΔF is a function of I_1, I_2, and I_3, and it is the Gaussian assumption which implies that, for the time being, only the first invariant plays a role. For simple extension and "compression" ($\lambda < 1$) we have Eq. 3.2a with $D = dw/d\lambda$, i.e.

$$D = \lambda - 1/\lambda^2, \tag{3.10}$$

whereas there remains some uncertainty with the front factor G. The experimental results are usually correlated in a Mooney diagram, Fig. 17b, f/D vs. $1/\lambda$. The Gaussian ψ level estimation of Eqs. 3.2 and 3.9 would give a constant value in this diagram. Typical experiments, however, show after some slope approximately described by a straight line,

$$f/D = 2(C_1 - C_2/\lambda) \tag{3.11}$$

(Mooney Rivlin equation, $C_1 - C_2 = G/2$), a steep upturn for larger deformation (smaller $1/\lambda$).

This behavior is often discussed in terms of a "hindrance" of fluctuation for crosslinks, Refs. 59, 60, or for chains, Refs. 61 and 62. In the zero stress state, the crosslink fluctuation is of order the mean chain end-to-end distance between the crosslinks (Ref. 63). The free room for movement of crosslinks or chains is gradually narrowed for increasing deformation, and ρ level arguments should be used to describe the effects. One expects a short range orientational correlation between chain parts (nematic coupling), especially in the environment of the crosslinks, and trapping of entanglements for large-M_c networks.

Thermodynamically we obtain a typical van der Waals situation (Ref. 64). Besides the entropic contributions – the larger the deformation is the more k's are switched on in Eq. 1.18 – we find increasing possibilities for energetic contributions. The nematic coupling leads to additional attractions between segments (tendency to "condensation" or even crystallization), and the decreasing free room leads to effective repulsion between the segments. The former corresponds to the a term, and the latter to the b term of the van der Waals equation for small molecule fluids, $(p + a/V^2)(V - b) = RT$, or rearranged,

$$p = RTV^{-1}\left(\frac{b^{-1}}{b^{-1} - V^{-1}} - \frac{a}{RT}V^{-1}\right). \tag{3.12}$$

The reciprocal self volume of molecules (b^{-1}) is mapped to a maximal deformation for networks, λ_m or D_m, and we obtain, from the analogy to Eq. 3.12,

$$f = GD\left(\frac{D_m}{D_m - D} - a'D\right). \tag{3.13}$$

This equation with three adjustable parameters (G, D_m, a') permits a reasonable representation of the upturn behavior in Fig. 17b. The slope in the Mooney Rivlin diagram near $\lambda = 1$ is now $G(D_m - a')$ corresponding to the second virial coefficient of the fluid. This means that this slope has contributions from both attraction and repulsion, also near $\lambda = 1$.

3.3 Flory Huggins formula

The mixing entropy of two different ideal gases,

$$\Delta S_m = -R\sum_B n_B \ln x_B, \tag{3.14}$$

with n_B the number of moles of component B, $B = 1, 2$, x_B the mole fraction, is calculated in two steps. (i) The (phase) volume is divided into a large number of equal cells, and the two sorts of particles (1, 2) are randomly placed. (ii) The Gibbs identity factors, $N_1!$ and $N_2!$, with N_B numbers of particles, must be taken into account for the Boltzmann probabilities to assure that a permutation of 1 particles (and of 2 particles) does not give a new macro state.

For polymer mixtures, Flory, Huggins, Staverman & van Souten (Refs. 65–67) and possibly others introduced the following formula for the ideal mixture (Flory Huggins formula for the *combinatoric entropy*),

$$\Delta S_m = -R\sum_B \frac{n_B}{N_B} \ln \phi_B, \tag{3.15}$$

with ϕ_B volume fraction, N_B number of segments *per chain*, and n_B mole number of B segments in the system.

The reasons for the modification of Eq. 3.14 are as follows. (1) From Sec. 3.1 we know that chains or larger chain parts are to be considered as identity carriers. Eq. 3.15 is exactly the ideal gas mixing entropy (Eq. 3.14) for $n_B \rightarrow \nu_B \equiv n_B/N_B$ i.e. the units are the chains. [For the time being, the picture of interpenetrating chain coils (or spheres) does not play an essential role here.] Only the universal gas constant R calls the segments to mind: Eq. 3.15 is usually considered to be an equation for a given mole number of segments (and not chains). (2) According to Fig. 1d the placement unit on a lattice is one segment. Paying attention to the chain connection by the placing possibilities, then, in case of statistical independence of segments (mean field approximation), Eq. 3.15 is obtained only if the factorials $\nu_B!$ are used to express Gibbs identity. That is different units (thermodynamic "species" or "entities") are used for random placement (segments) and for Gibbs identity (chains).

In Eq. 3.15 managed in such a way, n_B is therefore the mole number of segments whatever the precise meaning of this term may be. This number is divided by a large number N_B, the number of segments per chain. Therefore, the ideal combinatoric entropy is small for long chains. This corresponds to the action of only one k by small deformations of coils in Eq. 1.18. The volume fraction ϕ_B in the logarithm of Eq. 3.15, of course, does not depend on whether segments or chains are taken as identity carriers.

In polymer science, the volume fraction ϕ_B is often preferred to the mole fraction x_B to escape from the thermodynamic definition of the segment and, more important, to decouple the ψ level from the ρ level description. This decoupling can be realized by the requirement of constant partial densities in the mixture. Then we have $\rho_B = \phi_B \rho_{0B}$ with ρ_B the density of B segments in the mixture, and ρ_{0B} in the pure component. Eq. 3.15 is a courageous ansatz for the combinatoric entropy in the ψ level description.

The ρ level is taken into account by adding a χ term to the free energy. It contains energetic and entropic contributions from the ρ level, weak energetic contributions of the ψ level (long range interaction between segments as described in Sec. 1.4, screened at high polymer concentration) and, possibly necessary, corrections to the ideal segment and identity definition. For a binary mixture we put

$$\Delta G_{\mathrm{m}} = RT \sum_{B} \frac{n_B}{N_B} \ln \phi_B + \phi_1 \phi_2 \chi(\phi, p, T, \ldots), \tag{3.16}$$

$(B = 1, 2)$, χ is called the *Flory Huggins parameter* of briefly the χ parameter.

3.4 Discussion

Let us consider the thermodynamic consequences of the Flory Huggins formula Eq. 3.16 for the case that χ does not depend on ϕ. [This corresponds to the Porter ansatz for the free excess energy of small-molecule mixtures, $G^{\mathrm{E}}/RT = Ax_1x_2$]. For one mol segments (1 and 2, e.g. monomeric units) in molecular mixture we have

$$\Delta G_{\mathrm{m}}/RT = (\phi_1/N_1) \ln \phi_1 + (\phi_2/N_2) \ln \phi_2 + \chi\phi_1\phi_2. \tag{3.17a}$$

For constant partial volume, we can put $n_B/N_B = \phi_B/v_B$ with v_B volume ratio of chain over segment. Then, equivalently to Eq. 3.17a,

$$\Delta G_{\mathrm{m}}/RT = (\phi_1 v_1) \ln \phi_1 + (\phi_2 v_2) \ln \phi_2 + \chi\phi_1\phi_2. \tag{3.17b}$$

From Eq. 3.17a we can calculate the chemical potentials μ_B per one chain B as

$$\mu_B/kT = \ln \phi_B + (1 - (N_B/N_A))\phi_A + N_B \chi_{\mu B} \phi_A^2 \tag{3.18}$$

where $B = (1, 2)$, $A = (2, 1)$, $\phi_B = 1 - \phi_A$, and

$$\chi_{\mu B} \equiv \chi + \phi_B d\chi/d\phi_B = \chi \tag{3.19}$$

for constant χ (not depending on $\phi = \phi_A$, e.g.).

Now we can calculate the miscibility gap occurring for large enough χ, see Fig. 5a. (More thermodynamic details will be described in Chap. 9). The binodal curve $\chi(\phi)$ follows from the equality of chemical potentials for any component in the two phases. For a *symmetrical mixture* defined by equal chain length for both components, $N_1 = N_2 = N$, we find for the binodal, $\chi(\phi)$,

$$N^{-1} \ln [\phi/(1 - \phi)] + \chi(1 - 2\phi) = 0. \tag{3.20}$$

The limit of local diffusion stability is called spinodal and is to be calculated from $(\partial^2/\partial\phi^2)\Delta G_{\mathrm{m}} = 0$. From Eq. 3.17a one finds this curve $\chi(\phi)$ as

$$\chi = (2N_A\phi_A)^{-1} + (2N_B\phi_B)^{-1}. \tag{3.21}$$

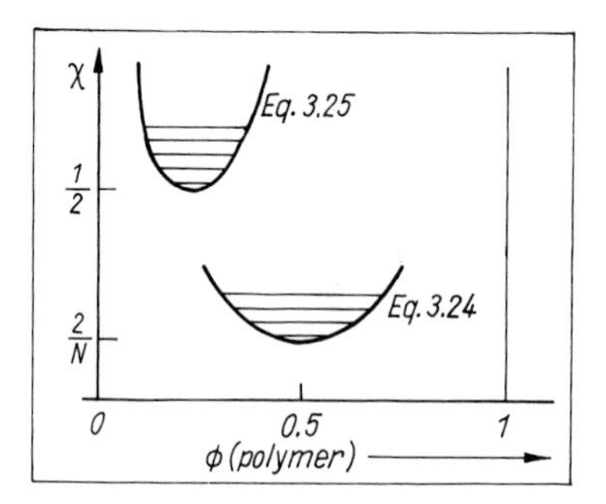

Fig. 18. Mixing gaps according to the Flory Huggins formula: Eq. 3.24 is for symmetrical polymer–polymer systems, Eq. 3.25 for a polymer solution with a small-molecule liquid.

The critical point (χ_c, ϕ_c) in a polymer mixture of two monodisperse components is the extreme of $\chi(\phi)$ (binodal or spinodal) curve. Using Eq. 3.21 we find from $d\chi/d\phi = 0$

$$\begin{aligned} \phi_C(=\phi_{AC}) &= \sqrt{N_B}/(\sqrt{N_A} + \sqrt{N_B}), \\ \chi_C &= (\sqrt{N_A} + \sqrt{N_B})^2/2N_A N_B. \end{aligned} \tag{3.22}$$

Special cases are the symmetrical small-molecule mixture ($N_A = N_B = 1$),

$$\phi_C = \tfrac{1}{2}, \chi_C = 2, \tag{3.23}$$

the symmetrical polymer mixture

$$\phi_C = \tfrac{1}{2}, \chi_C = 2/N, \tag{3.24}$$

and a A polymer solution in a small molecule solute ($N_B = 1$)

$$\phi_C = N_A^{-1/2}, \chi_C \approx \tfrac{1}{2} + N_A^{-1/2}. \tag{3.25}$$

The receding of the critical point to small polymer concentration ($\phi_A \ll 1$) for a small-molecule solute ($N_B \to 1$), Fig. 18, is really observed in many experiments. This is an indication for the tendency that longer chain molecules really lower the miscibility and that the identity problem (segment or chain) is at least approximately grasped for the three input possibilities of the Flory Huggins theory: segment for placement, chain for Gibbs identity, and unified (R or k) segments, not depending on N, as the units for the mol.

The motive for mixing in molecular scale is, of course, lowering the free energy. From

$$\Delta G_\mathrm{m} = \Delta H_\mathrm{m} - T\Delta S_\mathrm{m} \tag{3.26}$$

we see that a positive combinatoric mixing entropy is a motive for mixing, strong for small-molecule mixture because it is of order $R \ln 2 \approx 0.7\ R$ for $\chi = 0.5$. But this motive is heavily reduced for polymer–polymer systems by a factor N^{-1} small for long chains (compare Eqs. 3.14 and 3.15). Whether this is really the reason for the widespread incompatibility in polymer–polymer systems depends on the χ parameter and, therefore, mainly on the ρ-level properties, more precisely on the thermodynamic ρ-level activities of the segments.

Even small-molecule Lennard Jones mixtures with energy and diameter differences of 20% can have positive and negative mixing enthalpies and excess entropies of order several tenths of RT and R, respectively (Sec. 9.2). Since the ρ-level scale, one nanometer, is of order the Kuhn step even for flexible polymers, a comparably good packing of different segments would require considerable forces applied to the segments. But voids of order 1 nm are impossible there because of the high statistical weight they would occupy. We can imagine that, without special stereometric mutual fitting or without specific attraction between the different segments 1 and 2, it is the repulsive force – generally responsible for structure in ill-condensed matter – that pushes the segments into the virtual voids. Repulsion means, as a rule, $\Delta H_m > 0$, and from Eq. 3.26, $\Delta G_m > 0$. This means larger χ which implies the tendency to phase demixing (Fig. 5a) and, therefore, to incompatibility. Since the repulsive forces steeply depend on the distance between segments relatively large χ are prompted.

Generally, the imperfection of polymer packing causes some break up of the structure. The technical term is *free volume*. It amounts to a few per cent of the total volume, it is larger for higher temperatures, it can easily be redistributed, and enlarges the mobility (higher $\ln\omega$). The free volume is also assumed to be responsible for the larger compressibility and thermal expansion of polymers as compared to small-molecule liquids. As for polymer mixtures, "differences" of free volume are often discussed.

In the ρ level we have to do with short chain parts (segments) and a few chain ends. It is not easy to imagine that, if local mixing of segments is possible, the N-damped mixing entropy Eq. 3.15 would play the main role. Instead we should think about entropic mixing motives similar to the small-molecule case of Eq. 3.14 but hindered by the chain continuity. This concept is – not quite correctly – called noncombinatoric entropy.

In summary, there are several ρ-level contributions to the χ parameter. Without specific conditions we expect a tendency towards large values

($\chi \approx 1$ and larger), for specific conditions towards small values including $\chi < 0$, and in general entropic contributions of order some tenths of R per mol monomeric units. Unfortunately, because of the fine balance of thermodynamic optimization (two elephants on a letter balance, to use a parable by Koningsveld), these tendencies do not act additively. If differences are of interest, they act at least in quadratic terms (Sec. 9.2) with the burden of cross terms.

3.5 Thermodynamic systems: spatial and temporal aspects

The system concept is one of the most important concepts of thermodynamics. The relation between system and environment is quite different from the relation between particle and field in mechanics or electrodynamics. Contrary to a widespread opinion one should have the feeling that in concentrated polymer systems a single chain is not a very practical thermodynamic subsystem.

The definition of a *thermodynamic system* includes four aspects:
(i) (space) It must be said what belongs to it and what does not, the latter is the environment.
(ii) (system aspect) What are the contacts between system and environment.
(iii) (time) What is the relevant time scale; often meant inclusively since experiments give responses for all fluctuations with correlation time $\leqslant$ experimental time.
(iv) (size) It should contain so many (n) units (particles, segments, . . .) that observables can be established which are typical for the macroscopic scale. This does not mean that the observables must not depend on size (as for small systems, see Sec. 3.8), but the average, mean value of the (usually nonlocal) observable x should well be defined relative to the fluctuations; e.g. since the relative fluctuation is of order $1/\sqrt{n}$, $\Delta x/x \approx 1/\sqrt{n}$, one can require $n \gtrsim 100$ units for $\Delta x/x \lesssim 10$ per cent.

Let us make some further comments to the definition. The spatial aspect is illustrated in Fig. 19. In Fig. 19a the system is given by geometry, the part of space given defines the system. In this case the borders are called walls. If the particles can penetrate then their number in the system can fluctuate. In Fig. 19b a simply connected system is given by the black particles, where "black" means that they belong to the system. Here we can imagine a volume fluctuation of the system. This system can become disconnected in the case

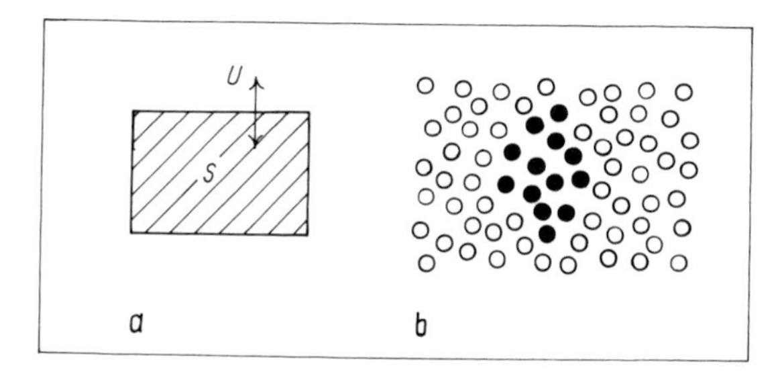

Fig. 19. Thermodynamic system: spatial aspects.
a. Geometric definition. S system, U environment, ↕ contacts through the "walls". b. "Chemical" definition.

of diffusion. The original sense is preserved only for small times $t < \tau_D \approx L^2/D$ with L its size and D the diffusion coefficient.

A contact can be characterized by the exchange of extensive variables. A system is called open if particles can be (better: are) exchanged. Mechanical work (or volume) and heat can also be exchanged. This corresponds, by the Zeroth Law, to a dynamic equilibrium with equal intensive variables in the system and in the environment: the chemical potential μ, the stress tensor (or pressure p), and the temperature T, respectively.

The time aspect is often related to the equilibrium and not to the system. Then the variables that cannot fluctuate within the experiment time because of too large correlation (fluctuation, relaxation, reaction, . . .) times characterize the kind of inhibition of the equilibrium. But in case of several dispersion zones with extremely varying relaxation times (depending on temperature, pressure, concentration, . . .) the inhibition problem is rather difficult to handle, and it can be of advantage to delegate alternatively the time restrictions to the system. In a deeper sense, calling the variables observables, this is also a question of what a thermodynamic experiment is (see Sec. 3.7, below). For instance, when for decreasing temperatures the relaxation time emigrates from the experiment time then we observe a thermal glass transition from equilibrium into a special kind of non-equilibrium, related to the dispersion zone of the system under consideration. This phenomenon will be described in Chap. 7.

3.6 Subsystems

The subsystem concept is a key to the operations and to the way of thinking in thermodynamics (see Ref. 45). It is related to the spatial aspect of the system definition, and a subsystem means, roughly, a part of the system that is separated from the whole only by the observer's mind and not necessarily by actual walls. More precisely, the term *subsystem* means a part of the larger system, homogeneous or not, which part is large enough to define reasonable thermodynamic observables and is only separated from the larger one by our definition, i.e. the natural interaction with the latter is the only one. Of course it is a main task to substitute the term definition by a reasonable real or gedanken experiment.

The subsystem definition permits various sizes and does not insist on the equality of nonextensive quantities in subsystems of different size. But, of course, a minimal size in the sense of the system aspect (iv) of Sec. 3.5 is required: $n \gtrsim 100$ units. The case that structures or correlations are larger than this "minimal subsystem" is discussed in Sec. 3.8 – small systems. Rescaling of subsystem size is an important operation in thermodynamics.

A single chain or chain part in a concentrated two-component (A, B) polymer system is not a good subsystem: The relation chain-environment is too difficult (e.g. there is no so-called μ space statistics as for A chains in dilute solutions), and rescaling would only be possible along the chain. The system curvature would imply to have always small systems for chain parts.

Consider now a homogeneous system with a cooperative molecular motion having a large correlation length ξ_a (not necessary from a formula like $\exp(-r/\xi_a)$) and a correlation time τ smaller than the system time according to aspect (iii) of Sec. 3.5. Consider the case that ξ_a^3 is larger than the volume of the minimal subsystem. Starting from a large subsystem of size $\lambda \gg \xi_a$ and decreasing the scale λ we would find, near the start, that all nonextensive quantities are equal, independent of subsystem size. But arriving at the ξ_a scale, $\lambda \approx \xi_a$, we expect deviations. The smallest subsystem having the same nonextensive quantities as larger subsystems is called a *natural subsystem* (see Fig. 20). It is the smallest thermodynamic representative for the whole system. Its volume is

$$V_a = (4\pi/3)\xi_a^3. \tag{3.27}$$

It is minimal with the intention of constructing a complete spectral density in it, but it is not a minimal subsystem as defined above ($1/\sqrt{n}$ small enough).

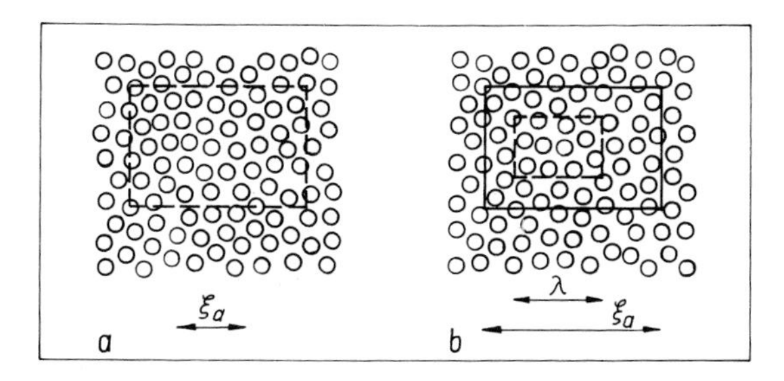

Fig. 20. Natural subsystem.
a. Correlation volume of size ξ_a^3 smaller than minimal subsystem (- - -). b. Correlation volume (——) larger than minimal subsystem of size λ (- - -): Then —— defines the natural subsystem of size ξ_a. [Do not compare sizes and particle numbers of part a and b.]

Natural subsystems and their relaxation times are the "points" $(\boldsymbol{r}, t)$ for macroscopic thermodynamic fields such as $T(\boldsymbol{r}, t)$ or $S(\boldsymbol{r}, t)$ for noncomplete equilibrium situations.

There are two problems with the subsystem concept. (1) What about a surface contribution? The answer is no for natural (and larger) subsystems in a homogeneous situation because there is no extra structure or correlation that could be attached to an interface. The answer is yes (or should be discussed) for a small subsystem (smaller than a structure or correlation). (2) What about statistical independence? The answers are opposite, we must exchange yes and no. The system of size λ defined by the broken line of Fig. 20b is, of course, a (small) subsystem according to the definition, but we hve no *complete* statistical independence from the natural subsystem since the correlation length ξ_a is larger than λ. Relations of such kind will be found to be important for a discussion of the glass transition.

A proposal for a relation between experiment and subsystem – as an object of the experiment – will be dared in the next section.

3.7 FDT again: What is a thermodynamic experiment?

A colleague of mine, Dr. Milan Marvan from Prague, once said to me: "Thermodynamics is the science of what's an experiment for macroscopic properties". We have to distinguish two kinds of thermodynamic experiments: (1) Intensive variables as temperature, pressure, and chemical potentials – equivalence class indices from the equivalence relation "mutual equilibrium"

(Zeroth Law) – can be measured from outside. To make a temperature measurement a thermal contact is made, we have to wait for mutual thermal equilibrium (i.e. the heat exchange ensures equal temperatures), and we read the system temperature from the thermometer.

(2) What about the conjugate (by the fundamental form of the First and Second law, TdS or SdT) extensive variable, the entropy S? This is done by a susceptibility measurement similar to the creep experiment of Fig. 10c. In a calorimeter we can measure the heat (or entropy) response to a temperature disturbance, $C_p = (\Delta Q/\Delta T)_p$ for isobaric conditions, or

$$C_p = T(\partial S/\partial T)_p = k\overline{\Delta S^2}, \tag{3.28a}$$

see Eq. ii of the Introduction. The second sign of equality is from the integral form of the fluctuation dissipation theorem FDT, Eq. 2.59. The heat capacity C_p is the retardation intensity of an entropy compliance. The entropy is obtained by integration of C_p with the measure $d \ln T = dT/T$. Equilibrium provided one can obtain absolute values for S from the Third (Nernst) Law.

But we should be astonished that such complicated constructions as T and S can directly be grasped, with no probability sensor and transmission line, when we think about the abstract definitions from statistical mechanics, e.g.

$$S = k \ln W_B,\ T = (\partial U/\partial S)_{V,N_B} \tag{3.28b}$$

with W_B the Boltzmann probability and U the internal energy. Or if we think about the construction that the temperature defined by the equivalence class construction of the Zeroth Law is always sharp whereas, from the standpoint of linear response with temperature modulus and spectral density $\Delta T^2(\omega)$, we have a finite temperature fluctuation, e.g.

$$\overline{\Delta T^2} = kT/C_V = \int_{-\infty}^{\infty} \Delta T^2(\omega)\, d\omega. \tag{3.29}$$

How is the temperature defined when during a fluctuation the system is off its average state?

The usual construction producing thermodynamic variables from mechanics is the *principle of local equilibrium*, PLE. This is more than the statement that any subsystem is in local equilibrium with its environment in a situation that permits global deviations from the equilibrium, e.g. with a gradient $\partial T/\partial r \neq 0$. The construction contains four steps. (1) A subsystem larger than natural is mechanically simulated in a computer during the proper time scale. We *determine* from the recordings for any time t

$$\begin{aligned} &\text{the volume } V(t) \text{ by geometry (see Fig. 19b),} \\ &\text{the energy } E(t) \text{ from mechanical energy,} \\ &\text{the particle number } N(t) \text{ by counting.} \end{aligned} \tag{3.30}$$

(2) We *measure* (or calculate from computed mean values for many states) the equation of state in a large system, homogeneously added from many such subsystems so large that all observables have sharp values,

$$\mathscr{X}_{\text{tot}} = \mathscr{X}_{\text{tot}}(\mathscr{U}, \mathscr{V}, \mathscr{N}) \tag{3.31}$$

where $\mathscr{X}_{\text{tot}}$ stands e.g. for entropy or temperature. (3) All extensive observables ($\mathscr{U}$, $\mathscr{V}$, $\mathscr{N}$) of the large system are reduced to the subsystem size by multiplying with $N/\mathscr{N}$ (scaling according to Fig. 21a). Since the fluctuation of the total system is small we obtain a sharp raster of thermodynamic values, e.g. $\mathscr{X}_{\text{tot}}(\mathscr{U}, \mathscr{V}, \mathscr{N})_{\text{scal}}$ of Fig. 21b, of course for equilibrium mean values. Varying $\mathscr{U}$, or $\mathscr{V}$, or $\mathscr{N}$ then $\mathscr{X}$ changes and we obtain a new horizontal (constant) line for the scaled values. (4) Now the subsystem simulations (Eq. 3.30) are pursued with time. Each value $E(t)$, $V(t)$, and $N(t)$ at time t is identified with the scaled values of $\mathscr{U}$, $\mathscr{V}$, and $\mathscr{N}$ from step (3). Then the off-equilibrium values of the subsystem are *defined* by

$$X(t) = X(E(t), V(t), N(t)) \stackrel{\text{def}}{=} \mathscr{X}_{\text{tot}}(\mathscr{U}, \mathscr{V}, \mathscr{N})_{\text{scal,eq}}\,. \tag{3.32}$$

The PLE can now be expressed as follows: the fluctuating thermodynamic variables of a subsystem as a function of determined variables are defined by the rescaled function thermodynamically measured for a large system. This principle permits the temperature – originally defined without fluctuation by the Zeroth Law – to be included in the stock of fluctuating thermodynamic variables.

But the question remains open whether or not the real susceptibility experiment acts according to this principle. The nontriviality of the thermodynamic experiment is demonstrated by the thermal glass transition. Starting from statistical mechanics, Eq. 3.28, the entropy is to be determined from the probability W_{B}. The time aspect can be best accounted for indirectly, discussing nonergodicity or restriction of the phase space. In a real experiment of $C_p = \dot{Q}/\dot{T}$, over the thermal glass transition, the result does depend on the rate of fluctuations, see Fig. 21c. If they are slower than the experimental time (for $T < T_g$), then they do not contribute to C_p, and if they are

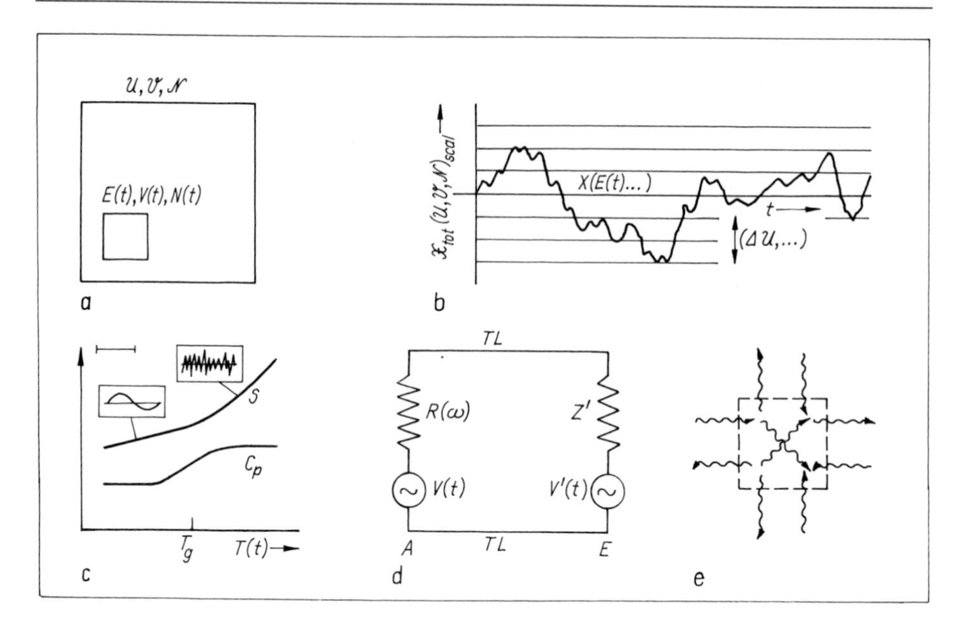

Fig. 21. Thermodynamic experiment.
a. Subsystem with mechanically simulated quantities E, V, N and total system with thermodynamic variables $\mathscr{U}$, $\mathscr{V}$, $\mathscr{N}$. b. Principle of local equilibrium, PLE. Definition of actual thermodynamic variables for a fluctuating subsystem. c. Heat capacity $C_p = (\dot{Q}/\dot{T})_p$ and entropy S at a thermal glass transition. The inserts are amplifications of fluctuation for $T < T_g$ and $T > T_g$ related to the bar ⊢⊣ representing the experimental time (as defined e.g. be the cooling rate). T_g glass temperature. d. Nyquist's scheme for the derivation of the FDT. E emitter, A absorber, TL transmission lines. e. Quantum cloud mediating the (self) experiment.

faster then they do. That means that the thermodynamic experiment measures time average values over time intervals not too short compared to the fluctuation (correlation) time. On the other hand, in the PLE, a susceptibility is modelled by changing the raster values corresponding more to the ensemble average.

This time aspect can be grasped by the fluctuation dissipation theorem FDT and the inclusion of the time aspect in the system definition: A system responds only with respect to variables that fluctuate within the time scale of experiment. Therefore we make the *experiment hypothesis* that the FDT is the equation for the thermodynamic experiment. (The relation to quantum mechanical experiments is discussed in Ref. 68).

This hypothesis will shortly be justified with Nyquist's derivation of the FDT (Ref. 48, see also Ref. 69). His derivation gives a generally valid result (except the zero-point contribution) and refers to thermal voltage noise $V'(t)$ of a "Brownian" complex resistance Z', the emitter of Fig. 21d. An essential element is Nyquist's transmission line for the signal transport from the object (Z', V') to the apparatus ($R(\omega)$, $V(t)$) where the signal is transformed to a pointer. All the emitted voltage waves can be absorbed by a proper adjustment (balance) of the line. This reminds us of a black radiator – a set of oscillators – that delivers, in the classical (non quantum mechanical) limit the oscillator energy factor kT that mediates between the fluctuation x^2 and the response ΔJ, see Eqs. 2.55–2.58.

The relation to a thermodynamic experiment is given by a little dictionary:

Nyquist experiment	subsystem experiment
emitter	subsystem
transmission line	cloud of quanta
absorber	environment, or the subsystem itself

We have constructed the following picture. The set of line oscillators corresponds to a quantum cloud interpenetrating the subsystem (Nyquist cloud). This is a set of $\hbar\omega$ quanta that is relatively independent, and the experiment corresponds to a registration of correlations of this cloud, differing for different activities. This cloud mediates the experiment (kT factor in the FDT), and is a representative for the system properties that can directly be experienced (by an "observer", or by the system itself acting as an apparatus: self-experiment, see Fig. 21e). [Perhaps, being a real object, the cloud can itself carry some properties, e.g. $k \ln \omega$ being some kind of "thermokinetic entropy" when related to a cooperative dispersion zone]. This picture underlines the primary role of frequency (ω identity of the FDT, see Sec. 2.6).

In the classical limit the FDT formulas Eqs. 2.55–2.58 do not contain the Planck constant $\hbar$. Then the quantum properties are only included in the concept, as sometimes observed in thermodynamics (size of phase space cells, Gibbs identity factorials $n!$), and are expressed indirectly by the identity acronym of thermodynamics, the Boltzmann constant, similar to the Avogadro or Dalton law for ideal gases. The time aspect of experiments comes automatically: times shorter than $1/\omega$ cannot be established by a noncoherent quantum cloud. The length aspect will be considered for natural subsystems (size ξ_a, see Eq. 3.27). A thermodynamic experiment can only detect correlative properties of the Nyquist cloud. The sensitivity to larger length scales $\lambda > \xi_a$ is systematically quenched by statistical independence. It will be said that a *characteristic length* ξ_a is needed and is sufficient for completing a linear response spectrum. Beyond this scale the sensitivity is lowered similar to ω^a for $\omega \to 0$ in the flow or transport zone (described by hydrodynamic modes, e.g.); examples are $G'' \sim \omega$, $G' \sim \omega^2$, $\varepsilon'' \sim \omega^a$ etc. In this zone, the sensitivity can be enlarged by a scattering experiment with scattering vector $\boldsymbol{q}$ that can evaluate the effect at $\omega = Dq^2$, see Sec. 2.8, esp. Eq. 2.66. The relation between linear response and scattering will be discussed further in Sec. 6.10.

With the experiment hypothesis we can transform the subsystem concept from an analytical tool to an experimental situation (at least to a gedanken experiment). Some identity questions (e.g. M_c for networks) can, in principle, be discussed on an experimental basis, one can discuss small subsystems of different size (Sec. 3.8), and subsystem thermodymamics can be divided into parts related to frequency deserts between dispersion zones (Sec. 3.9).

3.8 Small systems

As discussed in Sec. 3.6 (in homogeneous systems after reduction of extensive variables with the scaling factor $N/\mathcal{N}$) all subsystems larger than the natural ones have the same thermodynamic properties.

Small subsystems in the sense of Hill (Ref. 70) are subsystems smaller than the length scale of structures. Examples are given in Fig. 22a–c. This concept is of benefit if there is a spread between the size of minimal subsystems and of structure. In this book, a *structure* is a spatial correlation with a lifetime larger than the correlation time of the fluctuation covered by the subsystem (i.e. by the experiment considered). The usual structure concept is included by the case of infinitive lifetime such as for a polymer coil.

Small subsystems need not have, and this is the main issue, the same (reduced) thermodynamic properties as large subsystems. The larger structures (e.g. the many chains penetrating the walls of the small subsystem in Fig. 22a) induce constraints in the small subsystem that can result in size-dependent modifications of thermodynamic properties. This is easy to see in

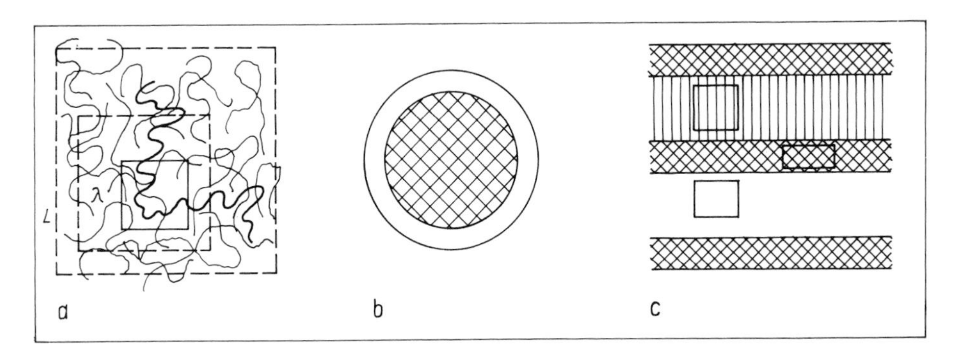

Fig. 22. Small subsystems.
a. Small subsystems of varying size ($\lambda < L$) in a homogeneous polymer sample. b. Droplet as a small subsystem. c. Small subsystems in different layers of a stack of lamellas for semicrystalline polymers, see Figs. 8c and d.

the examples of Figs. 22b (droplet) and 22c (layers) but should also be valid for the homogeneous case of interpenetrating coils, Fig. 22a.

The question of which kind of variables are concerned is the question of how important the corresponding contacts are. For the droplet it is the work (volume) contact that is modified by the surface tension. The pressure inside the droplet is higher because the surface curvature results in additional forces. Chemical potentials are established by particle exchange. We can imagine, referring to Fig. 22a, that the segment exchange can be modified by the many constraining chains going through the walls of a small subsystem. Therefore we should not exclude the possibility that even the chemical potentials in the equilibrium can depend on the size of small subsystems.

The experiment hypothesis of Sec. 3.7 gives a reasonable basis for a discussion of this question. If the subsystems are large enough to have their own Nyquist cloud then small systems are rather independent objects (and subjects) of thermodynamics, independent subsystems for experiments. In principle, all observables must be influenced by the larger structure and should depend on the size. The degree of modification, as mentioned above, depends on the constraint effects on different contacts (possibly depending additionally on the time scale considered). Moreover, the Gibbs identity (factor $\nu!$ or $n!$ in the Gibbs distribution) is now, because of the quantum basis for the FDT, an experimental issue: The smaller the small subsystem of Fig. 22a, the smaller the chain parts that are available as identity carriers for these small subsystems.

The concept of small subsystems is very important for polymers because 10-nm structures are so widespread (see Chap. 1) and permit minimal subsystems much smaller than 10 nm still containing enough segments for thermodynamics. A list of examples follows that are discussed in this book: (1) The properties of crystal lamellas are influenced by the interface properties. The constraints are the emanating chains (from crystal to interface). Figs. 8c,d, and 22c, Sec. 11.5. (2) If thick enough, the interface itself (between crystal and amorphous layers) is a small subsystem: interphase. Secs. 1.7, 8.5, and 11.5. A similar situation occurs for amorphous layers of small thickness between the crystallites or the interfacials. (3) The domains of incompatible block copolymer liquids are small systems constrained by chains going through interfaces and inducing elastic energies to realize constant densities in the domains. Secs. 10.6–7. (4) The droplet model can be used for nucleation. Secs. 10.4, 11.3. (5) The thermodynamic entities for large subsystems in melts, for instance, are the chains (ψ level); for networks it is the chain parts between the crosslinks; and for ϱ level discussions we have the monomeric units or, at best, segments. Secs. 1.1, 3.1–4, 9.1–3. (6) If chemical potentials are really influenced by the constraints then the polymer compatibility is also a question of spatial scale: granulated demixing. Sec. 10.9. (7) As the diffusion coefficient contains a thermodynamic factor it is possible to discuss a scale dependent diffusion. Sec. 10.5.

Further complications may arise from the time aspect of the experiment, or of the subsystem, and of the structure as defined above. This will be treated in the next section.

3.9 Functional subsystems. Thermokinetic structure

As discussed in Sec. 1.8, a multiplicity of relaxations is observed in polymers. Besides of local modes LM, we find two widely separated dispersion zones in amorphous polymers for $M > M_c$: the main transition MT and the flow transition FT, see Figs. 6b, 13c–d, and Fig. 23a here.

Having so distinctly separated dispersion zones suggests considering their modes separately from the thermodynamic aspect, too, in particular when time aspects are included in the system definition. Then a *functional subsystem* is defined to be a subsystem for which only the motion of one dispersion zone is considered. This situation is close to the experimental situation. Larger times (lower frequencies) can be quenched by the experimental time,

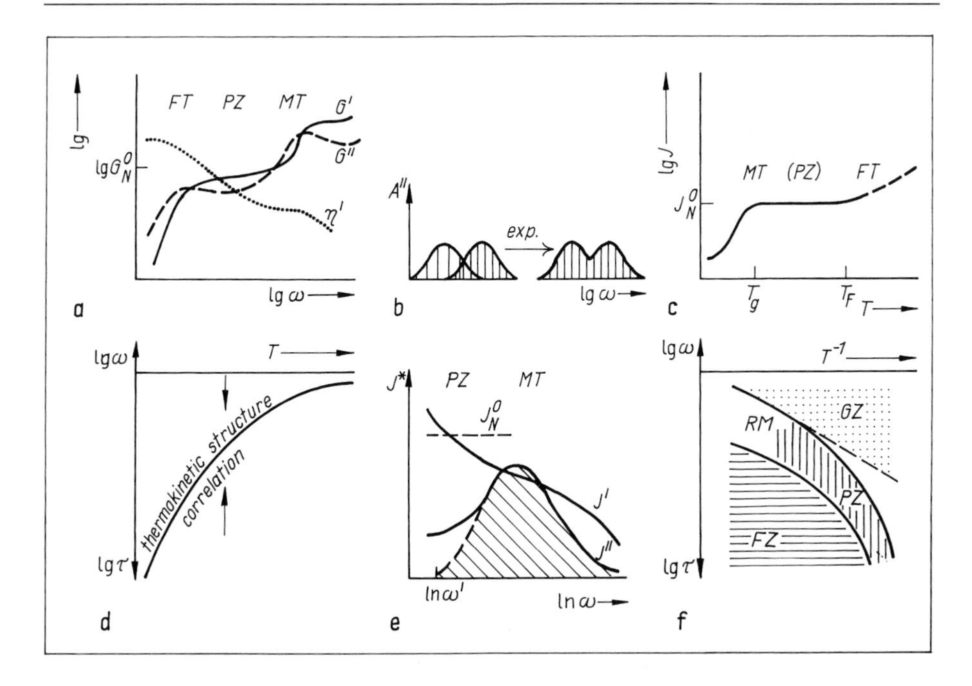

Fig. 23. Time aspects of thermodynamics.
a. Log–log shear modulus vs. frequency diagram of amorphous polymers. FT flow transition, PZ plateau zone, MT main transition. $G^* = G' + iG''$; η' real part of complex viscosity $\eta^* \equiv G^*/i\omega = \eta' - i\eta''$. b. The experiment (FDT) cannot distinguish between modes of same frequency ω from different dispersion zones. c. A thermodynamic experiment can successively separate functional subsystems. Example: shear compliance $J = (\partial x/\partial f)_\tau$ at constant time scale τ as a function of temperature T. MT, PZ, FT see Fig. 23a. d. Two aspects of any dispersion zone: thermokinetic structure (as seen from the high frequency side) and correlation (seen from below). e. Ferry's construction for the plateau compliance $J_N^0 \approx 1/G_N^0$ in the case of "spurios" relaxations (Andrade behavior) in the plateau zone. f. lg ω-T (or lg τ-T) windows for plateau-zone thermodynamics. FZ flow zone ("usual" thermodynamics), PZ plateau zone, GZ glassy zone. [RM Rouse modes (Secs. 5.3, 5.4) do not form a plateau zone.]

and shorter times (or higher frequencies) can be disregarded by a virtual filter applied to the quantum cloud of the experiment hypothesis, Sec. 3.7. [This filter is not an experimental possibility for overlapping dispersion zones, Fig. 23b, because the FDT cannot distinguish between different modes at the same frequency.]

The general scaling principle GSP, see Sec. 2.8, implies that the size of a natural functional subsystem is the larger the larger the typical time scale is.

The separation of functional subsystems can directly be demonstrated by a thermal experiment, e.g. by measuring the shear compliance at constant frequency (such as 1 Hz) from low to high temperatures with a corresponding heating rate (such as 1 or 10 K/min), see Fig. 23c. For a given time scale, here given by the frequency, or by the heating rate, the MT is at lower temperatures than the FT, see Fig. 37b, below. For $T < T_g$ the correlation time of fluctuation for both dispersion zones, if in equilibrium, would be much larger than the given experimental time scale. This means, from the FDT, that we find no response from either zone (they are frozen in) – small compliance. At the glass temperature T_g the experiment time becomes of order the equilibrium correlation time for MT. Their fluctuations start to work (they are "stimulated" at T_g) and we observe an increase of compliance. For $T_g < T < T_F$, in the plateau between MT and FT, the MT does work and the FT does not. I.e. the functional subsystem of MT has quick enough fluctuations for experimental response. The plateau compliance J_N^0 is the result of "switch on" the MT fluctuations. The (frozen-in) motions of the FT are felt as a structure, i.e. we can also define a *small* functional subsystem. At T_F the flow transition fluctuations are stimulated (their functional subsystem is switched on), and for $T > T_F$ both zones contribute to the compliance.

The steps observed in such a heating (or cooling) experiment at T_g and T_F (or at other dispersion zones) are called *thermal transitions*. The corresponding temperatures depend on the time scale of the experiment, they increase with shorter time scale as can be seen from the referred Fig. 37b. More details, especially nonequilibrium aspects of a thermal transition, will be discussed in Chap. 7.

Let us go back to the equilibrium again, Fig. 23d. Consider one dispersion zone in a lgω-T diagram. Consider a molecular movement far from this zone, i.e. with a typical frequency ω (at a given temperature) much lower or much higher than this dispersion zone, e.g. the motion in another dispersion zone, not drawn in Fig. 23d. Then, referring to our structure definition of Sec. 3.8, we have the following situation. Seen from the lower frequency motion, the dispersion zone of Fig. 23d defines some spatial correlation, whereas, as seen from the higher frequency motion, this dispersion zone defines a structure. Such a structure in equilibrium – with a larger time scale of the corresponding dispersion zone – will be called *thermokinetic structure*. For instance, the slow fluctuations of the FT are a (modifying) thermokinetic structure for the quick

MT motion. Inversely, the quick MT motions are a (prerequisite) correlation for the FT fluctuation, etc.

Suppose the plateau zone PZ would be empty – no activity – then in both diagrams Figs. 23a, c the plateaus are represented by horizontal lines, $G' = G_N^0 = \text{const}$ (plateau zone modulus) and $J' = J_N^0 = \text{const}$ with $G_N^0 J_N^0 = 1$. In reality, however, we always observe a certain slope which means that there are also some molecular motions in the plateau zone. Then the definition of J_N^0 values is not completely free of arbitrariness. Often the Ferry construction is used, see Fig. 23e. The dispersion zone at the higher frequency (MT here) is separated to the best of one's knowledge and belief, and then it is integrated according to the dispersion relations, Eqs. 2.35–37

$$J_N^0 - J_g = \frac{2}{\pi} \int_{\xi=\omega'}^{\infty} J''(\xi)\, d\ln\xi = J'(\omega') - J_g, \tag{3.33a}$$

or

$$G_N^0 - G_g = \frac{2}{\pi} \int_{\xi=\omega'}^{\infty} G''(\xi)\, d\ln\xi, \tag{3.33b}$$

putting ξ instead of $\xi - \omega$ in the denominator of Eqs. 2.36–37. This means that the plateau-zone level is determined by the next dispersion zone on the high-frequency side, i.e. by the quicker (and shorter) modes. This corresponds to the thermodynamic experiment also counting only the quicker fluctuations.

For small-molecule glass formers this construction can be applied to the steady state compliance, $J_e^0 - J_g \approx J_e^0$, although $G \to 0$ here.

An instructive example for the plateau zone is the Edwards Graessley correlation for $G_N^0 = 1/J_N^0$, see Ref. 71. They found that

$$G_N^0 l^3/kT \approx \text{const} \cdot (\nu L l^2)^2 \tag{3.34}$$

with ν the number of chains per volume (also for concentrated systems) and νL the chain density per area (meter^{-2}). The point is that the correlation contains the Kuhn step l as an indicator for the short range correlations of the main transition. The independence of G_N^0 from the chain length indicates that G_N^0 is not determined by the other neighbor, the flow transition at lower frequencies with larger-scale motion. [A formal G_N^0 construction from the FT dispersion zone is also possible, see Ref. 2, but without the experimental and correlative background.] If the plateau-zone responses do not, or do weakly, depend on time or frequency, then we have the usual thermodynamics there. The thermodynamic relations are restricted to this PZ, and the values of observables can differ from those in other zones such as in the glass zone (or glass state) or in the flow zone. Thermodynamics that is confined by dispersion zones to a restricted $\lg\omega$-T window (Fig. 23f) will be called *plateau-zone thermodynamics*. Thermodynamic processes in this window, e.g. in the plateau zone, feel other susceptibilities than in other windows (flow zone, glass zone).

Some caution is necessary for transport properties. In contrast with the constant viscosity in the flow zone, for instance, the shear viscosity $\eta' = G''(\omega)/\omega$ of the plateau zone decreases with increasing frequency (shorter times, see Fig. 23a). This seems to be in conflict with the increased "solidification" by the entanglement, as indicated by the finite modulus G_N^0. This decrease corresponds to the higher mobility of shorter modes in polymers. There is no conflict with the constant modulus because the short modes cannot percolate ("flow") through the entanglements, which is a thermokinetic structure for all the faster modes.

4. Theoretical physics code

The reader should not be too full of expectation to this chapter. The coarsening of this book does not allow the reader to apply the following methods without any further assistance. But more and more polymer papers use terms and concepts of theoretical physics. The aim of this chapter is, therefore, to give some general ideas about associations, origins, purposes, restrictions, and starting points for the approximations of several theoretical methods. This could be useful for the nonspecialist reader. The professional theorist is recommended to skip this chapter.

The usual starting point for the calculation of thermodynamic equilibrium values is the sum of states. For many polymer problems the classical form, the Gibbs distribution, is sufficient:

$$\sum_n e^{-E_n/kT} \cong \frac{1}{v!h^f} \int e^{-H(p,q)/kT} \, d^f p d^f q \approx g(T) \cdot Q, \tag{4.1}$$

with $H(p, q)$ the Hamilton function (Hamiltonian), i.e. the system energy in canonical coordinates q and momenta p of the f degrees of freedom. The separation of a kinetic part $g(T)$ refers to the kinetic energy of the segments and side groups. [It should not be forgotten that polymers in different equilibrium phases (and interfaces) or at different states can have quite different typical conformations.] The configuration integral (including the different conformations) is then

$$Q = \frac{1}{v!} \int_{(V)} \exp\left(-U(q)/kT\right) d^f q. \tag{4.2}$$

The term configuration is used here in the physical meaning (distribution in f-dimensional phase space) and does not mean the succession of chemical units along the chains (chemists's configuration). The very problem of Eq. 4.2 is the practical execution of the integration over a complicated potential energy function $U(q)$ of many important dimensions. The Gibbs identity factor $1/v!$ is usually related to the number of chains v, and there is no discussion of identity questions during the calculation. For mixtures $1/v_A!v_B!$ is used.

The free energy is then

$$F(V, T) = kT(\ln Q + \ln g(T)). \tag{4.3}$$

The symbol T in Eqs. 4.1–3 means the sharp temperature of a large heat bath fed by the Zeroth Law into the subsystem. Therefore, it is not easy to get a temperature fluctuation (first part of Eq. 3.29) from the Gibbs distribution.

The starting point for (classical) dynamics is Newton's equation for the segments, monomeric units, or groups. Written in Hamilton's canonical form, we have

$$\dot{q} = \partial H/\partial p, \quad \dot{p} = -\partial H/\partial q. \tag{4.4}$$

Usually, a distribution $f(p, q, t)$ of phase points in the $2f$-dimensional phase space is considered for an ensemble of equal systems. One point is one system. Then the dynamics is described by changes of $f(p, q, t)$. From Eq. 4.4, the Liouville equation can be derived,

$$\frac{\partial f}{\partial t} = -iLf, \quad f = f(p, q, t), \tag{4.5a}$$

with the Liouville operator

$$L = i\sum_{f}\left(\frac{\partial H}{\partial p_i}\frac{\partial}{\partial q_i} - \frac{\partial H}{\partial q_i}\frac{\partial}{\partial p_i}\right). \tag{4.5b}$$

The motion of the points corresponds to the flow of an incompressible liquid in the phase space if the forces are conservative. This does not mean that the problem of finding $f(t, p, q)$ will be simple.

For long chains it is often assumed that the system can be separated in ρ and ψ level coordinates.

4.1 Analytical methods

1. Order parameter and mean field

Order parameters are ψ level concepts. They are used for phase transitions (e.g. Landau theory, spinodal phase decomposition) or, more generally, for describing of the order in ψ level: orientation, gradients, diffusion etc. In pure substances the density deviation from the average (or the density difference of two phases), $\psi = \rho - \rho_m$, for mixture concentration or volume fraction differences

$$\psi(\boldsymbol{r}, t) = \phi(\boldsymbol{r}, t) - \bar{\phi} \qquad (4.6)$$

can be defined to be the *order parameters*. Being in the ψ level, the position $\boldsymbol{r}$ is "smeared out" in a ρ level scale, in contrast with the ρ level density Eq. 1.19 that used Dirac deltas. If the underlying quantity obeys a conservation law (e.g. mass), then the movement is restricted by the continuity equation, $\partial\psi/\partial t = -\operatorname{div} \boldsymbol{j}_\psi$. The theorists often compare complicated situations with magnetic systems (spins of a lattice). Then the order parameter refers to the magnetization,

$$\psi = \boldsymbol{M}(\boldsymbol{r}, t) - \bar{\boldsymbol{M}}. \qquad (4.7)$$

The *mean field h*, in a sense, is the conjugated ψ-level quantity: ψh is a ψ level energy or energy density. Examples are the Weiss field for the magnetic case or a long-ranged Kac potential for the van der Waals equation. The mean field is used to express the action of environment on a particle or a group of particles (Fig. 24a) in the ψ scale. Describing the particle arrangement by an order parameter we obtain a term ψh for this action. The corresponding ρ level term is the intermolecular potential. Since this potential depends on the mutual orientation and distance of molecules – exactly expressed by ψ in the ψ level –, mean field energies can also be expressed by terms such as ψ^2, ψ^3, $\ldots$, $(\nabla\psi)^2$, $\ldots$. The idea is that short-range potentials can be transformed into the ψ range by $h(\psi)$ so that we have $\psi h \approx a\psi\psi$ in the lowest order. Gradient terms are used for spatial inhomogeneities in the ψ level.

2. Effective Hamiltonian

In this spirit, the Hamiltonian is written down in terms of ψ, whereas h is now restricted to external fields. Thus, the problem is to find the right type of ψ and an interesting approximation

$$H(p, q) \cong H(\psi(\boldsymbol{p}, \boldsymbol{r}, n), \text{rest}) \qquad (4.8)$$

hoping that the rest is not too important. If successful, this coarsening from the ρ to the ψ level is much easier to handle than the canonical form, $H(p, q)$.

One method for this coarsening is, after spatial Fourier transformation, the integration over the large scattering vectors corresponding to the ρ level (Wilson method). Then, omitting the particle number dependence (n),

$$H(\psi(\boldsymbol{p}, \boldsymbol{r}), \text{rest}) \cong H_{\text{eff}}(\psi(\boldsymbol{r}, t)). \qquad (4.9)$$

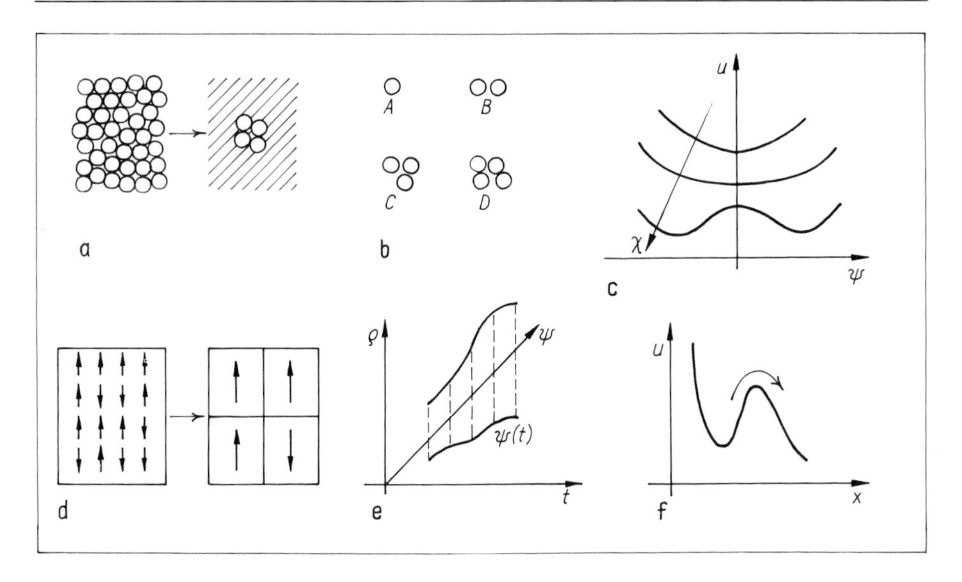

Fig. 24. Theoretical methods.
a. Mean field. b. Series expansion of the Boltzmann factor exp $(-u/kT)$. Typical Mayer clusters for virial coefficients (A, B, C, D, . . .) in the ρ level. c. Field energy configuration ($u(\psi)$, parameter χ) in the ψ level, see also Fig. 58b, below. d. Kadanoff scaling. e. (Mori Zwanzig) projection of ψ coordinates from $\rho\psi$ space. f. Potential function $U(x)$ for the Kramers equation, Eq. 4.19a.

A form often desired for such an effective Hamiltonian is the Landau Ginsburg form,

$$H_{\text{eff}} = \int d^3\boldsymbol{r}(a\psi^2 + b\psi^4 + \cdots + g(\nabla\psi)^2 - h\psi) \tag{4.10}$$

with a (often $a = \alpha(T - T_c)$), b, . . . the parameters, g the gradient term constant, and h the control field. Sometimes ψ^4 is a short hand symbol to express complicated spatial order relations like $\psi(\boldsymbol{r})\psi(\boldsymbol{r}')\psi(\boldsymbol{r}'')\psi(\boldsymbol{r}''')$ and corresponding integrations.

In many cases H_{eff} of Eq. 4.10 is considered to be the "mechanical" part of the free energy or of another thermodynamic potential. Then the equilibrium optimum can be found by thermodynamic variation methods (e.g. functional derivation). The effective Hamiltonian of Eq. 4.9 with time dependence can be applied to dynamic problems when it is accompanied by a kinetic ansatz for the ψ's.

3. Field theory methods

The original ρ level Hamiltonian depends on the particle coordinates (p, q), which implies that configurations in real space are the main point of interest, see Sec. 1.1. The effective Hamiltonian depends on the order parameter ψ field, and one looks for "field configurations".

In this sense the integration measure $dpdq$ of the Gibbs distribution or dq of the configuration integral (Eqs. 4.1 and 4.2, respectively) should be substituted by the measure $d\psi$. The position $\boldsymbol{r}$ as the basis for the integration $d^3\boldsymbol{r}$ in Eq. 4.10 is then indicated by an index $\boldsymbol{r}$, $d\psi_{\boldsymbol{r}}$ (or, after a spatial Fourier transformation, by an index $\boldsymbol{q}$ of $d\psi_{\boldsymbol{q}}$, where $\boldsymbol{q}$ is the scattering vector). Of course, $\boldsymbol{r}$ (or $\boldsymbol{q}$) is understood to be smeared over the ρ level. When permitting complex order parameters, $\psi_{\boldsymbol{q}} = \psi'_{\boldsymbol{q}} + i\psi''_{\boldsymbol{q}}$, the measure is now, instead of $dq^f = \Pi dq_i$,

$$D\psi_{\boldsymbol{r}} = \prod_{\boldsymbol{r}} d\psi_{\boldsymbol{r}} \quad \text{or} \quad \prod_{\boldsymbol{q}} d\psi'_{\boldsymbol{q}} d\psi''_{\boldsymbol{q}}. \tag{4.11}$$

Instead of Eq. 4.1 and 4.2, we then have expressions like

$$\Delta F = -kT \ln \int \prod d\psi'_q d\psi''_q \exp\left(-H_{\text{eff}}(\psi)/kT\right) \tag{4.12a}$$

or

$$Q = \int D\psi_r \exp\left(-\Delta F/kT\right) \tag{4.12b}$$

or, referring to the comment after Eq. 4.10, simply

$$\Delta F = H_{\text{eff}} \tag{4.12c}$$

for ΔF the contribution to the free energy. The "path integrals" in Eqs. 4.12a–b gain their power by the following Gaussian formula. Let H_{eff} be (as a rule: after sensible manipulations, see Ref. 73) of quadratic form,

$$(\psi, A\psi) = \sum_{ij}^{k} \psi_i A_{ij} \psi_j \ (\geqslant 0) \tag{4.13}$$

where the indices i, j can mean quite a lot of different things, depending on the problem, and A_{ij} is a positively definite $k \times k$ matrix. Then

$$\int_{-\infty}^{\infty} \cdots \int_{-\infty}^{\infty} \frac{d^k\psi}{\pi^{k/2}} \, \mathrm{e}^{-(\psi, A\psi)} = (\det A)^{-1/2}. \tag{4.14}$$

That is, the mathematics is quite simple: one has to calculate the determinant of the matrix A. The problem is the manipulation of the path integrals with the aim of finding the proper H_{eff} in the form Eq. 4.13.

In the ideal case the matrix A is diagonal. Then the ψ represent some kind of "free" modes as a basis for perturbative methods.

Usually the logarithm (deexponentialization) in Eq. 4.12a is realized by a series expansion with a small parameter (perturbation). This induces some structure, e.g. Mayer clusters for calculating virial coefficients in the ρ level (Fig. 24b), or Feynman graphs in field theory. We can draw pictures for the discussion of the problem. Then, in the ρ-level configuration integral Eq. 4.2, the pictures show spatial configurations. In the ψ-level path integral, the pictures show field configurations such as nematic potentials, tube potentials for chains within, or Landau-type potential curves for phase transitions of different kinds, see Fig. 24c.

4. Renormalization

Usually, far from thermodynamic stability limits such as spinodals, the fluctuation of the order parameter is smaller than the mean value $\bar{\psi}$

$$\overline{\Delta\psi^2} \ll \bar{\psi}^2 \; (\Delta = \text{fluctuation}). \tag{4.15a}$$

Then the integration in Eq. 4.10 is simplified because the effective Hamiltonian or free energy Eq. 4.12c has a ψ-level density,

$$\Delta F/V = a\bar{\psi}^2 + b\bar{\psi}^4 + \cdots + g(\nabla\bar{\psi})^2 - h\psi. \tag{4.15b}$$

Often the mean field ansatz is restricted to this case.

The Ginsburg criterion Eq. 4.15a is fulfilled the better the longer the range of (effective) interaction is. This is evident from the larger spatial scale of the ψ level as compared to the ρ level.

Thus, for concentrated polymer systems the chain interaction is distributed over the whole chain length, a fact which, of course, enlarges the range of any effective potential. Since this range does not prevent the polymer system from thermodynamic instabilities, the mean field approximation (Ginsburg criterion) also breaks down for these systems in the vicinity of spinodals, but it does so later and nearer to the spinodal than expected from small-molecule system experience.

The methods for calculating free energies from Eq. 4.10 in case the Ginsburg criterion breaks down (e.g. for the original ansatz) are characterized by *renormalization*. This term means that all intermediate (e.g. diverging) stages of perturbative or iterative calculations can be controlled with a finite (small) number of subtractions or redefinitions. The application of this method to polymers is described in Ref. 74a. Three examples will be given here.

(i) The microphase separation of block copolymer systems can be described by a free energy like in Eq. 4.15b leading to a second order phase transition, see Ref. 74b. Beyond the validity of the Ginsburg criterion, the effective Hamiltonian can be renormalized by redefinitions of the parameters a, b, . . . and by adding some new terms (such as ψ^6) coming from a "series development" of fluctuations (Brazowski method). Then the new terms remain mean field because the phase transition is changed, as expected for supercrystal formation, to a weak first order (so-called fluctuation induced) transition. This is discussed in Ref. 75.

(ii) The ε method is simply described in Ref. 76. The theorists have found that there is usually a minimal, higher space dimension (d_c, $d_c > 3$, critical dimension) where the Ginsburg criterion is nowhere violated for an H_{eff} of a given structure. The higher dimension can simulate a longer-ranged field because the number of nearest neighbors is steeply increasing for higher dimensions. Let us assume that the critical dimension is $d_c = 4$. Then the critical point singularities there are mean field properties (such as for the van der Waals or Flory Huggins equation). The idea of the renorm group method is as follows. We put for d(real) (=3 e.g.)

$$d(\text{real}) = d_c - \varepsilon, \tag{4.16}$$

and all formulas are expanded into an ε series. For $\varepsilon \to 0$ all the deviations from mean field would vanish. Finally, ε is taken as for the real space dimension ($\varepsilon = 1$ for d(real) $= 3$ and $d_c = 4$).

The formulas applied are asymptotic equations for the limit of a large correlation length ξ e.g. near the spinodal. They are obtained from a systematic and successive coarsening of the space scale in the ψ level (Kadanoff coarsening, see Fig. 24d for the magnetic case). Dimension analysis can be helpful to discuss the parameter combinations they depend on. This coarsening (or decimation) gets mathematical group properties, which leaves the structure of formulas invariant but changes the parameters. Useful equations are obtained from the requirement that the observables must be invariant against the coarsening. For $\xi \to \infty$ the coarsening arrives at a fixed point. This corresponds to the critical (or spinodal) point, and the asymptotic equations correspond to thermodynamic equations near the critical point.

The critical exponents (see Sec. 4.2.5 below) can then be calculated from the ε series (Wilson).

(iii) The polymer-magnetism analogy is described in Ref. 4. The chain dimensions in good solvents, especially the excluded volume exponent in Eq. 1.39 (Sec. 1.4) can also be calculated by renormalization methods. We have large fluctuations here because, as discussed in the comment to Eq. 1.18, the coils are very sensitive to small forces. If this sensitivity is characterized by the number of chain segments (N degrees of freedom have a one-k contribution to ΔF), then it diverges for $N \to \infty$. Therefore, $\xi \to \infty$ for the critical state of a special spin system corresponds to $N \to \infty$ for a coil in good solvent, if an analogy can be established between the two systems.

5. Direct correlation function

The direct correlation function ($c(r)$, see Sec. 1.1, Eqs. 1.24, 1.28–29) is a ρ-level concept. It is short-ranged and is similar to a Mayer f function

$$f(r) = 1 - \exp(-\varphi(r)/kT) \tag{4.17}$$

with $\varphi(r)$ the intermolecular potential (between argon atoms in theoretical physics).

For polymer systems the method of direct correlation functions (see e.g. Ref. 77) tries to obtain information about $c(r)$ from inter- and intramolecular interaction of chain segments. [The situation is complicated as discussed in the comment between Eqs. 1.29 and 1.30]. Then the problem is to enlarge the spatial scale up to the ψ level. The Ornstein Zernicke equation (Eqs. 1.24 and 1.29) transforms $c(r)$ into the larger $g(r)$ scale. This is perhaps large enough to describe the scale of spatial smearing for the ψ-level description. A larger scale can be reached for polymers because the correlation along the chains are longer-ranged. But it seems to be difficult to calculate ψ-level properties from fine details of the direct correlation function.

6. General Langevin equation (GLE)

This method is a refinement (Mori, Zwanzig) of the analysis of Brownian motion. For all details the reader is recommended to refer to Ref. 49, while some mathematical relations are explained in the Appendix 1.

The main concept is the separation of a few interesting variables (ψ, mostly mean field or ψ level, often called "observables") from the rest (mostly ϱ level). Starting point is the Liouville equation Eq. 4.5. From the many q, p

variables a "set of dynamical variables" ($\psi(t)$, indices surpressed) is selected, the rest (the overwhelming number of coordinates) is projected in the background, see Fig. 24e. As proved by Mori, this projection can be executed accurately, and one obtains from Eq. 4.5

$$\frac{d\psi}{dt} = i\Omega\psi(t) - \int_0^t dsK(t - s)\psi(s) + f(t) \tag{4.18}$$

where both Ω ("frequency") and K ("memory") can uniquely be calculated from the Liouville operator using the selected projection. The "force" f is determined, and K is influenced by the background variables.

The general Langevin equation Eq. 4.18 reminds us of the linear response, Eq. 2.6. Additionally, we find a perturbation term $f(t)$ that renders the name of the equation: Langevin had tried to transform mechanical models in the ψ or mean field level by using stochastic forces $f(t)$ to animate the Brownian motion.

The Kramers equation for describing the overcoming of a potential barrier $U(x)$, see Fig. 24b, is an example for a differential form of the Langevin equation,

$$M\ddot{x} + M\gamma\dot{x} + dU/dx = f(t), \tag{4.19a}$$

$$\overline{f(t)f(t')} = 2\gamma MkT\delta(t - t'), \tag{4.19b}$$

with M the particle mass and γ the "coupling" constant to the heat bath, the same γ also defines the damping.

The general equation Eq. 4.18 can be solved,

$$\psi(t) = L(t)\psi(0) + \int_0^t dt'L(t')f(t - t') \tag{4.20a}$$

with a function L given by (as Fourier Laplace transform with z the complex frequency)

$$L(z) = [zI - i\Omega + K(z)]^{-1}, \tag{4.20b}$$

where I is the unit matrix. Thus L corresponds to the compliance or the modulus of linear response and must not be confounded with the Liouville operator.

The connection between response and correlation, the heart of the FDT, is here mediated by the Langevin force $f(t)$. Embracing the drift in L, we have $\overline{f(t)} = 0$. The fluctuations are "generated" by $f(t)$. In the spirit of the FDT, the f correlation is put as white noise,

$$\overline{f(t)f(t')} = 2\chi kT\delta(t - t'), \tag{4.21}$$

to avoid any additional characteristics of the heat bath. The susceptibility is denoted by χ, and kT comes from FDT-like considerations.

Now the correlation function can be calculated by $\psi^2(t) = \overline{\psi(t)\psi(0)}$, and the spectral density as a function of imaginary frequency z is

$$\psi^2(z) = ikT\chi/(z - \Omega + iK(z)). \tag{4.22}$$

This is the goal of the approach: If a reasonable projection can be achieved, the response $L(z)$ and the correlation $\psi^2(z)$ can generally be calculated in closed form (Eqs. 4.20b and 4.22). Similar to Eq. 1.6 for Brownian motion, a diffusion equation can also be connected to Eq. 4.18. Such equations are called Fokker Planck equations and are an important tool for the practical calculations.

The transition to the Fokker Planck equation and its evaluation usually requires drastic approximations. Sometimes the index for the set of dynamic variables is a ψ level position $\boldsymbol{r}$ or the corresponding scattering vector $\boldsymbol{q}$. Then, for isotropic cases, $\psi \rightarrow \psi_q(t) = \psi(q, t)$, and it is tried to give the Langevin operator, L in Eq. 4.20b, the "Brownian" form of Eq. 2.65, but now with the diffusion coefficient being a function of q and z,

$$L(q, z) = [z + q^2 D(q, z)]^{-1}. \tag{4.23}$$

This means that the models for Ω and K are constructed in the spirit of a generalized diffusion coefficient $D(q, z)$. In a dispersion zone, however, (see Sec. 6.10, below) $D(q, z)$ is a rather complicated function and the diffusion character of L can be lost.

A general problem is the selection of ψ from microscopic variables. When selecting velocities the kT of Eq. 4.21 is implied in the equipartition theorem of statistical mechanics. If the $\boldsymbol{r}$ in $\psi(\boldsymbol{r})$ stands for (minimal) subsystems, then kT follows from the FDT. If ψ is not a thermodynamic observable, then the kT in Eq. 4.21 should be subject to discussion; practically it can be absorbed in the susceptibility χ.

A special development of GLE is the *mode coupling method*. The memory $K(t - s)$ in Eq. 4.18 is connected with the "modes" $\psi(q, s)$ itself, e.g., symbolically, $K \sim \psi^2$. The result is a nonlinear, collective scheme, e.g. for a correlation function

$$G(z) = \frac{1}{z + i/d(z)} \tag{4.24a}$$

and the subsequent coupling of d with G,

$$d(z) = d_0 + \lambda \int_0^\infty dt\, e^{izt} G^2(t) + \text{higher order terms.} \tag{4.24b}$$

The coupling parameter λ depends on the state, e.g. $\lambda(T)$, and can, in principle, be calculated from partition functions of the reference system (by closing a hierarchy), successively if higher orders are important. The basic idea of the mode coupling theory (see Ref. 178) is that, as for the example of Eqs. 4.24a–b, a fluctuation of a given dynamical variable decays predominantly into pairs of hydrodynamic modes.

7. Random phase approximation (RPA)

This approximation (Ref. 78) was introduced into polymer science by Edwards and de Gennes (Ref. 78a and 78b). It is connected to a local variant of the FDT. Let us imagine fictitious external fields (u_A, u_B) that couple to order parameters at ψ level points $\boldsymbol{r}$, $\boldsymbol{r}'$ and create spatial ψ level inhomogeneities. Assume that this corresponds to an experimental situation, and let the effect be connected with a second order contribution to the free energy

$$\Delta F \approx \frac{1}{2kT} \int d\boldsymbol{r}' d\boldsymbol{r}\, S_{AB}(\boldsymbol{r} - \boldsymbol{r}') u_A(\boldsymbol{r}) u_B(\boldsymbol{r}') \tag{4.25}$$

where $S_{AB}(\boldsymbol{r} - \boldsymbol{r}')$ is the corresponding correlation function [e.g., for ψ being the partial density S_{AB} is a density-density correlation function; neglecting indices we have $S = G - \bar{n}$ in terms of Eq. 1.20. The name "random phase approximation" is derived from the breaking of higher correlations by randomization of phases for the case of electron systems. This closing of a hierarchy of equations leads to the second order of the approximation.] The order parameters are defined via a system susceptibility χ_{AB} related to the u action,

$$\psi_A(\boldsymbol{r}) = \sum_B \int d\boldsymbol{r}' \chi_{AB}(\boldsymbol{r} - \boldsymbol{r}') u_B(\boldsymbol{r}'), \tag{4.26}$$

analogously for ψ_B etc. This convolution is Fourier transformed into a product that can be rearranged for explicit u. Using the FDT for the structure factor (see e.g. Eqs. 2.60 or 2.64a),

$$S_{AB}(q) = kT\chi_{AB}(q) \tag{4.27}$$

and identifying it with S_{AB} from Eq. 4.25 we obtain, e.g. for u_A

$$u_A(q) = \sum_B \chi_{AB}^{-1}(q)\psi_B(q) = kT \sum_B S_{AB}^{-1}(q)\psi_B(q). \tag{4.28}$$

Hence, from a Fourier transform of Eq. 4.25, we obtain in the second order a typical RPA formula:

$$\Delta F/kT \approx \frac{1}{2}\int dq\{S_{AA}^{-1}(q)|\psi_A(q)|^2 + S_{BB}^{-1}(q)|\psi_B(q)|^2 + 2S_{AB}^{-1}(q)\mathrm{Re}(\psi_A^*(q)\psi_B(q))\}. \tag{4.29}$$

The reciprocal structure factors, $S^{-1}(q) = 1/S(q)$, stemming from the rearrangement of Eqs. 4.25–27 in favour of u are characteristic for RPA. The relation to the FDT (if kT is not absorbed in χ) is justified only if the positions $\boldsymbol{r}$, $\boldsymbol{r}'$ can be identified with subsystems and when the u's correspond to a situation where they can be part of an $f du$ term in the First Law fundamental form for these subsystems.

Typical approximations can be discussed, for instance, in case of scattering from concentration fluctuation in a binary polymer system. The three partial structure functions (S_{AA}, S_{BB}, S_{AB}) can be substituted by one S if the density is constant (see Ref. 79, "Babinet principle"). Applied to chain segments, with $\boldsymbol{r}$ being their position, the Debye structure factor, or even its approximation Eq. 1.15b can be used. Eq. 4.28 suggests that the deviations from the "ideal" structure factor should be considered as an energy term V, $u = u_{\text{ideal}} + V$. Hence

$$S_{\text{collective}}^{-1} = S_{\text{noninteract.}}^{-1} - V(q). \tag{4.30}$$

In the Flory Huggins approximation we have $RT\chi$ for interaction, with χ the Flory Huggins χ parameter (and not a susceptibility). Delegating the identity question (of what a segment is) to the FDT, and reducing by RT, we obtain from Eq. 4.29

$$\frac{1}{S_{\text{coll}}(q)} \approx \frac{1}{\phi_A N_A} + \frac{1}{\phi_B N_B} - 2\chi + \frac{1}{18}\frac{a^2}{\phi_A \phi_B} q^2 \tag{4.31}$$

where the a term (with a being the structure length) is the Tyndall effect of the ρ level. Eq. 4.31 (Ref. 80) links the correlative fluctuation with an energy, corresponding to a coarsening of Eq. 1.31-type equations from the ρ to the ψ level. The q^2 term corresponds, in a way, to the Ornstein Zernicke approximation, see Sec. 10.2.

8. Directory

Table III gives a memo for the theoretical methods mentioned. As a rule, no

Table III. Short hand characteristics of several theoretical methods

Method	Characteristics
Gibbs distribution	Handling of spatial configuration
Mean field	Simplification of particle or ϱ level environment
Direct correlation function	Linking the $\boldsymbol{r}$ point of the ψ level with a ϱ level structure
Path integral	Handling of field configurations
GLE (General Langevin Equation)	Generalization of diffusion and diffusion coefficient
Mode coupling	Collectivity
RPA (random phase approximation)	Linking reciprocal structure factors with energy
Renorm group	Handling large ψ fluctuations, calculating (critical) exponents

method can supply an exact solution for a complicated situation. The trend is to use combinations of two or more methods.

One can ask for the reason of the broad application of the scattering vector $\boldsymbol{q}$ instead of the position difference in space $\boldsymbol{r}$. Small r, i.e. the ρ level, is usually not a subject for series expansion. This can be seen from Fig. 2b. Small q, however, correspond to large spatial scales. Expanding the scattering effects for small q one can catch some global labels of ρ level details for a ψ-level description, similar to the Born approximation in particle scattering or Tyndall effects in light scattering. The ψ-level functions are usually smooth at small q, see Fig. 1e. A further advantage of the q picture is the mathematical property that a spatial convolution is a product in the q space. Moreover, RPA shows a way to connect the q picture with thermodynamic potential parameters. The cooperativity and ψ collectivity cannot easily be disentangled in direct calculations with the Gibbs distribution.

4.2 Scaling and crossover

The term scaling is used in different situations with different semantic meanings. The general application is best described by the dictionary formulations "to arrange in a graded series" or "the size of a sample (as a model) in proportion to the size of the actual thing". This section is to describe some typical examples.

1. General scaling principle (GSP) again

This term was first mentioned in the Introduction and again in Sec. 2.8, Figs. 16c–f. General scaling means a dispersion law inside a dispersion zone, and it can also be applied to compare two different dispersion zones with similar molecular mechanisms (for a material in a given state): The shorter the mode length the shorter the corresponding time. This means that we assume that typical molecular motions are the slower the larger their spatial measure is.

2. Absence of a typical length

The absence of a typical length is often called *self-similarity*. If we have, for instance, a spectrum of typical modes (e.g. flow or hydrodynamic modes, Rouse modes) inside a larger mode length interval with no typical length within, then all responses are *power functions* of ω or τ. All (and only) curves of the form

$$y = ct^p, p \text{ given,} \tag{4.32}$$

can be identified by redefinition (rescaling) of the axis labels. This is possible because we need not consider a special mark at the axes mapped from a length to y or t, e.g. by a dispersion law. Self-similarity is governed by a *Theorem* (proved in e.g. Ref. 39): Power laws are the only functions that are scale invariant in the following sense

$$J(ct) = M(c) \cdot J(t). \tag{4.33a}$$

This means that Eq. 4.33a is equivalent to

$$J = c^p J(t) \quad \text{or} \quad J = J(1)t^p. \tag{4.33b}$$

Observe the symmetry between the parameter c and the variable t. A counter example was discussed in Sec. 1.4. Nanoheterogeneity in semi solutions destroys (or restricts) scaling with M or c.

3. Reduction of variables

The reduction of variables is also called scaling. A well known example is the

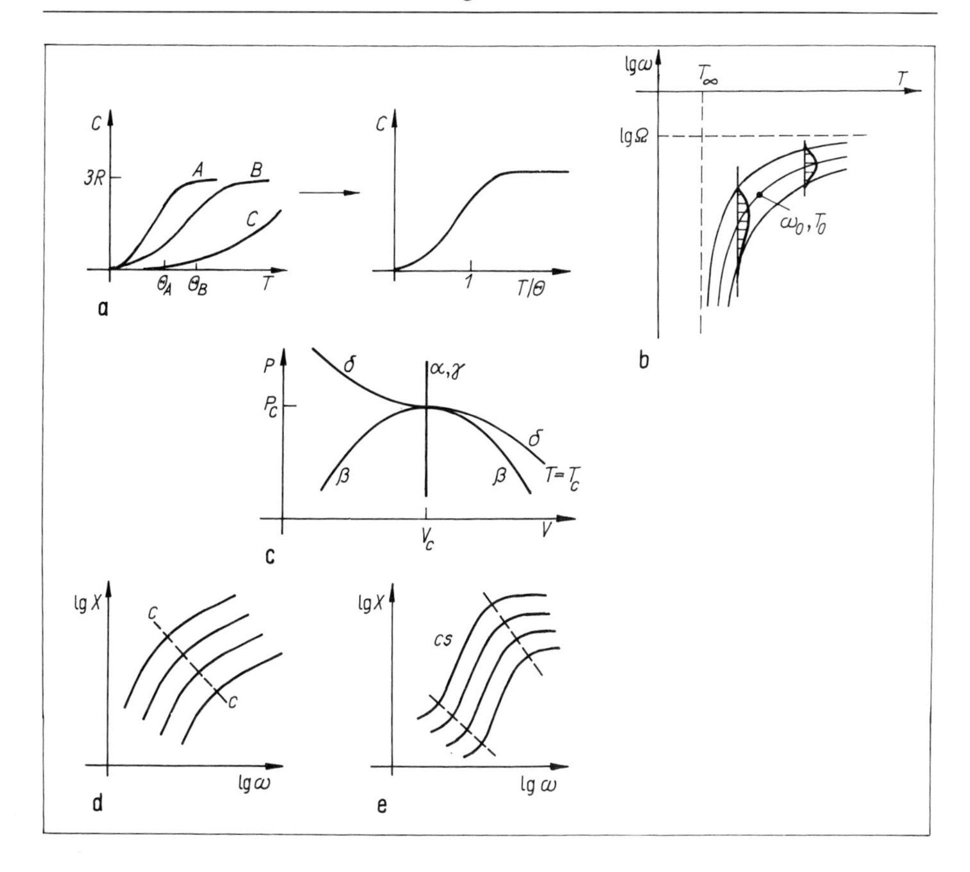

Fig. 25. Scaling.
a. Reduction of Debye's law for different substances (A, B, C). b. WLF scaling of a spectrum (hatched). c. The different ways for the critical exponents in a *PV* diagram of critical state (*P* pressure, *V* volume; the same picture is used for the definition of critical exponents in a binary mixture, with $P\phi$ or $T\phi$ instead of *PV* or *TV*, ϕ volume fraction of one component). d. Formal definition of a crossover (C . . . C) for a hypothetical property *X* as a function of frequency ω and one parameter (not specified). e. (Formal) confined scaling region (c.s.).

reduced representation of the Debye law for the heat capacity at low temperatures,

$$c(T, \theta) \;=\; c(T/\theta) \sim (T/\theta)^3 \tag{4.34}$$

with θ the Debye temperature, see Fig. 25a. Another example is the reduced van der Waals equation, the thermodynamic correspondence principle.

The fit of a spectrum $f(\omega)$ with logarithmic broadness $\Delta \lg \omega$ into a set of hyperbolas with common asymptotes ($\lg \Omega$, T_∞) in a $\lg \omega$–T diagram is called *WLF* (or VFT) *scaling*. From Fig. 25b and properties of hyperbolas we learn that

$$f(\omega, T; \Delta \lg \omega, \omega_0, T_0, \Omega, T_\infty) = f(a_T \omega, b_T \Delta \lg \omega) \tag{4.35}$$

with a_T, b_T the *shift factors* for, e.g., the maximum and for the broadness of the spectrum, respectively,

$$\lg a_T = -\lg \left(\frac{\Omega}{\omega_0}\right) \cdot \frac{T - T_0}{T - T_\infty} \left(= \frac{c_1^0 (T - T_0)}{c_2^0 + (T - T_0)} \right), \tag{4.36}$$

$$b_T = (T - T_\infty)/(T_0 - T_\infty), \tag{4.37}$$

where

$$c_1^0 = \lg (\Omega/\omega_0), \quad c_2^0 = T_0 - T_\infty \tag{4.38}$$

are the so-called WLF constants that are related to a reference point (ω_0, T_0) onto one hyperbola. Guiding the whole spectrum by the hyperbola set means that its broadness is inversely proportional to the temperature difference to T_∞ (the *Vogel temperature*), $\Delta \lg \omega \sim (T - T_\infty)^{-1}$.

4. Reduction and dimensional analysis

The linkage between these two tools is demonstrated in de Gennes' book "Scaling principles . . . " (Ref. 4). Dimensional analysis is used to see how important parameters should be combined for the reduction of variables. This is completed by physical ideas to estimate the kind of functions, see also Sec. 4.2.2. Let us consider an example from the book. Calculate the change of a Gauss-chain end-to-end distance r by application of a pair of forces $(f, -f)$ to the coil (ends). From Eq. 1.8 we see that with no force

$$R_0^2 = \overline{r^2} = N b^2. \tag{1.8}$$

With force we expect a new r, $r \to r_f$. Which are the parameters it can depend on? The answer is

$$r_f(kT/R_0, f). \tag{4.39}$$

From dimensional analysis we see that the dimensionless ratio r_f/R_0 can only depend on a dimensionless combination of parameters. That is

$$r_f/R_0 = g(f R_0/kT) \tag{4.40}$$

with g a function to be determined. Provided that there is no typical length between segment length b and coil radius R_0 in the ψ level, as is true for Gauss chains, and if f is too small to define a new length by coil deformation we obtain a power law from Eqs. 4.32–33,

$$r_f/R_0 \sim (fR_0/kT)^\kappa. \tag{4.41}$$

Moreover, the absence of a length implies $r_f \sim N$ (or a linear region between r_f and f). This means $\kappa = 1$ and therefore

$$r_f \sim kT\ R_0^2 f. \tag{4.42}$$

This relation can, of course, also be obtained from the free energy Eq. 1.18, $f = -\partial\Delta F/\partial r$. The calculation of the sum of states has thus been substituted by scaling arguments.

5. Critical scaling (Widom scaling)

Many thermodynamic variables become singular in the vicinity of critical points. Because of the absence of a typical length within the long-ranging fluctuations of critical opalescence – no length between the ρ level and their correlation length ξ – the singularities are described by poles with power laws. The equation of state can be reduced and one obtains (Ref. 81) homogeneous functions (in the sense of Euler), e.g. for the potential:

$$G(p, t) \approx |h|^{1/\delta} g(t/|h|^{1/\beta\delta}) \tag{4.43a}$$

with t, p, h being small variable differences to the critical point,

$$t = T - T_c, \quad p = P - P_c, \quad h = p - bt, \tag{4.43b}$$

where $b = (\partial p/\partial T)_V \approx$ const and $g = g(x)$ is a continuous function of the argument $x = t/|h|^{1/\beta\delta}$. In the magnetic case, h would be proportional to the field.

The exponents (α, β, γ, δ) are called *critical exponents* (or indices) and describe the singularities on the ways indicated in Fig. 25c,

$$C_V \sim t^{-\alpha}, \quad v_{\mathrm{coex}} \sim p^{1/\beta}, \quad (\partial p/\partial v)_T \sim t^\gamma, \quad p \sim v^\delta, \tag{4.44}$$

with $v = V - V_c$. Because of the network of thermodynamic relations (existence of entropy or of the potential $G(p, t)$) only two of the four exponents are independent; the exponents α and γ can be calculated from β and δ by the "scaling laws"

$$\alpha + 2\beta + \gamma = 2, \quad \beta\delta = \beta + \gamma. \tag{4.45a}$$

The correlation length varies with $\xi \sim |t|^{-\nu}$ along the critical isochore. Neglecting fine details one finds

$$\nu \approx \gamma/2. \tag{4.45b}$$

6. Crossover

The more or less broad transition of curves or surfaces from one (scaling) region to the next one is called crossover, see Fig. 25d. Usually the (power) law is changed, and different parameters are observed on the two sides. A crossover is usually connected with a typical length, it has nothing to do with phase transitions and, for dynamics, it often corresponds to a boundary of a dispersion zone.

7. Confined scaling

Typically the dispersion zones for polymers, in a lg ω–T diagram, are stripes or bands, not too broad when seen from a large scale of many logarithmic decades, see Fig. 27c, below. The term confined underlines the fact that there are boundaries of the dispersion zones or the "plateau zones" between them. Examples are Rouse or Rouse Zimm modes of polymer coils (Sec. 5.1–2, below): The small-scale boundary is given by the segmental motion with a characteristic length in the ρ level, while the large-scale boundary is a crossover to hydrodynamic modes of a Newton liquid (see Eq. 2.18) or, for entanglements, to the rubbery plateau zone. This crossover is characterized by the coil radius in the former, or the entanglement spacing in the latter case. For lower temperatures the Rouse scaling in concentrated polymer systems is mapped into the main transition. The lower spatial boundary is then the confined flow zone after the glass transition (length scale $\xi_a \approx 2\,\text{nm}$) making the segments mobile), and the upper boundary is the entanglement spacing again hindering this flow. Such boundaries will be called scaling limits.

There are several possibilities for scaling limits: true structures (as segments or mean distances between crosslinks of a network), or thermokinetic structures (e.g. entanglements), or correlation lengths (lower ones as the glass transition length for the confined flow, or upper ones as the correlation length for the dispersion zone considered).

In this book, the term *confined scaling* will mostly be restricted to the latter case, i.e. when the large scaling limit is the correlation or a characteristic length ξ_a of the dispersion zone under consideration. So, for the glass transition, the scaling is confined between the particle diameter (≈ 0.5 nm) and the characteristic length (expressed as a diameter, $2\xi_a \approx 4 \ldots 5$ nm for polymers). As the modes close to the upper scaling limit are slow, they have aspects of a thermokinetic structure, if one applies this concept of Sec. 3.9 also inside – across – the dispersion zone. The glass transition as a whole is a correlation that describes a thermokinetic structure only for local modes at much higher frequency (β relaxation). The term correlation in the own dispersion zone means peculiarities of the particle (or segment) motion that cause them to be distinct from completely random motion. In the confined scaling region the scale of peculiarities must scale with the length, or time (*length-time scaling* of correlation), because there is no further typical length within. According to the general scaling principle, a larger length corresponds to a larger time. The correlation will generally be weaker at larger distances, and the characteristic length signalizes the crossover to completely random motion.

[If the picture of Fig. 21e is correct for a thermodynamic experiment, and if the minimal subsystems are smaller than the natural subsystem (of order ξ_a^3) for glass transition, then the small-scale self experiments may, in essence, only collect the typical frequencies from inside, i.e. the higher frequencies. This means that there are reasons for a length-time scaling of statistical independence. Consider the situation from a given mobility inside confined scaling. Then the larger scale with lower frequency is more correlative, and the shorter scale with higher frequency is statistically more independent.]

4.3 Computer simulation

The aim of computer simulation is to model complicated (mostly ρ level) molecular situations that are too complicated to be imagined otherwise. Two methods are applied: The classical Monte Carlo MC method tries to calculate the configuration integral Eq. 4.2 and related distribution functions. One starts from a random q space (phase space) configuration for which the energy $U(q)$ and $\exp(-U/kT)$ are calculated. There are algorithms (e.g. Metropolis algorithm Ref. 82) with the property that few configurations are rather representative for the configuration integral. The second method – molecular dynamics MD – stakes on the numerical integration of Newton's equation Eq. 4.4 for a larger number of particles with interaction. The elementary time step must be small, e.g. 0.01 ps, and it is of advantage to use periodic and stabilizing boundary conditions (see e.g. Ref. 83). Periodic boundaries are to minimize artificial walls of subsystems. One can calculate $E(t)$, $V(t)$ and $N(t)$ of the subsystem and can use the PLE and related methods to calculate thermodynamic variables and their fluctuations, see Sec. 3.7.

Modern computers allow to handle about 1000 interacting particles or segments over several orders of time compared to the elementary step. To compare computer simulations to analytical methods in polymer science we closely follow the judgment of Binder, Ref. 84. Disadvantages of analytical methods are: prejudice cannot be excluded by the selection of seemingly important modes or degrees of freedom, the (necessarily few) parameters are intended to be adjustable, and the necessary mathematical simplifications and approximations cannot be easily surveyed in any case (e.g. the next order cannot be handled). It is possible that a parameter adjustment is also satisfactory with irrelevant models. The advantages of computer simulation are: the number of degrees of freedom is not so limited, pretests can be made on a microscopic basis, one is not so restricted to ideal cases, and the reliability of models can be tested by a broad parameter variation. The latter can be a help to find the truly essential conditions, informative parameters, configurations etc. for the phenomenon under consideration. Moreover, many experiments cannot be analyzed without the aid of parallel computer simulations.

Of course, analytical methods also have advantages and computer simulations also have disadvantages. One point is what is called understanding. Analytical methods are closer to words than computer simulations.

The main problem for present day computer simulation of relaxation is the large amount of time (compared to the elementary step) necessary e.g. to establish a WLF guided dispersion zone, especially when the elements are not particles or segments but larger units as for the flow transition.

II Relaxation

Chapters 5 and 6 in this second part of the book are devoted to linear response on external perturbations. We know from the fluctuation dissipation theorem that relaxation and fluctuation are equivalent phenomena in the linear region. These chapters could, therefore, also be named "polymer dynamics". The restriction to the linear region does not mean that only small molecular oscillations etc. can be described. The liner response is generated by the full amplitude of highly nonlinear and chaotic motion of molecules, chains and their segments in equilibrium. The term "linear" only means that, compared to natural fluctuation, the external action does not deform the correlation function and spectral density of fluctuation.

In Chap. 7 we consider the nonequilibrium situation that is generated when, during cooling, a part of fluctuation becomes too slow to contribute to the experimental response. In a sense the thermokinetic structures of the equilibrium at $T > T_g$ become real structures at $T < T_g$, when related to the experimental time given by the cooling rate. These real structures influence the action of the remaining real-time fluctuation and, on the other hand, can relax themselves (structure relaxation), e.g. during annealing at T not too far from T_g.

Some typical examples of relaxation are described in Chap. 8.

5. Three simple models

This chapter deals with three concepts of extraordinarily frequent use in polymer papers: activation energy, Rouse (or normal, also chain) modes, and reptation. The basic concepts are introduced in their simplest version followed by a discussion and criticism. A modern technical presentation is not in the scope of this book and can be found, for Rouse modes and reptation, in the literature (e.g. Ref. 6). A systematic description of polymer dynamics is dared in Chap. 6.

5.1 Barrier model. Arrhenius mechanism

Consider a particle in a potential $\varphi(x)$ according to Fig. 26a where two valleys are separated by an energy barrier of height ε_A (*activation energy*). The x coordinate can for example be an angle describing a conformation change of the polymer chain. Consider now an ensemble of such situations. The particles have a distribution of energies that is characterized, in equilibrium, by the Boltzman factor

$$w(x) \sim \exp\left(-\varphi(x)/kT\right). \tag{5.1}$$

(The phase space factor is neglected.) Kinetically, a particle at low energy has a high oscillation frequency (ν_0) given by the curvature of $\varphi(x)$ in the valley, see Fig. 26b. Having an energy distribution, however, also means to consider higher energies. We have a certain frequency ν for overcoming the barrier. The success/attempt ratio is a normalized frequency that can be related to the Boltzmann probability by

$$\begin{aligned} \frac{\nu}{\nu_0} &= \frac{\text{frequency per time}}{(\text{frequency per time})_0} = \frac{w(\nu)}{w(\nu_0)} = \frac{w}{w_0} \\ &= \exp\left(-\varepsilon_A/kT\right). \end{aligned} \tag{5.2}$$

A mechanism obeying

$$\ln \nu = \ln \nu_0 - \varepsilon_A/kT \tag{5.3}$$

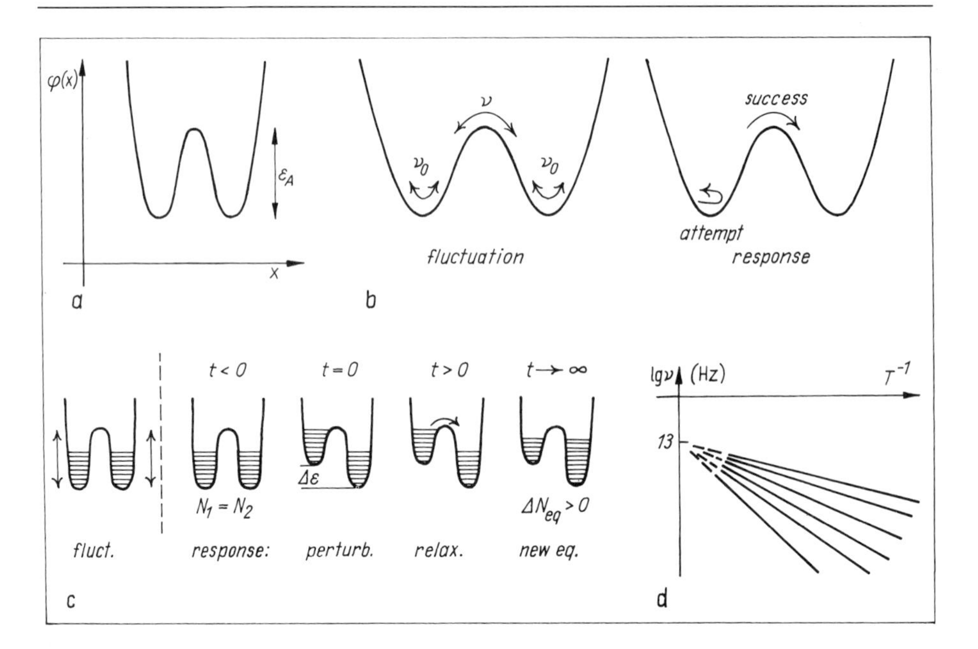

Fig. 26. Barrier model.
a. Activation energy ε_A for the potential function $\varphi(x)$. b. Fluctuation and response pictures are equivalent according to the fluctuation dissipation theorem FDT. c. Calculation of response. d. Heijboer's Arrhenius-plot correlation for local modes of several polymers.

is called an *Arrhenius mechanism*. This is a straight line in a log ν–T^{-1} plot (Arrhenius diagram) with the intercept ln ν_0 for $T^{-1} = 0$ ($T = \infty$) and the slope $-\varepsilon_A/k$.

It will be shown in the following that the experimental response of this process is an exponential decay (standard or Debye relaxation, Lorentz curve, see Eqs. 2.22–2.27 and Fig. 11). The thermal fluctuation of the environment can be imagined as a fluctuation of the valley depths. According to the FDT, Eq. 2.55, the response can be calculated from the system answer to the step perturbation (Fig. 26c): switch on of a valley difference ε at time $t = 0$.

The new equilibrium can be calculated from the new Boltzmann factors now being different for the two valleys. Characterize the energy distribution by horizontal lines: occupation numbers (N_i, $i = 1, 2$). In the old equilibrium ($\Delta\varepsilon = 0$, $t \leqslant 0$) we had

$$N_1 = N_2, \quad \Delta N \equiv N_2 - N_1 = 0. \tag{5.4}$$

The new equilibrium ($\Delta\varepsilon > 0$, $t \to \infty$) is

$$\frac{N_1}{N_2} = \frac{w_1}{w_2} = \frac{\exp(-(\varepsilon + \Delta\varepsilon/2)/kT)}{\exp(-(\varepsilon - \Delta\varepsilon/2)/kT)} \approx 1 - \Delta\varepsilon/kT, \tag{5.5}$$

where the approximation is for

$$\Delta\varepsilon/kT \ll 1. \tag{5.6}$$

The transition behaviour is calculated from a balance: change = access − loss. A kinetic equation of this structure that uses transition probabilities is called master equation. Then

$$d(N_2 - N_1)/dt = w_{12}N_1 - w_{21}N_2, \tag{5.7}$$

where the access to 2 is given by the transition probability from $1 \to 2$ times the occupation number of 1, and so on. The transition probabilities are calculated by success-to-attempt ratios again,

$$w_{12} = \nu \exp(-(-\Delta\varepsilon/kT)) > w_{21} = \nu \exp(-(\Delta\varepsilon/kT)), \tag{5.8}$$

with the ν from Eq. 5.3. Expanding with respect to the small parameter of Eq. 5.6 we obtain in the first order

$$d\Delta N/dt = \nu\Delta N + \nu\Delta N_{eq} \tag{5.9}$$

where ΔN_{eq} is the occupation difference in the new equilibrium that can be calculated from Eq. 5.5. The solution of the differential equation Eq. 5.9 compatible with $\Delta N \to \Delta N_{eq}$ for $t \to \infty$ is

$$\Delta N = \Delta N_{eq}(1 - \exp(-t/\tau)), \quad \tau = 1/\nu. \tag{5.10}$$

Since ΔN is an extensive variable it can be considered to be the relevant compliance. That is, Eq. 5.10 is the standard retardation for $N_1 \neq 0$, $N_2 \neq 0$, and dynamic loss factors are Lorentz curves as proposed.

Heijboer (Ref. 85) analyzed the Arrhenius plots for secondary relaxation in many polymers. For the maximum of shear loss factors ($\nu(T_m)$ or $T_m(\nu)$) the following correlation was obtained for local modes,

$$\frac{E_A}{\text{kcal/mol}} = \left(0.060 - 0.0046 \lg \frac{\nu}{\text{Hz}}\right) \cdot \frac{T_m(\nu)}{\text{K}}. \tag{5.11}$$

This implies that the limit frequency $\nu_0 = \nu(T = \infty)$ has the same order for these polymers

$$\lg(\nu_0/\text{Hz}) = 0.060/0.0046 = 13.0(\pm 1) \tag{5.12}$$

corresponding to the oscillatory frequency of monomeric units (or particles) in condensed systems, see Fig. 26c.

Approximate values for activation energies are

$$\begin{array}{l} 20\,\text{kJ in polymer solutions (high temperatures),} \\ 25\ldots 60\,\text{kJ in concentrated polymer systems.} \end{array} \qquad (5.13)$$

The latter values are also valid for secondary relaxations (local modes) in the glassy state, $T < T_g$. For solutions, where "high temperature" means considerably above the glass temperature, the barrier height can be estimated to come 50/50 from the conformation barrier and from the influence of the segment environment (viscosity).

5.2 Discussion

The barrier models were qualified in the 1930's by Eyring and Kramers, see Ref. 86 and 87. The point is to consider lengths and angles (phase space) in addition to energy, e.g. a simple energy mountain over two coordinates, x_1 and x_2. The barriers turn into saddles that can be connected with a statistical weight (transition state). The activation variables in the exponent get a thermodynamic meaning that is expressed e.g. by an activation entropy S_A. The universal Eyring formula for the attempt frequency, $\nu_0 = 2kT/\hbar$, (from $\hbar\omega \approx kT$) was modified by Kramers to include properties of the potential function $\varphi(x_1, x_2)$. He analyzed the activation on the basis of (his) generalized Langevin equation, Eqs. 4.19a–b.

In polymers, activation models with a few x coordinates correspond to secondary relaxations SR, e.g. the motion of side groups SG, or local modes LM of the chain itself, possibly with a mutual influence of both (Ref. 85). These are straight lines in an Arrhenius diagram (Fig. 27a), but experience shows that their spectrum is much wider than the 1.2 logarithmic decades of the Lorentz curve. This is linked with a distribution of activation energies of spatially separated processes, parameterized by an interval $\Delta\varepsilon_A$, due to the disorder of real ($T < T_g$) or thermokinetic ($T > T_g$) structure. Below a certain temperature (T_Δ, from $\varepsilon_A/kT_\Delta \approx 1.2$), the broadness of the spectrum is then determined by $\Delta\varepsilon_A$,

$$\delta \ln \omega \approx \Delta\varepsilon_A/kT \sim 1/T, \quad T < T_\Delta, \qquad (5.14)$$

see Fig. 27b. The proportionality to T^{-1} is only valid for $\Delta\varepsilon_A$ being independent of T. This condition is usually not fulfilled because $\Delta\varepsilon_A$ is heavily

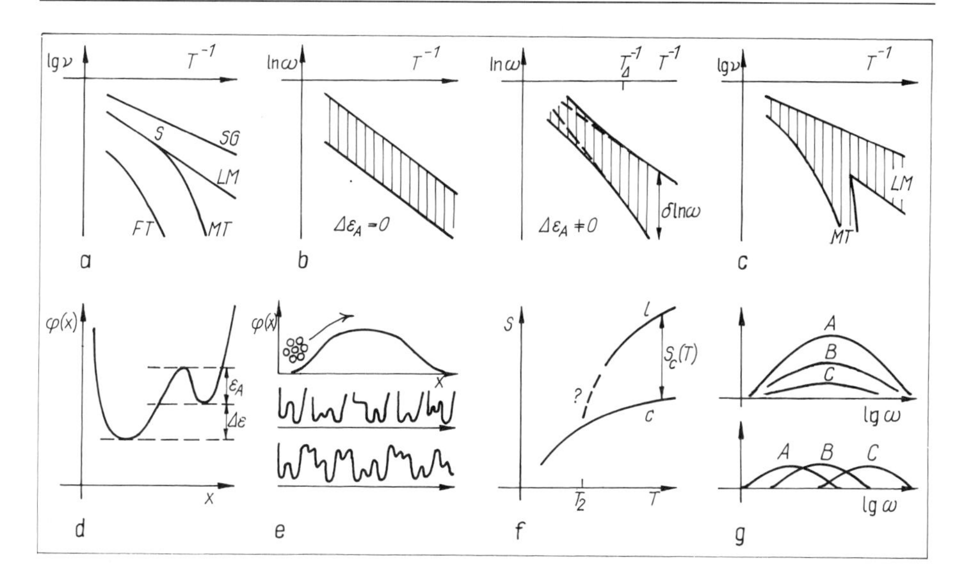

Fig. 27. Discussion of the barrier model.
a. Arrhenius diagram of an amorphous polymer like polystyrene. SG side group motion, LM local mode (LM + SG: secondary relaxation), MT main transition, FT flow transition. S splitting point between LM and MT. b. Spectral width of local mechanisms with (on the right) and with no (left) distribution of activation energy $\Delta\varepsilon_A$. c. Spectral width of LM and MT, schematically. d. The two elements of the energy landscape: activation barrier ε_A and energy step $\Delta\varepsilon$. e. One-dimensional simplification of an energy landscape. Upper part: wrong picture of a "thermodynamic" energy barrier. Middle part: With singularities from steep particle (or segment) repulsion. Lower part: Accessible landscape along ways avoiding the singularities by using the next saddles. This landscape slightly depends on temperature. f. Entropy extrapolation crisis at Kauzmann temperature T_2. S_c configurational entropy, l liquid, c crystal. g. Arrangement of different activities A, B, C for a broad spectrum of local modes (upper part) and a cooperative process (lower part), schematically, see also Fig. 16e.

influenced by the density due to steeply repulsive intermolecular potentials (see Ref. 88). Roughly, a 5% enlargement of the interparticular distances is sufficient to transform the broad spectrum into a Lorentz curve. Moreover, the real mechanism is heavily influenced by the molecular environment. Packing effects are often more important than overcoming a given barrier, and a compliance of the environment is to be taken into account.

Activation is a simple model to generate arbitrarily large time ratios (ν_0/ν)

in short range molecular arrangements. But the main transition MT and the flow transition FT cannot be described by Arrhenius processes, not even approximately. At low temperatures the relaxation times grow successively larger than calculated by activation. They are represented by strongly downward curved lines in an Arrhenius plot (see Fig. 27a again) that may be extrapolated to finite asymptotic temperatures, e.g. to $T_\infty > 0$ and $T_{\infty F} > 0$, the Vogel temperatures for MT and FT, respectively. In case of WLF scaling, Sec. 4.2.3, the broadness of the spectrum diverges according to Eq. 5.14, i.e.

$$\delta \ln \omega \sim (T - T_\infty)^{-1}, \tag{5.15}$$

see Fig. 27c.

The question of a finite Vogel temperature, $kT_\infty > 0$, is one of the most essential problems in relaxations of complex systems. It cannot be explained by an ε_A enlarged to a "thermodynamic" order of smallest subsystems because of two reasons. (1) A large ε_A value, $\varepsilon_A < \infty$, would mean that $T \to 0$ is still the only possibility for $\lg \nu \to -\infty$. [There is also no sense in interpreting a tangent at the curve as a large "apparent" activation energy because their intercept would be larger than the segmental frequency, $\lg (\nu_0(T)/\mathrm{Hz}) \gg 13$.] (2) According to Simon, Ref. 89, there is no thermodynamic sense in describing the freezing-in at the thermal glass transition by such a thermodynamic activation barrier because this would be in conflict with the reversibility concept used in the Nernst Simon form of the Third Law, see Chap. 7.

The way out of this problem was a more serious treatment of cooperativity. Goldstein (Ref. 90) pointed towards the necessary enlargement of the number of coordinates x in $\varphi(x)$. Hence the energy hypersurface is called *energy landscape*. It is characterized by a large number of barriers, valleys, and saddles of different height and broadness. The thermokinetic structure of the dynamic glass transition is reflected by some disorder (amorphicity) of the landscape. But there are delicacies in this structure because it is generated by a larger-scale optimization process in the equilibrium.

The cooperativity of this conception will be described in Sec. 6.2, below. It includes two aspects: space and time, or thermokinetic structure and correlation functions. The space aspect is characterized by a relation between energy differences of different parts of the landscape, and the time aspect is characterized by a certain sequence of molecular activation steps, i.e. a sequence of sufficiently random success/attempt processes at different but correlated regions of the landscape. We have therefore two essential elements, energy steps $\Delta\varepsilon$ and barrier heights ε_A, see Fig. 27d. The energy landscape of one cooperativity region must be described by a broad distribution of both, $\Delta\varepsilon$ and ε_A (Fig. 27e), reflecting the larger spatial correlation (volume V_a) and broad spectra (broadness $\delta \ln \omega$).

There is no sense in separating the time and space aspect, they are tightly interwoven. It is not possible to find the full spectral broadness in parts smaller than the cooperativity region, and the full correlation length can only be established by experiments in times that correspond at least to the low-frequency end of the spectrum. The thermodynamic reflection of this situation is the natural functional subsystem of a glass transition, see Sec. 3.9. These subsystems are real representatives of the cooperativity, including their fluctuations.

[Of course, by measuring the mean temperature fluctuation ΔT of one mol glass-forming substance with a thermometer, we find $\Delta T \approx 10^{-12}$ K. But all measurements that have any reference to kinetic aspects — that means all measurements of glass transition phenomena — reflect the fluctuation in these substems, e.g. ΔT is of order 1–10 K.]

With respect to such a picture, Adam & Gibbs (Ref. 91) introduced the concept of cooperatively rearranging regions CRR for the (range of) cooperativity: volume V_a, particle number N_a. This concept is compatible with material homogeneity, i.e. a picture of static clusters is not relevant here. The point is that the size of a CRR is assumed to increase with decreasing temperature and that this increase is connected with the curvature in an Arrhenius plot. Thus the Vogel temperature $T_\infty > 0$ is necessarily connected with $N_a(T) \to \infty$ at $T = T_\infty$ (see also Sec. 6.4, below, especially Eq. 6.19).

The Adam & Gibbs approach is based on the entropy-temperature activity pair, i.e. on the statistical weight of the energy landscape in the lower part of Fig. 27e. In the sense of Eyring, this corresponds to an activation entropy S_A and they put

$$\ln (\nu/\nu_0) \approx -S_A/R. \tag{5.16}$$

The problem now is to connect $S_A \to \infty$ for $T \to T_\infty$ with $N_a \to \infty$. Since S_A is an extensive variable that grows larger with increasing CRR's, there should be no principal difficulties. To make formulas they factorize S_A in a thermodynamic and a kinetic part.

The thermodynamic factor reflects, at the end, the statistical weight of the energy steps ($\Delta\varepsilon$) of the energy landscape. Fig. 27f shows the entropy of liquid and crystal of a typical glass former. In the liquid, as compared to the crystal, additional entropy is generated that is called configurational entropy,

$$S_c(T) = S(l) - S(c) = \int_{T_2}^{T} \Delta C_p(T) d \ln T \tag{5.17}$$

with

$$\Delta C_p = C_p(l) - C_p(c) \tag{5.18}$$

being the differences of heat capacity between liquid and crystal. The integration starts at T_2 (*Kauzmann temperature*) that is defined by extrapolation to $S_c(T) \to 0$ from the high temperature side. $S_c(T) < 0$ would mean an "entropy crisis" because it is hardly imaginable that the entropy in the crystal is much larger than in the liquid (Ref. 92). [Observe that at the melting point $S(l) > S(c)$ because of thermal stability. It should further be mentioned that a general connection of metastability with glass transition is dangerous because there are thermal glass transitions also for stable substances, see Refs. 93–95.]

Linearizing the situation near T_2 would result in $TS_c \sim T - T_2$. The same relation would follow from $\Delta C_p \approx \text{const}/T$, for which there is some experimental evidence for good glass formers (Ref. 96). Then, if one will connect $\ln \nu \to -\infty$ with $S_c \to 0$ at T_2, and Adam & Gibbs do so, then $S_c(T)$ must be put at the denominator of the thermodynamic factor, $S_A \sim S_c^{-1}(T)$.

The kinetic factor is assumed to be given by activation, $\overline{\varepsilon_A}/kT$. Referring to the energy landscape, $\overline{\varepsilon_A}$ reflects some average of the elementary activation steps ε_A. Therefore we have $S_A \approx$ (kinetics) $\times$ (thermodynamics),

$$S_A = \frac{\overline{\varepsilon_A}}{kT} \cdot \frac{s_a^*}{S_c(T)}. \tag{5.19}$$

Since the kinetic factor is dimensionless and not extensive, the thermodynamic factor must be corrected by an entropy s_a^* in the numerator and must be made extensive (increasing with size of CRR, $s_a^* \sim N_a$).

Adams and Gibbs put $s_a^* = k \ln W_B$ with a Boltzmann probability for the statistical weights of all the activation steps in a CRR. They put, therefore,

$$N_a = N_A s_a^* / S_c(T) \tag{5.20}$$

where the Avogadro constant N_A relates S_c to one mol. That means that their "derivation is based essentially on the assumption of independent and equivalent subsystems, i.e. on the premise of weak interaction of a cooperativity region with its environment." A linear correlation between lg ν (or lg τ, ln η, etc.) and $1/TS_c(T)$, based on Eqs. 5.16 and 5.19, is called *Adam Gibbs correlation.*

In total we have

$$\ln(\nu/\nu_0) = Q/R(T - T_2) \tag{5.21a}$$

which corresponds to a WLF curve for Q = const if the Kauzmann and Vogel temperatures are the same,

$$T_2 = T_\infty . \tag{5.21b}$$

This equation is confirmed for several good glass formers (Ref. 97).

Unfortunately, the size of cooperativity cannot be uniquely calculated from this approach. Q = const means that there were no crucial difference between the weights s_a^* and the thermodynamic Δc_p. When supposing that s_a^* does not become singular at T_∞ (which is difficult to connect with $N_a \to \infty$ for $T_\infty > 0$) then we obtain from Eq. 5.20,

$$N_a \sim (T - T_\infty)^{-1}, \quad \xi_a \sim N_a^{1/3} \sim (T - T_\infty)^{-1/3}. \tag{5.22}$$

A more symmetrical partitioning of the singularity at $T \to T_\infty$ between s_a^* and S_c would be, for instance, $s_a^* \sim N_a^{1/2}$. This is suggested for the case when s_a^* represents a fluctuation of an extensive variable in a CRR, related to the fluctuation formula $\Delta C_p = \overline{\Delta S^2}/k$. Then we would have $\xi_a \sim (T - T_\infty)^{-2/3}$.

There are several serious attempts to understand the glass transition in the time of frequency picture only. For instance, Ngai (Ref. 98) asked the question if the slow motions ($\tau > 100$ ps) can be described as new excitations with very low energies, $E/k < 0.1$ K. Relaxation of given few levels systems would be modified by such a random background in the direction of broader spectra. Many universal results can be calculated from such a picture (coupling model) which are in good accordance with natural phenomena (e.g. Ref. 129, p. 442). A final specification, however, seems to call for spatial aspects to be included as well.

Finally let us consider the question of what the indications for cooperativity are at all, besides $T_\infty > 0$. The experiment (FDT) cannot directly distinguish between a distribution of isolated local modes with different activation energies ε_A and a cooperative process (see Fig. 23b).

One possibility for detecting cooperativity and their lengths is the arrangement of activities across the dispersion zone (see Secs. 2.7 and 2.8). If, for a distribution of local modes LM, the dispersion for one local mode is much smaller than for the distribution ($T < T_\Delta$), then, of course, all activities have the same maximum location and the same functional form (Fig. 27g) provided that the different activities of the local mode are not differently influenced by the parameter ε_A.

A collection of different processes that can respond differently to different perturbations, however, would result in a different location and form of different activities, see lower part of Fig. 27g. A cooperative process can be expected to belong to this case. The experimental experience is that for the main transition MT (Ref. 52–54), and for the flow transition FT (Ref. 99) different moduli (and separately, different compliances) have different forms and are located at different frequencies. The fact that they belong to the same primary spectrum is indicated by the common guiding of the activities in an Arrhenius diagram. In contrast, the activities for the so-called γ relaxation in polyethylene – an Arrhenius process – are conform similar to the upper part of Fig. 27g.

Therefore we can confirm the indication from $T_\infty > 0$ that MT and FT are cooperative processes, whereas the secondary relaxations are expected to be more or less local.

Another possibility to distinguish between local and cooperative processes might be the dependence of typical relaxation times on the label or probe size used to investigate the relaxation, see e.g. Ref. 100. Some other evidences for the characteristic length of glass transition are indicated in Refs. 101 and 150a.

5.3 Rouse modes

Rouse modes describe the relative segment motion of single Gauss chains in solution or melt approximating the interaction to solution molecules or to segments of other chains by a mean-field friction coefficient ζ. In contrast to the mechanical models of Sec. 2.1 the spatial aspect of chain segment distribution occurs explicitly. The time scale is confined between local modes (or glass transition) and chain flow, free for $M < M_c$ and hindered by entanglements for $M > M_c$.

The model (Refs. 102, 103) is explained in Fig. 28a. Consider a chain of Z monomeric units; z_k units are comprised into one bead, N_k beads then make the chain, $N_k z_k = Z$. Let z_k be large enough so that the distance between neighboring beads is subject to Gaussian coiling, i.e. the distance is larger than the persistence length or the Kuhn step. A bead is, therefore, a ψ level concept. Then, in the linear region, the "elastic" force between the beads is given by the entropy change of the deformed coil: Eqs. 1.18 and 4.42 describe the corres-

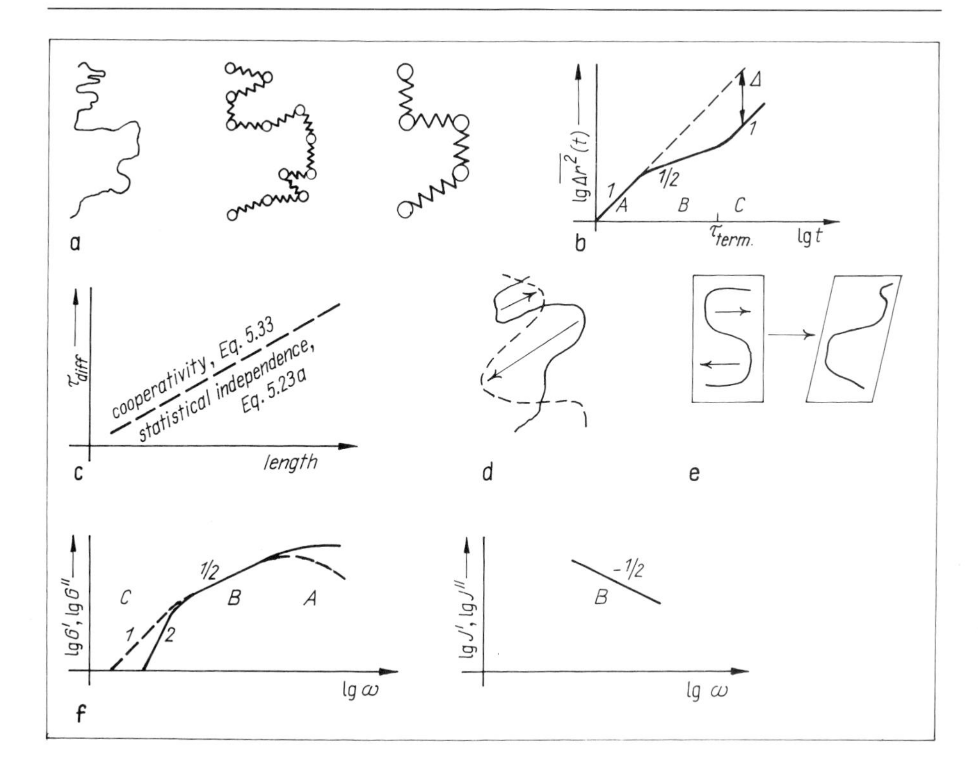

Fig. 28. Rouse modes.
a. Coarsening of the bead spring model. b. Mean square displacement of a bead. A segmental free diffusion, B Rouse diffusion (length scale: from minimal coiling up to the mean end-to-end distance of the chain), C total Brown diffusion of the chain. c. Length-time scaling of statistical independence for Rouse modes. d. Normal mode. e. Shear activity of normal modes. f. Rouse behavior of shear modulus $G^* = G' + iG''$ and shear compliance $J^* = J' - iJ''$.

ponding entropy springs. This force is relatively small because the deformation is connected with only one degree of freedom as discussed in Sec. I.

The bead size of this *bead-spring model* can be freely chosen between the minimal size introduced above and a maximal size that must not be exceeded for modelling a Gauss behavior of the whole coil. That means that the bead size (z_k) does not correspond to an additional length and we expect scaling with power laws.

According to this model, the motion has three aspects: a thermal (Brown, Langevin), a geometrical (Gauss), and a mechanical one (Rouse, Ref. 104).

Because of the small spring forces, any bead moves freely in small scales (as compared to the corresponding mode length, see below). It can be described by a free Brownian diffusion, Eq. 1.5. Substituting the diffusion coefficient by a Stokes friction on the bead we have

$$\overline{\Delta x^2} = (kT/3\pi\eta R)\tau \tag{5.23a}$$

with R the bead radius and η the viscosity of the environment. Supposing the bead friction, i.e. the bead radius R, is proportional to z_k (the number of monomeric units of a bead) then the friction can hypothetically be related to one monomeric unit (extrapolation into the ρ level). Then

$$\overline{\Delta x^2} \sim (kT/z_K\zeta_0)\tau \tag{5.23b}$$

where ζ_0 is called *monomeric friction coefficient* (unit Pa · s).

The model still contains both longitudinal and normal (lateral, transverse) components, relative to the chain direction. It is sometimes called "normal mode model" when lateral activities are considered.

The mean distance σ between two neighboring beads of z_k monomeric units is given by the Gaussian formula

$$\sigma = a z_K^{1/2} \tag{5.24a}$$

with a the structure length (≈ 0.7 nm for vinyl polymers, see Tab. I). Generally, for large arbitrary numbers $z_p (< Z)$ we have

$$\sigma_p = a z_p^{1/2}. \tag{5.24b}$$

From the mechanical point of view the molecular motion for times much larger than pico seconds corresponds to a strongly damped oscillation (creep case). The acceleration can, therefore, be neglected. Consider the Cartesian coordinates x_i (or x_i, y_i, z_i) of the ith bead. Then Newton's equations are

$$\text{const}_D \cdot \dot{x}_i + \text{const}_F(2x_i - x_{i-1} - x_{i+1}) = 0 \tag{5.25}$$

for $i = 1, 2, \ldots, N_K$, the indices D and F stand for the damper ($\sim \zeta_0$) and the spring force, respectively. On the ψ level the scale can be coarsened by making the discrete i values to a continuous variable along the chain contour (also denoted by i), $x = x(i, t)$. Then, from Eq. 5.25,

$$\frac{\partial x}{\partial t} = c \frac{\partial^2 x}{\partial i^2} \tag{5.26}$$

where the constant c expresses the force ratio spring/damper, F/D.

Using boundary conditions with regard to the larger mobility of the end beads (neglected in Eq. 5.25), such as

$$\frac{\partial x}{\partial i}(1) = \frac{\partial x}{\partial i}(N_K) = 0 \tag{5.27}$$

then Eq. 5.26 has the eigensolutions

$$x_p(i, t) = \text{amplitude} \cdot \cos(\pi p i / N_K) \exp(-t/\tau_p) \tag{5.28}$$

with the mode number $p = 1, 2, \ldots, N_K$ as eigenvalue. The corresponding mode times can be calculated from

$$\tau_p^{-1} = c(\pi p / N_K)^2. \tag{5.29}$$

A complete solution for the problem can be obtained from a transformation from Newton into Langevin equations. More simply, it is obtained by the equivalence of the length and time scale of all the three aspects, Eqs. 5.23b, 5.24b, and 5.29 for each mode p, and gauging the constant of proportionality, i.e. putting $\overline{\Delta x^2} \approx \sigma_p$ for $\tau = \tau_p$ in Eq. 5.23b. Defining a *Rouse rate* (or Rouse mobility) by

$$W \equiv 3kT/\zeta_0 a^2 \tag{5.30}$$

(dimension s^{-1}), then the mode time becomes

$$\tau_p \approx z_p^2 / W \sim 1/Wp^2 \tag{5.31}$$

with $z_p \sim 1/p$ the number of monomeric units per mode p. According to Eq. 5.31 the mode time increases with the mode length in correspondence with the general scaling principle. The largest mode, $\sigma = R_0$ for $p = N_K$, has a mode time proportional to the square of the chain length

$$\tau_{\text{terminal}} \sim N^2 \sim Z^2 \sim R_0^4. \tag{5.32}$$

The free diffusion behavior of beads (Eq. 5.23b) is modified if the mode amplitudes get in the order of mode zero distance, $\sigma_p = a z_p^{1/2}$, so that the mode p takes part in the activity of larger modes $p' > p$. Then the free Brownian diffusion changes into the Rouse diffusion given by the mean square of displacement,

$$\overline{\Delta r^2}(t) \approx a^2 (Wt)^{1/2}, \tag{5.33}$$

as can be seen from Eqs. 5.23b, 5.29 and 5.31. In this case the bead diffusion is modified by the springs, see Fig. 28b.

This is an interesting property of the Rouse model (length-time scaling of stastical independence, see Fig. 28c): The diffusion can be dissected into Rouse modes (index p). Each mode is characterized by statistically independent free Brownian diffusion $\overline{\Delta x_p^2} \sim t$ (Eq. 5.23b) up to an amplitude typical for each mode. But taking part at larger modes, their total diffusion contribution is given by the Rouse diffusion Eq. 5.33 that is different from the free Brownian diffusion.

A Rouse mode is schematically shown in Fig. 28d. Casting it in a box (Fig. 28e) we see that the normal (lateral) part of the Rouse modes is shear active. The mean shear angle fluctuation is proportional to the mean square displacement (Eq. 5.33). The latter is an extensive variable, and, therefore, according to the FDT, $\overline{\Delta r^2}(t)$ corresponds to a shear compliance, see Eq. 2.37 and Fig. 12f above. Thus we obtain for the linear shear response of Rouse modes

$$\begin{aligned} J(t) &\sim t^{1/2}, \quad J'(\omega) = J''(\omega) \sim \omega^{-1/2}, \\ G(t) &\sim t^{-1/2}, \quad G'(\omega) = G''(\omega) \sim \omega^{1/2}. \end{aligned} \tag{5.34}$$

This behavior of shear compliance (J^*) and shear modulus (G^*) is sketched in Fig. 28f.

5.4 Discussion

In polymer solutions with a small-molecule solute of high mobility the 1/2-power case of Sec. 5.3 is called the free draining case. The monomer friction coefficient is proportional to the viscosity of the solute, $\zeta \sim \eta_s$. This case can be modified by hydrodynamic interaction of chain parts due to the evading flow of the solute, see Fig. 29a (Zimm model, Ref. 105). This interaction causes the exponent to increase from 1/2 to 2/3 which means $G'' > G'$ and $J'' > J'$, see Fig. 29b. The enhanced dissipation increases the liquid-type character of the system.

In small spatial scales (ρ level, large q) the model breaks down because neither monomeric units can serve as beads nor covalent bonds as soft springs. One must expect deviations from Stokes friction law and we should discuss the intra and intermolecular origins of friction. All the deviations from the concept of monomeric friction coefficient, especially from $\zeta_p \sim z_p$, are collected by the concept of "internal viscosity" (Refs. 106, 107) which leads to observable effects that are discussed in Ref. 4.

The free draining model is accepted for melts at high temperatures. There is no additional hydrodynamic interaction because the environment of the

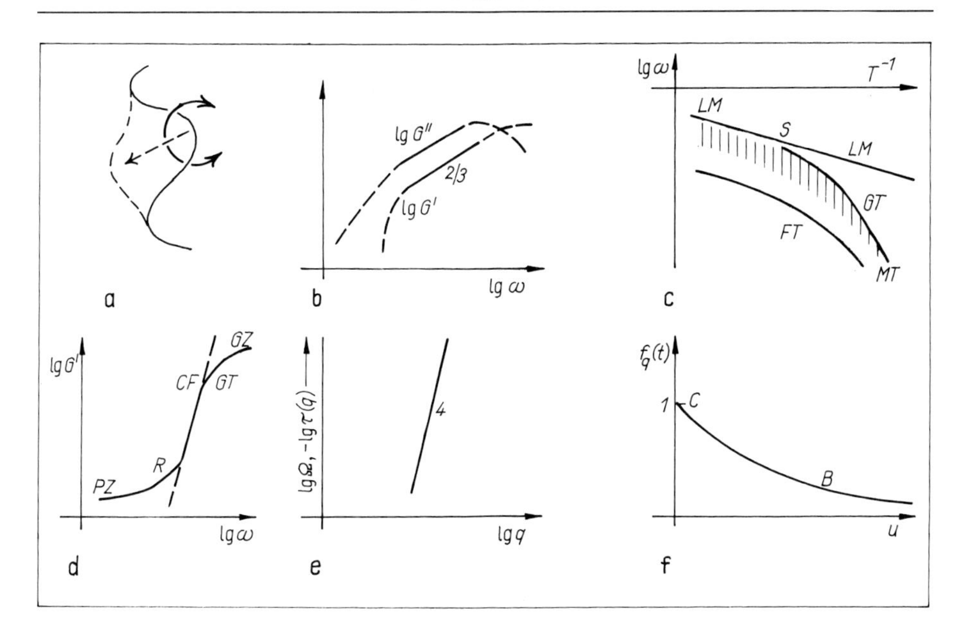

Fig. 29. Rouse modes continued.
a, b. Zimm correction for the hydrodynamic interaction (compare to Fig. 28f). c. Integration of Rouse modes (hatched) into the main transition MT below the splitting region S. LM local modes, GT dynamic glass transition, FT flow transition (in entangled polymer melts). d. Fine structure of main transition in entangled polymer melts. GZ glassy zone, CF confined flow ($d \lg G'/d \lg \omega \geqslant 1$), R Rouse like shear exponent, PZ plateau zone, e. First cumulant Ω ($\sim$ reciprocal Rouse-mode time $\tau^{-1}(q)$) as a function of scattering vector q. f. Structure-corrected decay of Rouse correlation in dimensionless coordinates, u see Eq. 5.36. The regions B, C correspond to the different regimes of Fig. 28b, B = Rouse regime.

chain has the same mobility as the test chain itself; it is a chain in its own solute. It should be underlined that all chains are completely equivalent in their fluctuations. This fact is reflected, in the ψ level model, by statistical independence of Rouse motions for neighboring chains.

We ask how the Rouse modes in the melt are integrated into the relaxation diagram for entangled polymers, Figs. 27 a or c above. We begin with a length scale discussion at a high temperature $T > T_s$ above the splitting point S in Fig. 27a. The highest frequencies of the Rouse spectrum correspond to segmental motions, i.e. they are local motions and follow to local modes LM, see Fig. 29c: Rouse modes are below the LM. Below the splitting point temperature ($T < T_s$) the coupling of local modes generates the cooperative motions of the dynamic glass transition with characteristic length in the nm scale, increasing with decreasing temperatures. Therefore, the local modes at the high-frequency end of the Rouse modes are substituted by glass-transition motions. The WLF dependence of the latter should now determine the time scale for $\zeta_0(T)$.

The onset of dynamic glass transition modifies the Rouse modes in a strange way. For $T > T_s$ we observe a shear exponent $d \lg G'/d \lg \omega = 1/2$, i.e. $G' = G''$, the border line between solid and liquid like behavior. Naively we would expect a trend to more rigidity at low temperatures. But a gradual enlargement of the shear exponent up to 1 . . . 1.5 (different in different polymers) is observed for $T < T_s$ indicating a larger fluidity for low temperature ($\tan \delta > 1, G'' > G'$, see e.g. Refs. 108 and 109). This phenomenon will be called *confined flow*. (In Newtonian flow we would have $d \lg G'/d \lg \omega = 2$, see Eq. 2.18.)

The main transition in entangled polymer melts therefore shows a *fine structure*, Fig. 29d. Starting from a high frequency we have the following succession: The cooperative glass transition (length scale $\approx$ 2 nm), then the confined flow with large shear exponent, followed by a "hindering zone" with rather low shear exponent (like as for Rouse modes) – here is also the maximum of $J''(\omega)$ – and finally the onset of the plateau zone with a typical length scale of $d_E \approx 7$ nm.

For high temperatures corresponding to short times ($\lessdot \mu$s) the Rouse modes can directly be observed by coherent dynamic neutron scattering (Ref. 110). The scattering function $S(q, t)$ describes the temporal decay of the Rouse correlations between the segments. To get away from the structure, $S(q, t)$ is divided by the static structure factor, $S(q) = S(q, 0)$, a function similar to Fig. 1e,

$$f_q(t) \equiv S(q, t)/S(q, 0) \leqslant 1. \quad (5.35)$$

In the Rouse regime, i.e. in the scaling region of Rouse modes (free selection of bead size within the scaling limits near a and R_0, or d_E for entangled systems), the length can be reduced by the structure length a, and the time by the Rouse mobility W; the dimensionless variable $f_q(t)$ can depend only on the dimensionless combinations aq and Wt. They are connected by the Rouse diffusion equation, Eq. 5.33. Therefore, $f_q(t)$ can depend only on the combination

$$u \equiv q^2 a^2 (Wt)^{1/2}, \quad f_q(t, W) = f(u). \quad (5.36)$$

In this scaling region the typical correlation time depends on the scattering vector $\boldsymbol{q}$. According to Eq. 5.36 we find

$$\tau^{-1}(q) = W(aq)^4. \tag{5.37}$$

This equation reflects the z_p dependence in Eq. 5.31, and, for the terminal limit, the R_0 dependence in Eq. 5.32. The initial slope of $f_q(t)$ is often used for the adjustment of experimental results. It is called first cumulant and is defined by

$$\Omega = -\lim_{t\to 0} \frac{d}{dt} f_q(t). \tag{5.38a}$$

Having the dimension of a reciprocal time it is given by

$$\Omega = \frac{W}{36}(aq)^4 = \frac{1}{36}\tau^{-1}(q) \sim \frac{u^2}{t} \tag{5.38b}$$

in the Rouse regime (see Fig. 27e). The factor 1/36 comes from a settlement of the open proportionality constants, i.e. from gauging of ζ_0. The proportionality $\Omega \sim u^2/t$ shows that Ω is a measure for a special (Rouse) kind of diffusion in a dispersion zone. [In a transport zone we would have $D \sim \overline{\Delta x^2}/t$.]

The function $f_q(t)$ starts from 1 at $t = 0$ and decreases with increasing t, thus indicating the successive decay of the correlations. The essential parts are described by formulas like (Refs. 111 and 112, τ_1 cf. Eq. 5.32)

$$\langle \exp iq(\boldsymbol{R}_m(t) - \boldsymbol{R}_n(0))\rangle = \exp\left(-\left(\frac{q^2}{6}\phi_{mn}(t)\right)\right),$$

$$\begin{aligned}\phi_{mn}(t) &= \langle(\boldsymbol{R}_m(t) - \boldsymbol{R}_n(0))^2\rangle \\ &= 6D_{\text{ges}}t + |n - m|a^2 + \frac{4Na^2}{\pi^2} \\ &\quad \times \left(\sum \frac{1}{p^2}\cos\left(\frac{p\pi m}{N}\right)\cdot\cos\left(\frac{p\pi n}{N}\right)(1 - e^{-p^2 t/\tau_1})\right).\end{aligned} \tag{5.39}$$

The Rouse scattering originates in an interference of the cosine functions (Eq. 5.28) of the different Rouse modes, $p = 1, 2, \ldots$ For small u (large length, small p) the limit is given by the center-of-mass diffusion coefficient D_{ges} depending on R_0. This corresponds to the diffusion regime C of Fig. 28b, lowered by a factor $\Delta = \Delta(R_0)$ when compared to segmental diffusion. Therefore,

$$f_q(t) \cong \exp(-\Omega t), \tag{5.40}$$

with $\Omega = D_{\text{ges}} q^2 \sim \tau_q^{-1}$ for the C regime. For large u (small length, larger q, but within the Rouse regime B of Fig. 28b, before ρ level) the interferences are sharpened, of course, and one obtains a simple formula again,

$$f_q(t) \cong \exp\left(- \frac{2}{\sqrt{\pi}} (\Omega t)^{1/2} \right) \tag{5.41}$$

with Ω now from Eq. 5.38b. This is a "diffusion formula" corresponding to the special Rouse diffusion Eq. 5.33 in a disperson zone. Eqs. 5.40 and 5.41 are no power laws since the scattering vector represents an additional length.

5.5 Reptation: tube model

The tube model (Refs. 113, 114) – a linear macromolecule moves in a virtual tube constructed by neighboring entanglements – was invented to explain the large time difference between main and flow transition for amorphous polymers with large molecular weights, $M > M_c$, see Figs. 6a, b, above.

The pristine version of the model is depicted in Fig. 30a. The main concept is the *tube*: The movement of the chain in longitudinal direction (along the tube) should be faster than the movement in the transverse direction (perpendicular to the tube). The entanglement points along the test chain, represented by hooks in Fig. 30a, are considered to be obstacles against the transverse motion and are used for the construction of a hypothetical tube wall. The tube diameter is equal to the entanglement spacing, $d_T = d_E$; otherwise we had to do with an additional length. The chain motion (relative to the wall) is imagined to occur inside the tube, and we have five regimes with increasing length scale:

a. Brownian motion of segments below the Rouse scaling,
b. free Rouse modes for length $< d_T$ (Rouse wriggling),
c. Rouse modes in tube contour direction, perpendicularly restricted by the tube walls for length $< a\sqrt{ZZ_c}$,
d. reptation (i.e. a diffusive contour motion in the tube, like that of a snake), and
e. diffusion of the whole macromolecule.

The *tube contour* is explained in Fig. 30b, it is the length of the tube for the test chain.

In this picture, the flow transition is defined by the transition d-e, the

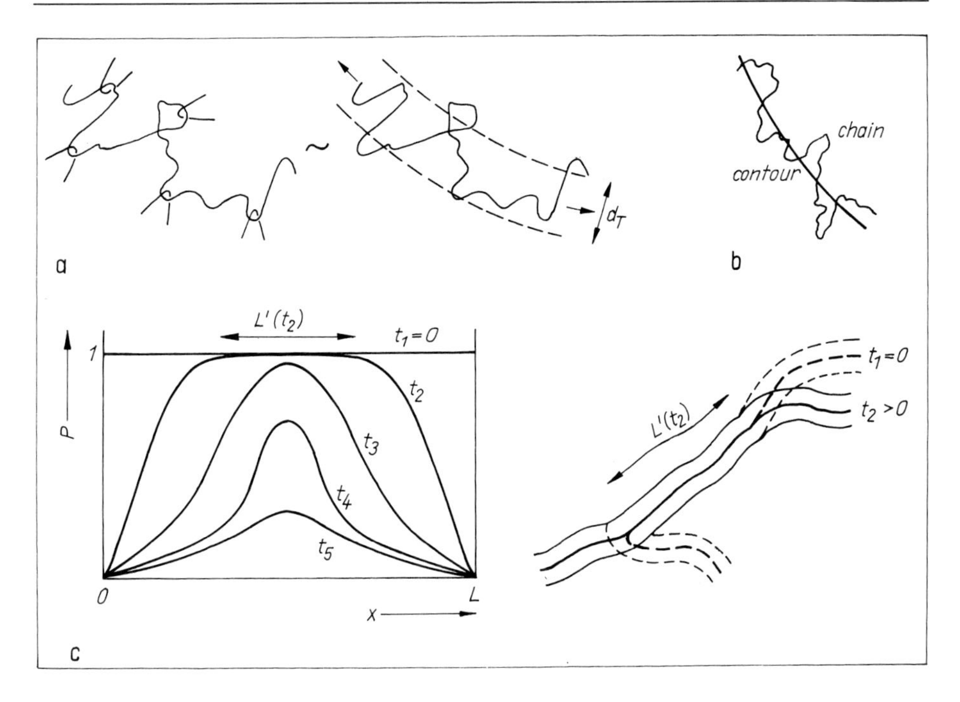

Fig. 30. Tube model.
a. Model construction. d_{T} tube diameter $= d_{\mathrm{E}}$ entanglement spacing of Fig. 6f. b. Tube contour, preaveraging of chain length in the d_{T} scale. c. Tube renewal by chain end fluctuation. $P = P(x, t)$ disentanglement probability density, i.e. the probability to remain in the original $t_1 = 0$ tube; $L'(t)$ is the contour length of the remaining chain fraction; x, $0 \leqslant x \leqslant L$, is a coordinate along the contour (internal space).

corresponding time is called *reptation time* τ_{R}. This time is a disengagement time and can be estimated as follows (Ref. 115).

The chain mobility μ_{T} in the tube is defined by $N \sim Z$ Stokes spheres,

$$\mu_{\mathrm{T}} \sim 1/N. \tag{5.42}$$

The diffusion coefficient of the tube contour in the tube (D_{T}) is defined by the Einstein relation between mobility and diffusion,

$$D_{\mathrm{T}} = kT\mu_{\mathrm{T}} \sim 1/N. \tag{5.43}$$

The diffusion is gauged (triggered) by an elementary step D_1 (local mode for $T >$

T_s, T_s = splitting point temperature, see Fig. 29c, above), e.g. reduced by N

$$D_T = D_1/N, \tag{5.44}$$

$$D_1 = \text{numerical factor} \times a^2/\tau_0$$

with a the structure length and τ_0 the time scale of the local mode. Since the length of the tube contour, L_T, is smaller than the extended chain length – due to Gauss coiling in tube diameter dimension – we obtain, from $d_T = d_E$,

$$L_T = aZZ_e^{-1/2} \sim NN_e^{-1/2} \tag{5.45}$$

with the index e for entanglement, $N_e \approx N_c$.

The observables of interest, self-diffusion D and viscosity η, are calculated by application of two Brownian diffusion equations. The typical time is the reptation time τ_R. For in-tube diffusion, the typical length is the contour length. Therefore,

$$D_T = L_T^2/\tau_R. \tag{5.46a}$$

From Eqs. 5.44, 5.45 we see that

$$\tau_R \sim N^3. \tag{5.46b}$$

The typical length for self-diffusion is the chain coil radius, R_0. Therefore,

$$D = R_0^2/\tau_R \sim N^{-2}. \tag{5.47}$$

From Kubo formulas we know that η is proportional to the typical time (if this is the mainly varying variable). Therefore,

$$\eta \sim \tau_R \sim N^3. \tag{5.48}$$

The term disengagement time for τ_R means the following. For monodisperse melts the tangling and loosening of entanglement points result mainly from fluctuative movements of the own chain ends. Consider the tube of a test chain at $t = t_1 = 0$. This original tube is partially "dissolved" for $t_2 > t_1$ at the chain ends. This process can be described by a disengagement probability P as explained in Fig. 30c. There is a length $L'(t)$ for the middle chain fraction remaining in the original tube. This part is the shorter the larger the time is. The corresponding loss of memory can be described by an in-tube diffusion equation for P, with D_T being the diffusion coefficient and with obvious initial and boundary conditions. The disengagement time is the time when the original tube, by fluctuation of chain ends, is completely left. Then we have a new tube for the test chain, so that this process is also called tube renewal.

Starting from Fig. 30c, Doi and Edwards (Ref. 6) developed a viscoelastic material equation. They used the term "relaxation after deformation", equivalent to fluctuation in the linear range. The basic statement is: "Since only the segments in the deformed tube are oriented and contribute to the stress, the stress is proportional to the fraction of the polymer still confined in the deformed tube". This theory qualitatively describes also some nonlinear effects and is oriented along the stress, i.e. the shear modulus G.

There are some open problems with the shear compliance J. Without going into detail we remark that an essential criterion is the numerical value of the product $J_e^0 \cdot G_N^0 > 1$. This relation is not in conflict with the implication $G(\tau) \cdot J(\tau) \leqslant 1$ from Eq. 2.35 because J_e^0 (Fig. 12e) and G_N^0 (Fig. 23a) are related to different times (or frequencies), and therefore to different mode lengths. The compliance indicates the onset of flow, for longer modes, that cannot be described adequately by the stress reduction corresponding to $L'(t) \to 0$ for shorter modes.

5.6 Discussion

The proximity of theoretical ($\eta \sim M^3$, $D \sim M^{-2}$) and experimental exponents ($\eta \sim M^{3.4 \pm 0.2}$, $D \sim M^{-2.0 \pm 0.1}$) shows that some essential features of the flow transition can be described with the tube model. There is still no generally accepted improvement of the theory that enlarges the η exponent up to 3.4 without changing the D exponent. An unknown referee describes the problem with the experimental exponents as follows: "If chains diffuse with D scaling as M^{-2} then the time for a chain to diffuse a distance of order its coil size scales like R_0^2/D or M^3. If the viscosity and the stress relaxation time scale as $M^{3.5}$, then a paradox emerges because very long chains must diffuse many times their own size without relaxing stress!" Many elements are added to the pristine tube model and there are many modifications of it (see Ref. 115). Models without tubes or even without topological elements are also suggested. They are reviewed in Ref. 117 and use concepts such as packing, displacement, and others.

We shall mention some ideas that are not going beyond the tube concept. A second process for tube renewal is *constraint release*, see Fig. 31a and Refs. 22, 23. An entanglement point in the middle part of a test chain can be loosened by the movement of the host's chain end. This enlarges, of course, the mobility. Clearly constraint release is especially effective when the host

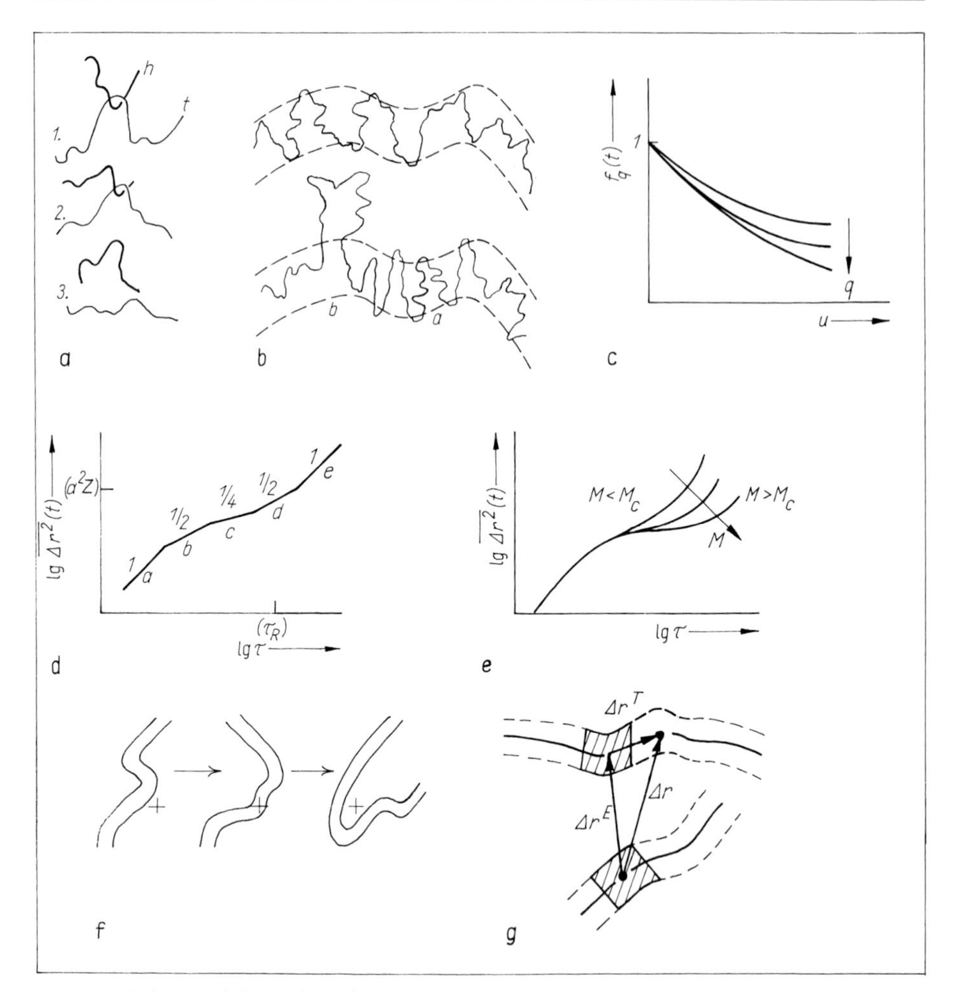

Fig. 31. Tube model continued.
a. Three stages of constraint release. *t* test chain, *h* host chain. b. Two elements of chain breathing. c. Effect of entanglement spacing on the reduced scattering function (cf. with the Rouse case in Fig. 29f), $u = q^2a^2(Wt)^{1/2}$, see Eq. 5.36. The parameter is the scattering vector q. d. Mean square displacement of a given chain segment as a function of time in a fixed-tube model. a, b, c, . . . different regimes according to Sec. 5.5, d reptation regime, τ_R reptation time, *a* structure length, *Z* degree of polymerization. e. Computer simulation of the first regimes for different chain lengths (Ref. 121). M_c critical molecular weight. f. Model of fluctuating tubes. g. Superposition of in-tube (T) and (lateral) tube element (E) motion. For the sake of clarity, the amplitudes in Figs. f and g are (probably) exaggerated.

chains are much shorter than the test chains, i.e. in polymer mixtures with different chain lengths. Estimations show (Ref. 115) that the contribution of constraint release is small in monodisperse melts.

The retraction of tethered arm chains is important for the flow transition in melts of star molecules. It can be discussed with the tube model for the arms and with changed boundary conditions for the Rouse modes. The mobility of stars is lower than of linear chains with the same molecular weight.

The contour length, being shorter than the extended chain length, can fluctuate (chain breathing, see Ref. 118). Reasons for this are the fluctuation of the in-tube Gauss coils (a in Fig. 31b), local off-tube loops between entanglement points (b in Fig. 31b), and local constraint release. This leads to increasing longer times because in an N^3 law the longer contours are more effective. Since the fluctuation effect disappears for very long chains, one would find an enhancement only in one half or one logarithmic decade in N, i.e. no scaling behavior with higher exponent than 3.0.

Now we shall proceed to an experimental test for the typical length of the model. The tube diameter d_T (= entanglement spacing d_E) should break the Rouse scaling of Fig. 29f. For small scattering vectors q and corresponding long times t two effects were expected and are verified by dynamic neutron scattering (Ref. 119). The new length splits the Rouse-scaled scattering curves at $d_T q \approx 1$, and beyond this the scattering function is nearly constant because the entanglement points can temporarily hinder a further decay of correlation functions, see Fig. 31c. The d_T values correspond to d_E calculated from the plateau modulus by Eqs. 1.41 and 1.42. The scattering curves can qualitatively be explained by a de Gennes formula (Ref. 120), combining the Rouse and reptation model in a simple way,

$$f_q(t) = 1 - \frac{d_T q}{36} + \frac{(d_T q)^2}{36} \, e^{u^2/36} \cdot \operatorname{erfc}(u/6) \tag{5.49}$$

with the Rouse u according to Eq. 5.36.

The mean square displacement of a given segment, as a function of time, was also calculated by de Gennes, see Fig. 31d. The curve is more complex than for the Rouse case (Fig. 28b) and we can see five parts corresponding to the five regimes a . . . e of Sec. 5.5. Quasi ab initio computer simulations (Ref. 121) up to the spatial d_T scale confirm the first three regimes and indicate the transition c-d to the reptation regime, see Fig. 31e. Despite their different conceptions, reptation and Rouse models are mathematically isomorphic in a sense: The diffusion equation Eq. 5.26 for Rouse beads corresponds to the in-tube diffusion equation for the disengagement probability for tube renewal Fig. 30c. This results in similar formulas for the corresponding spectra although the boundary conditions and the "particles" are rather different. The units for tube renewal and reptation are of size d_T and will be called *tube units*.

Both Rouse and reptation models describe, after all, motions of a *chain*, see Figs. 28d and 30c. Both underline the collective character in the unit motions, and one can ask, especially for low temperatures, for more cooperative elements. For instance, we can ask whether the chain end motion – without any doubt crucial for changing the topology – should be so distinguished in the spatial scale. Is it possible that all tube units move with a comparable amplitude (Ref. 122, Fig. 31f)? Models permitting flow like lateral tube motions will be called *fluctuating tubes*.

Such motions would influence regime d (Fig. 31d) of the pristine tube model where the tube obstacels are fixed in space. Lateral tube motions would be shear active (see Fig. 28e), but would not influence the diffusion because the "chemical" environemnt (segment in a tube) is not changed by lateral tube fluctuation. These attributes will be comprised by the term *microflow*. That means that we introduced a hypothetical mechanism with different contributions to η and D desired for a decoupling of η and D exponents in the pristine reptation model (3.0 and -2.0). Decoupling is a precondition for solving the paradox mentioned above.

Flow like lateral motions are also suggested by the experimental finding (Ref. 63) that even chemical crosslinks in networks show a considerable spatial fluctuation in the scale of crosslink distance.

Tube motions seem to be in conflict with the very idea of the tube model restricting the lateral chain motion. But this is only a question of reference: A chain motion can be restricted relative to the tube walls and can simultaneously share in the lateral tube fluctuation, if microflow motion of tubes is possible.

Then we would have an interesting space situation. Since chain parts of some ten different chains are contained in a tube unit, each tube unit belongs to about ten chains meeting there. For moderate chain lengths we have more tube units than tubes. The tubes are a kind of internal space for chain movement that is carried along by lateral tube motion in the normal, external space (Fig. 31g). As tube units are defined by topology and not by segments, an additional tangential in-tube motion of chains can be considered besides tube fluctuation. In other words: reptation is an internal motion that may be superposed by lateral external motion.

To construct a picture of fluctuating tubes imagine that all chains have a different color. Then, coarsening the picture, each tube unit has a certain mixed color. If we could see a movement of the colored regions, and not only color-changing regions, then we would have a model for fluctuating tubes.

If this picture were correct, the natural functional subsystem for the flow transition would consist of tube units (instead of particles) with a minimal coupling between them mediated by chain reptation in internal spaces (tubes).

6. Glass transition. Multiplicity in amorphous polymers

The three models presented in Chap. 5 cannot easily be united. This chapter is an attempt at a systematic representation of relaxation in amorphous polymers. The glass transition in simple glass formers such as glycerol is considered as the basis for understanding the complex relaxations. We shall here confine ourselves to the dynamic glass transition, i.e. to a discussion of typical dispersion zones in the equilibrium (e.g. $T > T_g$ for the main transition).

The experimental basis for polymer relaxation was founded by Tobolsky and Ferry in the Forties and Fifties (Refs. 2, 38, 123). The results were crowned by the scaled representation with the aid of the famous WLF equation (Williams, Landel, Ferry, Ref. 124), a relation in the ln ω–T or ln τ–T (or T^{-1}) diagram. WLF means a set of curves that guides the frequencies of maxima, inflection points etc. of different activities in their dependence on temperature. The Kubo formulas link WLF with the VFT equation (Vogel, Fulcher, Tammann Refs. 125–127) for the temperature dependence of viscosity found in the Twenties.

The reasons for the universality of the *leading* traces in the ln ω–T diagram and their connection to the multifarious activities *led* by them seem to be the central problems of the dynamic glass transition.

Many papers on glass transition have been published in the last 15 years (see the reviews in Refs. 128, 129, 129a). Highly sophisticated methods of theoretical physics are now being applied to solve the glass transition puzzles: mode coupling to understand the onset of dynamic glass transition at the splitting point (Refs. 130–132), abstract computer simulation with an attempt to model cooperativity by means of local topological conditions (Ref. 133), and generalization of experience with scaling methods for phase transitions (Ref. 134, 135) applied to T_∞. Although very fruitful for the development of science, they did not contribute too much to the central problems as defined above.

Also in the Eighties, there have been entirely new experimental results from dynamic scattering methods in the frequency region $10^8 \ldots 10^{12}$ Hz (e.g. Refs. 136–138) that are probably connected with essential features of classical glass transition problems.

As there are considerable differences in the conceptions of glass transitions, the terminology is going to be illustrated in advance by a longer quotation from Ref. 133, before the terms will be introduced in the further text without any reference to the following model.

"Supercooled liquids, by virtue of their high densities, possess strong constraints on the dynamics of the individual atoms or molecules. A particular molecule is trapped by its neighbors in a "cage", which may persist for long periods of time. To destroy the cage (required for viscous flow), very cooperative dynamical events in the vicinity of the molecule are required. This follows because the neighboring molecules that constitute the cage are themselves caged. The spatial extent[1] over which cooperative rearrangements must occur to relax a cage very likely increases as the fluid is densified. Long times are required for such cooperative rearrangements involving large numbers[2] of molecules. Hence, when describing the events leading to glassy dynamics in Euclidean space . . . , we say that dynamical constraints at molecular distances lead to a high degree of cooperativity in the dynamics at larger distances[3], . . ."

We shall use the following terms:

[1] This spatial extent is called cooperativity region, with volume V_a containing N_a relevant particles, see also Sec. 3.6, Fig. 20.

[2] This number (N_a) is called degree of cooperativity or, in short, cooperativity.

[3] Characteristic length ξ_a, $V_a \approx \xi_a^3$.

6.1 Glass transition in simple glass formers

Firstly we consider the classical behavior in the kilohertz region (a few ten kelvins above the glass temperature, T_g), i.e. the linear response in the equilibrium state at crossing the dynamical glass transition zone GTZ, see Fig. 32a. Typical examples are glycerol, ortho terphenyl, or the salt melt $(KCa)NO_3$. Fig. 32b shows some isothermal activities (Refs. 139, 140) across the glass transition zone schematically. The activities, represented in a linear scale, are quite similar, but a careful analysis results in significant differences in curve form and frequency position for a given temperature T (Refs. 142, 54). The variation of this parameter indicates the shift *along* the glass tran-

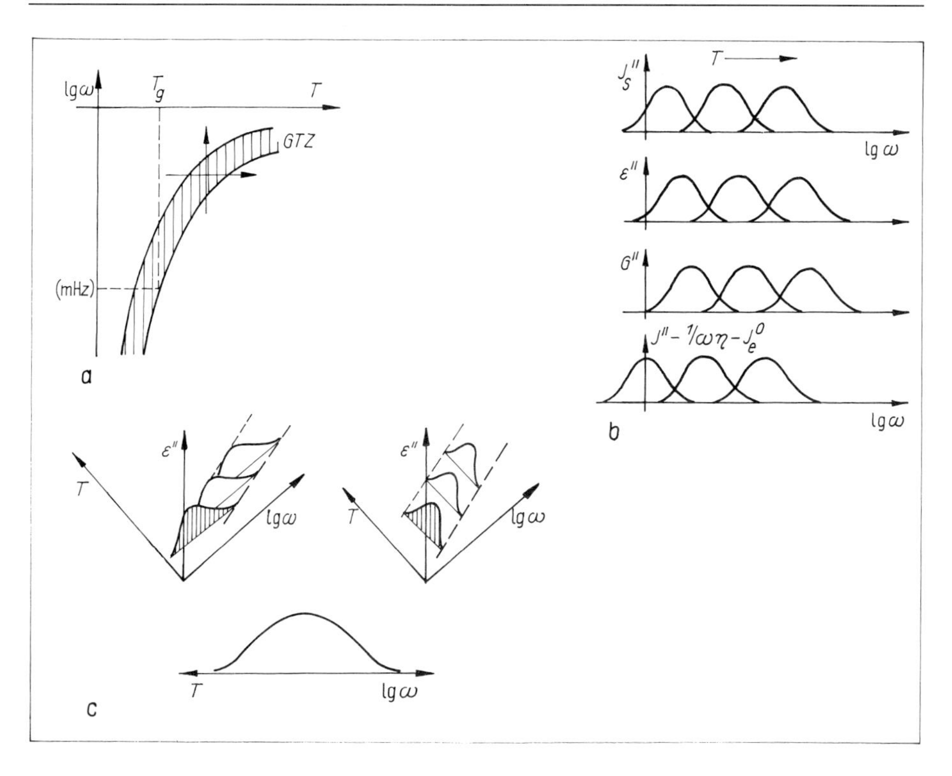

Fig. 32. Dynamic glass transition in simple glass formers.
a. Isothermal (↑) and isochronal (→) linear response experiments across the glass transition zone GTZ. b. Examples of loss parts for different activities (schematically). J_s'' entropy compliance, ε'' dielectric permittivity, G'' shear modulus, J'' shear compliance, parameter T temperature. c. Local time-temperature equivalence.

sition zone; increasing the temperature results in increasing mobility (lg ω, ω frequency).

Similar curves are obtained for isochronal experiments (ω = const, T varied). The term *across* the glass transition zone means that now higher temperatures correspond to lower lg ω when the crossing is "perpendicular" to the zone. The situation is sketched in Fig. 32c. For an isotherm and an isochrone, crossing in the GTZ, one unique curve is obtained for a given activity in a good approximation, if lg ω and T are counted in the opposite direction on the abscissa. This symmetry is called local *time-temperature equivalence*. The relation between T and lg ω on the abscissa is given by dT/d lg ω along the dispersion zone.

This equivalence is a consequence of the fact that the curves are approximately unchanged when shifted by one spectral width over lg ω or T. The interesting point is that it is the logarithm of frequency lg (ω/Ω) (and not ω, or ω^2) that is symmetrical to the temperature difference, $T - T_\infty$, or to pressure, concentration and so on), even if the temperature dependence is not Arrhenius'.

The curves are usually broader than the standard (Debye) relaxation, Eq. 2.26. Expressing the logarithmic curve width by the parameter β^{-1} of a KWW formula, Eq. 2.48, applied to the relaxation itself, then values around $\beta \approx 0.5$ are observed, depending somewhat on the temperature. The curves are not symmetrical over lg ω, they have a tail at high frequencies (corresponding to short times or low temperatures, see Fig. 32c). According to the general scaling principle this tail corresponds to short modes. For small-molecule substances it may be a small-scale molecular jiggling within the first shell. On the other hand the low-frequency end describes the gradual decay of cooperativity with larger spatial scales. Labelling the axis linearly in ω this part is, of course, the main fraction of the curves. It is not probable that such different things can be united precisely in a simple formula like KWW.

Now the scenario is described how the *dynamic glass transition* is discussed in this book (see Ref. 141): At high temperature the molecular relaxation in liquids is mainly determined by local rearrangements in the first shell. Decreasing the temperature we observe an increasing organization of the molecular motion corresponding to the completion of the liquid structure (more general, of a structure with an inherent disorder). The first step is the coupling of rather local modes. But then a crossover is observed (near the "splitting point", see Fig. 27a) to a typical cooperative process with universal features and specific symmetry. On decreasing the temperature further, the cooperativity increases which is accompanied by a universal, dramatic increase of the time scale (viscosity).

In principle, the crystalline state of a classical glass former is not closely neighbored to the liquid state. (There is no critical point between liquid and crystal.) The dynamic glass transition has, therefore, nothing to do with the crystalline state. It is a question of crystal nucleation and growth rate (generally: of the difficulties for mutual transformation) whether and where crystalline phases can intervene the dynamic glass transition. This means that undercooling (metastability) is not a precondition for the dynamic glass transition. But, on the other hand, the increasing organization of a liquid-like cooperative thermokinetic structure does have, of course, a tendency to metastability.

Obviously, there are complex liquids or materials which can offer different elements for the organization of cooperativity. This can lead to different dispersion zones and branching of dispersion zones, so that the phenomena can become very complex. Usually the cooperativity at T_g is not too large. A typical value is about 100 units. In contrast to critical phenomena with their very large correlations, we therefore observe much less universality for the glass transition. The general aspects are superposed on intermolecular individualism of the special glass former considered. In addition more or less universal branching tendencies are observed, and neither the flow nor the glass zone near the GTZ are relaxation deserts. Therefore, general laws for glass transition cannot be detected easily in a world of individual peculiarities, and it depends on the scenario what is considered to be typical or universal.

Because of the dramatic increase of the time scale τ, a cooperativity much larger than 100 units cannot directly be observed due to the intervention of practical time scales, e.g. minutes at T_g. Correspondingly we have a large field of speculation what can happen for $\tau \to \infty$ (T_∞ problem).

6.2 Comparison: phase transition and dynamic glass transition

A phase transition in a pure substance is characterized by a sharp borderline in a p-T diagram with, if any, a sharp critical point. Sharp means millikelvins or smaller. This corresponds, using the formula $\Delta T \sim N^{-1/2}$, to large phase zises (droplets, bubbles etc.) for first order phase transitions and to large correlation lengths for critical points or λ phase transitions. A large correlation length (critical opalescence) implies *universality*, i.e. retraining of intermolecular individualism, and leads to a classification into universality classes only depending on dimension, range of intermolecular forces, and number of spin or order parameter components.

A phase transition is described by order parameter(s) ψ and one can always write down an effective Hamiltonian in the ψ level. The general behavior can be explained by mean field methods (van der Waals equation, Weiß field, Bethe's quasichemical method etc.), additional singularities emerge from the necessity to include fluctuations of the order parameters (Sec. 4.1.5).

The temperature of a phase transition is explained by an energy-to-entropy ratio, $T_u = \Delta H/\Delta S$ for first order transition, a critical point is usually related to a typical *energy*, kT_c. The energy of intermolecular potentials, if short-ranged, is transferred into the ψ level by coarsening (decimation, see Fig. 24d) that renormalizes the interaction, i.e. the form of the effective Hamiltonian is usually the same for blocks of all sizes, but the parameters become running. Such energy parameters finally determine the fixed point of the method, kT_c.

The transition temperatures T_u or T_c are therefore determined by the equation of state, i.e. kinetic aspects do not play any role for their values. Of course, the other way round, the kinetics can heavily be influenced by a phase transition. A well known example is critical slowing down: near T_c the diffusion becomes very slow (soft mode) because the thermodynamic driving force tends to zero near the limit of material stability.

The Vogel temperature T_∞ is probably not a thermodynamic point that can be calculated from the Gibbs distribution, because it is kinetically defined by $\lg \tau \to \infty$, and by cooperativity $N_a \to \infty$. The latter property is less a property of the energy hypersurface (landscape) in the configurative part of the phase space than a question of how many coordinates (dimension) of the landscape give the correct representative of the whole system, i.e. give the natural subsystem. It seems to be difficult to analyze the size of cooperativity in the energy landscape.

If one still insists on considering the glass transition as a phase transition at T_∞ (Ref. 134) then one is confronted with the problem to enlarge the singularity region up to T_g or T_s (i.e. up to fifty or more kelvins) for inclusion of the representative effects of an actual glass transition into the description, see also Fig. 34, below.

The alternative is to look for an entirely new mechanism working in or near the ρ level and linking kinetics and temperature as symmetrical partners: $\ln\omega$–T or time-temperature equivalence.

It seems to be a matter of logical consistency to forget order parameters and the dominance of energetic aspects for a discussion of the dynamic glass transition in simple glass formers. This means: no effective Hamiltonian in the sense of Eq. 4.10. Instead, the problem is describing cooperative movements in the time scale of milliseconds, or hours (without using disappearing driving forces), although the molecular dynamics (of course, with

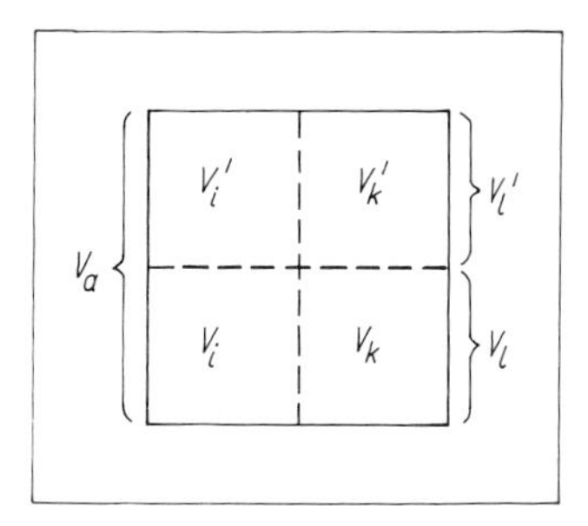

Fig. 33. Decimation schema for minimal coupling. V_a volume of the natural subsystem = size of cooperativity. V_i, V_k $(i, k = 1, 2, \ldots, n)$ parts of V_a. $V_l = V_i + V_k$ $(i \neq k)$ larger part of V_a.

Hamiltonian) is triggered by time steps of tenths of picoseconds. Thus quite abstract and unusual concepts are expected to occur in glass transition phenomenology.

How to substitute the order parameter techniques? If we assume that there is no additional length between the ρ level and the size of cooperativity ξ_a, assumed to be large enough, then we have the situation of confined scaling. Then the method is still decimation, i.e. division of the cooperativity volume V_a into smaller parts and analyzing the interaction between them, see Fig. 33. [Decimation must not be confounded with Landau's introduction of subsystems in large systems, see Ref. 45, where statistical independence is reasoned by no coupling between them.] The problem is now to find a mutual thermo-kinetic relation between the parts without energetic coupling. Such conceptions will be called *minimal coupling*. The questions are: What about mutual interpendence between fluctuation frequencies of the parts, and what is their dependence on temperature?

The situation can be illustrated by an enlargement of Kanig's bus model (Ref. 143). Consider an overcrowded bus and a passenger trying to get the way out from the back benches. Then all people have to move (turn) simultaneously, or accepting disorder, in some random succession, to act jointly with the others. Such a movement is called *cooperative*. The quality or state of being cooperative is called *cooperativity*. This is necessary for the diffusion of our "defect": the passenger considered. If the cooperativity is blocked, e.g. by a too small bus, then only local motion remains without defect diffusion [i.e. without the modes that are typical for glass transition]. If there are less or slimmer people in the bus (higher temperature), then the movement is easier and the degree of cooperativity becomes smaller. Consider fluctuation in the density (people per square or cubic meter) then there are regions of higher density where it is hard to penetrate (low frequency, low transition probability) and there are regions of lower density easier to come through (high frequency, large transition probability). We can imagine that this is not regulated by energetic interaction between the passengers but simply by the "geometric" situation: increasing the density in one region is accompanied by decreasing the density in another, and so on. This may be an example for minimal coupling between the different regions inside the bus of volume V_a.

Furthermore, we can imagine that, although the motion is cooperative in the bus scale, the local attempts of a passenger (attempts before success) are of higher frequency, and are statistically independent from those of a second passenger far away. This means we can define a *hierarchy* (maybe scaled) of statistical independence inside the cooperativity volume.

We should be careful with the terminology here: *collective* means that all particles are doing the same, as in a wave. In the bus model: the passengers are drilled by a conductor to organize moving kinks or inclinations. Whereas the term cooperative includes some randomness, disorder, individualism, stochastic elements (no conductor). Obviously, a glass transition is rather a cooperative than a collective effect since a broad spectrum of collective modes is hard to imagine in a small volume of the size of a few nanometers.

If such a conception is successful, then a dynamic glass transition in simple glass formers and a phase transition are two different things. Nevertheless it is an interesting question (Ref. 144) whether there are systems (e.g. spin glasses) that can compose the two things to a complex phenomenon, see Fig. 34b.

6.3 WLF scaling

As mentioned in Sec. 6.1 the loss curves of Fig. 32b can be approximately mastered i.e. be brought into line with each other by a shift along the lg ω (or lg τ) axis. This reduction: multiplication of ω with a shift factor a_T depending on the temperature according to Sec. 4.2.3, is called *temperature-time superposition*. WLF wrote in Ref. 124: "This treatment is quite independent of the nature of the relaxation spectrum and the time dependence of mechanical and electrical properties; it appears to be equally applicable to narrow and broad relaxation distributions". This is a very precise generaliza-

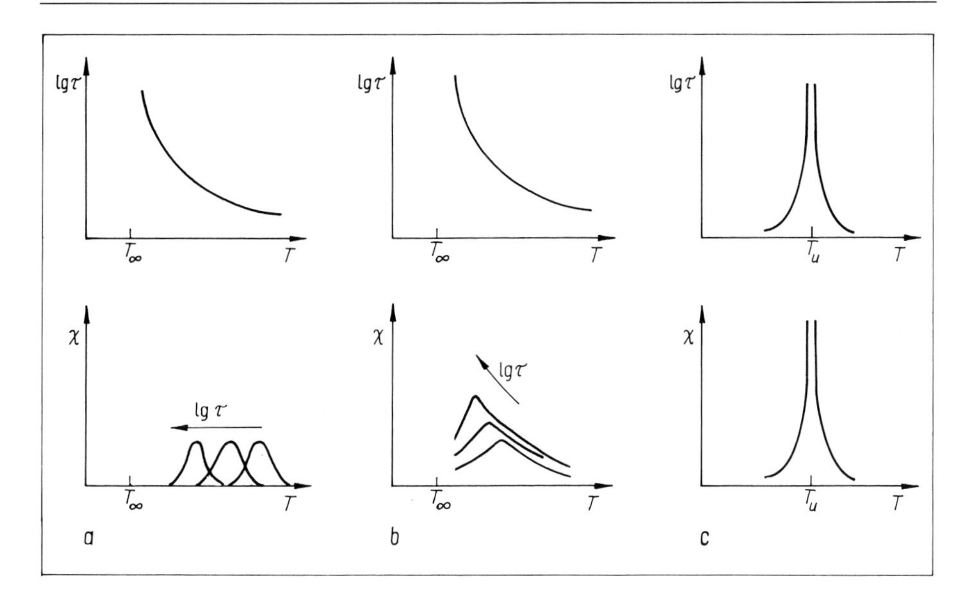

Fig. 34. Comparison: glass transition – phase transition.
a. Dynamic glass transition. b. Spin glass (schematically). c. Phase transition, without supercooling effects etc. τ typical fluctuation time (soft mode for phase transition), T temperature, χ typical activity (susceptibility) with parameter constant frequency ($\approx \tau^{-1}$).

tion of the experimental experience that indicates a certain independence of lg ω as a physical quantity, it points to an underlying lg ω–T physics ("thermokinetics").

The *WLF equation* for the shift factors, Eq. 4.36, is a global symmetry between temperature T and "thermokinetic entropy" lg ω: The equivalent equation

$$(T - T_\infty) \lg (\Omega/\omega) = (T_0 - T_\infty) \lg (\Omega/\omega_0) \tag{6.1}$$

is, in coordinates T and lg ω, a set of hyperbolas with common asymptotes, T_∞ (Vogel temperature) and lg Ω ($\Omega \approx 10^{12} \ldots 10^{15}\,\mathrm{s}^{-1}$ for simple glass formers). This set is parameterized by one parameter, e.g. by ω_0 (at a given temperature $T_0 > T_\infty$) or by T_0 (at a given mobility lg $\omega_0 <$ lg Ω), see Fig. 35a.

Although WLF assumed that the hyperbolas guide only one point of a relaxation curve, e.g. the maximum of the loss part, and that the spectra are otherwise form invariant (thermorheologically simple), there is now growing

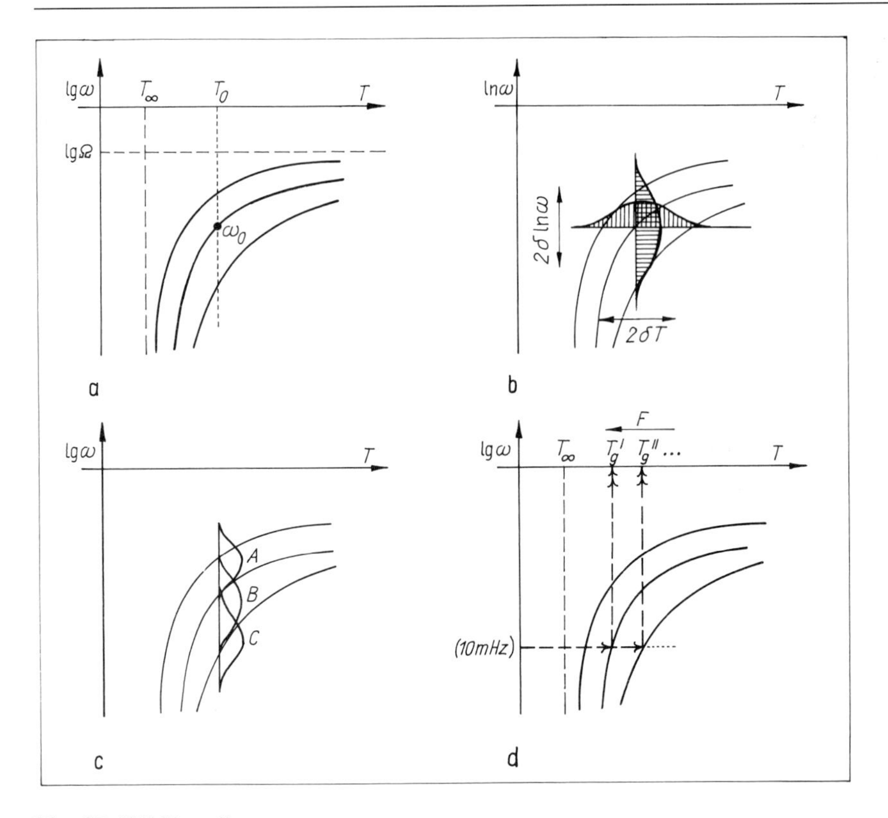

Fig. 35. WLF scaling.
a. The WLF equation is a set of hyperbolas with common asymptotes T_∞ and lg Ω. b. Spectral width δ ln ω (expressed by means of a loss curve for temperature fluctuation in the glass transition zone) and mean temperature fluctuation δT of the relevant subsystem. c. Activity arrangement across the glass transition zone for one substance, see also Fig. 27g. d. Fragility parameter definition Eq. 6.5 for different substances.

evidence that, for "ideal" examples, the other points of the spectra are also guided by the same set of hyperbolas (i.e. also their width, see Eq. 4.37 and Fig. 25b). Otherwise we must allow the awkward question why only a special point of one special activity is guided.

To remain at the coordinates T and lg ω we consider the pair of spectral widths δT and d lg ω (or δ ln ω, see Figs. 32c and 35b). Guiding these widths by the hyperbolas Eq. 6.1 means

$$\delta \ln \omega \sim (T - T_\infty)^{-1} \quad \text{and} \quad \delta T \sim (T - T_\infty), \tag{6.2a}$$

and we see from Figs. 25b and 35b that

$$\frac{\delta \ln \omega}{\delta T} = \frac{d \ln \omega}{dT} = \frac{\ln (\Omega/\omega)}{T - T_\infty} = 2.3 \frac{c_1^0(T)}{c_2^0(T)} \tag{6.2b}$$

where the derivative must be taken along a given hyperbola, e.g. for the spectral maximum. Eq. 6.2b means that this derivative is determined by the corresponding ratio of spectral widths – in agreement with the principle of local equilibrium, Sec. 3.7.

As for δT we remark that according to the FDT the temperature modulus ($K_T^* = K_T' + iK_T''$) is linked with the spectral density of temperatue fluctuation in the glass transition zone ($\Delta T^2(\omega)$, see Eq. 3.29),

$$K_T''(\omega) = \pi\omega\Delta T^2(\omega)/kT, \tag{6.3}$$

$$\delta T^2 = \int_{-\infty}^{\infty} \Delta T^2(\omega) d\omega = kT^2 \Delta(1/C_V). \tag{6.4}$$

$1/C_V$ is the reciprocal value of the extensive heat capacity for the relevant natural function subsystem (cf. Sec. 3.9 for this term). Therefore, δT is the mean temperature fluctuation of the natural functional subsystem defined by the cooperativity of the glass transition zone.

Two physical aspects of the set of hyperbolas can be considered.

(i) For a given substance we can ask whether and how the spectra for different activities (A, B, C in Fig. 35c) can be distinguished. [This is also the question of how individual molecular particularities can be detected in the general features of low glass-transition cooperativity, see Secs. 2.8 and 5.2.]

(ii) For a given activity, e.g. the entropy compliance $J_s''(\omega) = \Delta C_p''(\omega)$, for different substances we can ask which hyperbolas correspond to its maximum, see Fig. 35d. This is the conception of *fragility* of a substance (see Refs. 145, 141). If we define a *dynamic glass temperature* T_g as the temperature, where the maximum of $J_s''(\omega)$ lies at a (conventional) frequency of e.g. $\nu_0 = \omega_0/2\pi = 10\,\text{mHz}$, then the hyperbolas can be parametrized by the *fragility parameter*

$$F = T_g/(T_g - T_\infty). \tag{6.5a}$$

The fragility therefore characterizes the slope of the lg ω–T curves (or lg η–T curves, η viscosity). In terms of glass blowers low fragility (e.g. $F = 2$) corresponds to long glasses: they have a long time interval for blowing and forming the glass at a given cooling rate; high F values (e.g. $F = 5$) correspond to short glasses. The experience shows that for a given class of substances higher F tends to higher cooperativity N_a at T_g (see Ref. 146, for general arguments see Ref. 141).

For polymers WLF gave, as a rule of thumb, $T_g - T_\infty \approx c_2^g \approx 50\,\mathrm{K}$. This means that polymers are usually short glasses (with high F values, $F \approx 5 \ldots 7$).

It is the hyperbola form of the WLF equation that gives the instruction how to get a finite temperature asymptote for the extrapolation $\lg \tau \to \infty$, the Vogel temperature T_∞.

It is worth the while to consider the rather unique situation: except the asymptotes themselves, $\lg \Omega$ and T_∞, there is no characteristic frequency or time nor a characteristic temperature or energy kT in the interesting thermokinetic region $\{\ln \omega < \ln \Omega, T > T_\infty\}$. The use of glass temperature is only a conventional definition, albeit a very practical one, suggested by a practical time scale of seconds or minutes, or a practical cooling rate of some K/min. From physical arguments, T_g is mainly defined by T_∞, modified by fragility, and somewhat specified by the further convention to use the entropy compliance as the activity:

$$T_g = T_\infty F/(1 - F). \tag{6.5b}$$

6.4 Ideal dynamic glass transition

This section is an attempt to give a thermokinetic explanation of WLF scaling.

The global thermokinetic symmetry between T and $\ln\omega$ of Eq. 6.1 can be transformed into a local form by the following *Theorem* of differential geometry (proof in Ref. 147): Excluding a set of parallel straight lines a set of curves is a set of hyperbolas with common asymptotes if and only if the product of coordinate differences between any pairs of the curves is invariant against shifting along the curves,

$$\delta T \cdot \delta \ln \omega = \text{const} \Leftrightarrow \text{Eq. 6.1}, \tag{6.6}$$

see Fig. 36a.

The set of parallel straight lines can be excluded by physical arguments: There is no simple activation model that would give finite times for zero temperature. Therefore, the "hyperbolic" global thermokinetic symmetry between T and $\ln\omega$ can be expressed by a "local WLF equation",

$$\delta T \cdot \delta \ln \omega = \text{const.} \tag{6.7}$$

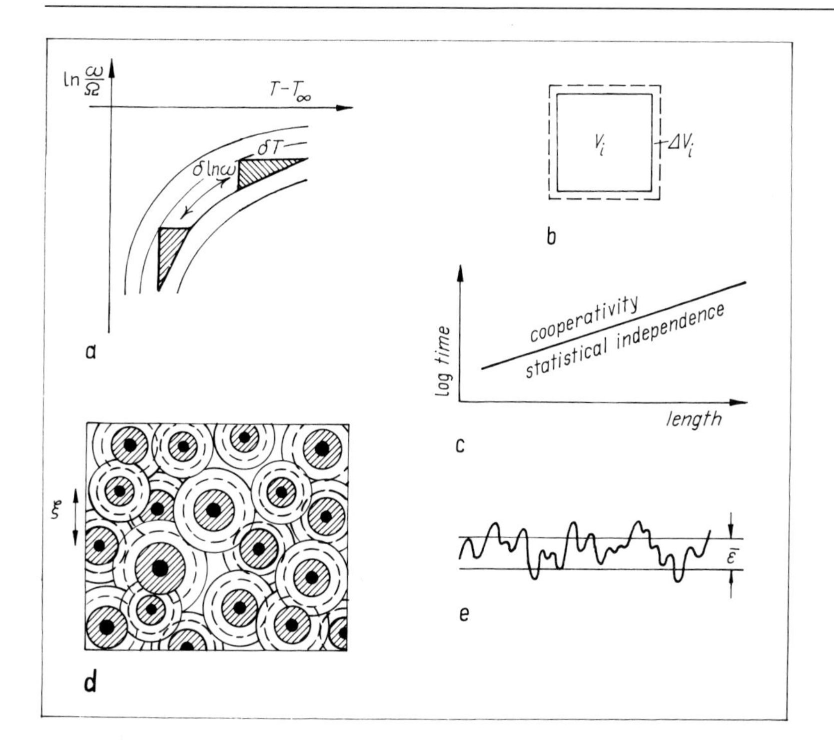

Fig. 36. Ideal glass transition.
a. The hatched area is invariant against shifting along the WLF hyperbolas in a ln ω–T diagram. b. Fluctuation of a part (block) of the cooperativity region. c. Scaled independence induces some hierarchy in the glass transition concept. d. Snapshot of the distribution of fluctuating local free volume. The black spots are the defects (or cages) with high concentration of free volume, the hatched areas around them are (scaled) regions of successively lower mobility. $\xi = 2\xi_a$ with ξ_a the characteristic length. e. The Vogel temperature as a mean roughness $\bar{\varepsilon}/k$ of the accessible energy landscape, see Fig. 27e.

Now we have to look for a parameter describing the shift. According to Adam and Gibbs (Ref. 91) we should use the size of cooperativity (N_a particles or units) as the main parameter. If we assume that N_a is the only important parameter, then we have the following strategy for the substantiation of WLF scaling. (i) Since

$$\delta T \sim N_a^{-1/2}, \tag{6.8}$$

Eq. 6.7 would be fulfilled if

$$\delta \ln \omega \sim N_a^{1/2}. \tag{6.9}$$

(ii) Look for a set of sufficient conditions for Eq. 6.9.

In this way the frequency is considered as a variable to do physics with. The conceptual basis is the fluctuation dissipation theorem (FDT, see Secs. 2.6 and 3.7) connecting the frequencies of external programs of dynamic experiments one-to-one with internal fluctuation frequencies (ω identity), and, moreover, representing the internal experimental situation (ω being a transition probability) for the definition of thermodynamic susceptibilities, see Figs. 21d and e. In this sense the frequency is not connected with a characteristic system energy: we need not have an effective Hamiltonian here.

In this book a dispersion zone that corresponds to the following two Assumptions, made in the spirit of Secs. 6.1 and 6.2, is called *ideal dynamic glass transition.*

Assumption 1 = minimal coupling (see Fig. 33). Let any block with volume V_i be connected to a typical frequency ω_i. Consider an ensemble of partitions (many sets $\{V_i; V_a\}$) so that the total distribution of all ω_i represents a reasonable temperature fluctuation spectrum of the cooperativity region. Consider then the fluctuation of a one block volume, $V_i \rightarrow V_i + \Delta V_i$, see Fig. 36b. This corresponds to $\omega_i \rightarrow \omega_i + \Delta\omega_i$. Then it follows from minimal coupling that $\Delta\omega_i$, and for the ensemble, that ω_i only depends on ΔV_i of the own block (and not on any renormalized energetic interaction),

$$\omega_i = \omega_i(\Delta V_i), \quad i = 1, 2, \ldots, n. \tag{6.10}$$

[An entropy variant of this approach will be given in Sec. 8.6.] If within each partition the minimal coupling is mathematically sharpened to statistical independence of ω_i for different i (for representative partitions of sufficent small blocks), then

$$\omega_a(\Delta V_a) = \omega_1(\Delta V_1) \cdot \omega_2(\Delta V_2) \ldots \omega_n(\Delta V_n). \tag{6.11}$$

The minimal coupling is then expressed by a geometric interdependence,

$$\Delta V_a = \Delta V_1 + \Delta V_2 + \ldots + \Delta V_n. \tag{6.12}$$

By the way, Eqs. 6.11–6.12 correspond to a "democracy" of ω_i and ΔV_i for different blocks i.

Assumption 2 = confined scaling. This means that there is no further length scale between the molecular diameter and the size of cooperativity (ξ_a). This is mathematically sharpened to: all the functions $\omega_i(\Delta V_i)$,

$i = 1, 2, \ldots, n; a$, should have the same form. Then Eqs. 6.11 and 6.12 form a system of functional equations that has only one solution,

$$\ln (\omega_i / M\{n\}) = \Delta V_i / v_0, \qquad (6.13)$$

where $M\{n\}$ is an unimportant constant describing the partition, and v_0 is a small volume that does not depend on the index i. [v_0/V_a being of order the fractional free volume – a few per cent.] Therefore, applied to the ensemble,

$$\Delta \ln \omega_i \sim \Delta V_i. \qquad (6.14)$$

This is an important result in our strategy. It is the logarithm of frequency (and not ω itself, or ω^2 etc.) which is the thermokinetic partner of a thermodynamic fluctuation.

Since, in the ensemble, ΔV_i is the fluctuation of an extensive variable we obtain from the statistical independence Eq. 6.11

$$\delta \ln \omega \sim N_a^{+1/2} \qquad (6.15)$$

for $\delta \ln \omega$ the mean fluctuation of $\ln \omega$ in the subsystem, as desired. We proved that ideal glass transitions with cooperativity as the only important shift parameter are WLF scaled.

Statistical independence inside the cooperativity region seems to be a logical contradiction. But the situation can be managed with the conception of independence scaling indicated at the end of the bus model (Sec. 6.2, see also Sec. 5.3, Fig. 28c, for a similar situation with Rouse modes), changing the yes-no property statistical independence into a continuous, relative property. Roughly speaking only independence of small blocks is needed for the derivation of Eq. 6.15. Consider a given spatial scale, e.g. $V_l = V_i + V_k$. Statistical independence between smaller blocks (V_i, V_k; high frequency) does not exclude cooperativity in a larger scale (V_l; low frequency). This situation can be repeated, or scaled: given a larger volume (lower frequency) $V = V_l + V_l'$ (Fig. 33), then V_l and V_l' can be statistically independent, at relatively high frequencies, but the sum, $V_l + V_l'$, at relatively low frequency, can be cooperative again.

The argumentation can be repeated for a given frequency or time scale. The situation will be called *scaled independence* of fluctuation and is schematically depicted in Fig. 36c. This independence is a scaling in space and time. Complementarily to the democracy of Eqs. 6.11–6.12, it is the scaled transition from statistical independence to cooperative motion that defines a scaled hierarchy in the glass transition. (See also the bus model in Sec. 6.2; hierarchy concepts for glass transition are introduced in Refs. 148, 149.)

The slow modes of cooperativity mark the low frequency flank and the end of the spectrum and define the thermodynamic concept of natural subsystems with volume V_a, see Sec. 3.6, Eq. 3.27. For $V > V_a$ the spectral density of cooperative fluctuation falls below the level of statistical independence between the larger Landau subsystems. It is difficult to detect the latter by linear response, the play is rather controlled by collective phenomena in the transport zone such as by hydrodynamic modes, see Sec. 6.10. The cooperative correlations are then condensed in the transport properties η, D, . . . and in the steady state compliance, J_e^0.

In this way, Eq. 6.8 introduces a characteristic length, $\xi_a \approx V_a^{1/3} \sim N_a^{1/3}$, into the glass transition approach. From the FDT we obtain V_a for the temperature activity at the glass transition temperature T_g,

$$V_a = kT_g^2 \Delta(1/c_v)/\rho\delta T^2, \tag{6.16}$$

if the cooperativity regions homogeneously cover the material. In Eq. 6.16, $\Delta(1/c_v)$ means the relaxation strength of the reciprocal *specific* heat at constant volume, and ρ the mass density. Using Eq. 6.16, this characteristic length (a "radius" ξ_a from $V_a = (4\pi/3)\xi_a^3$) is obtained from experiments to be of order 2 nm (Ref. 146). This size corresponds to length scales characterizing disorder in frozen-in glasses (Ref. 150), and to length scales that modify (see Ref. 101 and Ref. 150a) and finally block the typical motion in thin liquid layers.

Eq. 6.16 can be grasped from continuity arguments of the corresponding Landau Lifschitz fluctuation formulas for temperature fluctuation with regard to subsystem-volume scaling, $V\overline{\Delta T^2}(V) \to V_a\delta T^2$ for $V \to V_a$ when the natural subsystem size is reached from above. The Δ in Eq. 6.16 means that only the relaxation intensity (step) of the relevant functional subsystem is to be used. Different activities define different characteristic lengths, albeit in the same order of magnitude. The principle is given in the following way: Write down the general fluctuation formula from Ref. 45, e.g.

$$\overline{\Delta T^2} = kT/C_v. \tag{6.17a}$$

Then introduce the relaxation strength of the glass transition considered ($\Delta(1/C_v)$), express the extensive variable explicitly ($C_V = \rho V c_v$), and apply the formula to the natural subsystem ($\overline{\Delta T^2} \to \delta T^2$, $V \to V_a$). Then we arrive at Eq. 6.16, for $T = T_g$. Other examples: For pressure fluctuation we have

$$\delta p^2 = kT\Delta(B_s^{-1}), \tag{6.17b}$$

where B_s is the isentropic volume compliance, and for concentration fluctuation in polymer mixtures

$$\delta\phi^2 = \frac{kT}{\nu V_a} \Delta \left(\frac{\partial^2 G}{\partial \phi^2}\right)^{-1}, \tag{6.17c}$$

where ν is the number of segments (all components) per volume and G is the Gibbs free energy per segment. If $\xi_a < R_0$ we must consider the modifications for small subsystems (Sec. 3.8), i.e. G depends of the system size, $G = G(\xi_a)$.

For the shear stress and shear angle fluctuation of natural functional subsystems at the glass transition in simple glass formers ($G_e \rightarrow 0$) we have

$$\delta\sigma^2 = kTG_g/V_a, \tag{6.17d}$$

and

$$\delta\gamma^2 = kT(J_e^0 - J_g)/V_a, \tag{6.17e}$$

respectively. Of course, analogous equations can also be recorded for fluctuations of other extensive variables starting from $\overline{\Delta S^2} = kC_p$, $\overline{\Delta V^2} = kT(\partial V/\partial p)_T$ etc.

Although the V_a, determined from different activities, are not precisely the same (since different activities can have different sensitivities to different mode lengths), their temperature dependence is the same (because all of them are guided by WLF scaling that is only based on the $\ln\omega$ behavior). The formula for the temperature dependence of $V_a \sim N_a$ is an implication of Eqs. 6.1, 6.2, 6.7 and 6.8,

$$V_a \sim N_a \sim (T - T_\infty)^{-2}. \tag{6.18a}$$

Therefore the characteristic length is

$$\xi_a \sim (T - T_\infty)^{-\nu} \tag{6.18b}$$

with the exponent $\nu = 2/3$ (or $2/d$ for any dimension $d \geqslant 1$). Observe that this value of ν is similar to the critical exponent ν, Eq. 4.45b, and Eq. 10.15, below, although the mechanism of critical scaling is different from confined scaling for glass transition.

Considering the aspect of the general scaling principle (Sec. 2.8) in the two Assumptions we see that the distribution of frequencies corresponds to a distribution of mode lengths. Since a natural subsystem is a representative of the whole system this means that there is one spot (*defect*) in each cooperativity region, with especially short modes, high frequencies, high independence and low local density (i.e. large local free volume, see Fig. 36d). Two defects in one V_a are not possible on the average since then we would have an additional length: the mean distance between them would destroy scaling. The cooperative motion of this defect is governed by larger mode lengths and is therefore slower than the motion in it.

The conception of defect diffusion was introduced by Glarum (Ref. 151, see also Ref. 152). The defects can have very low energy barriers because small local enlargements of mean distances between the particles (a few percent) can considerably lower local potential barriers. Such defects will be called *cages* that will be further discussed in Secs. 6.8 and 7.6 below. Defects seem to be typical for thermokinetic structures.

The WLF asymptotes are "outside" the thermokinetic play, and their physical interpretation must therefore be deduced from the background. The frequency limit Ω is the extrapolated frequency of local elements. The very problem is the Vogel temperature T_∞. The corresponding asymptotic energy kT_∞ is the only energy that can enter ideal glass transition physics, since a marked energy $kT > kT_\infty$ would destroy WLF scaling, the global symmetry between T and $\ln \omega$ (see the last paragraph of Sec. 6.3). According to Sec. 5.2, we have to refer to Goldstein's energy landscape (Figs. 27d and e). As thermokinetics refers to temperature *and* frequency, we have to construct a characteristics of this landscape that includes energy barriers *and* energy steps. Both enter via Boltzmann factors. Therefore, the Vogel temperature corresponds to a mean roughness of the accessible energy landscape,

$$kT_\infty = \bar{\varepsilon}, \tag{6.19}$$

(see Ref. 153, precursors of this idea are suggested in Ref. 153a). Since the energy landscape is scaled by parameters of the intermolecular potentials (for polymers also by intramolecular potentials, see the Gibbs & DiMarzio paper of Ref. 153a), the Vogel temperatures does also scale with them (see Ref. 95 for a relevant correlation of T_g in simple organic glass formers).

Observe that $N_a \to \infty$ is essential for $T_\infty > 0$. If N_a would be finite, then we would have an activation process with a finite number of saddles etc. This would imply an Arrhenius behavior for sufficiently low temperatures with $T \to 0$ for $\lg \omega \to \infty$, despite of the large activation energy expected. That means that $\bar{\varepsilon}$ is a sharp characteristic of the whole sample (large natural subsystem) that cannot be understood as a local property.

From Eq. 6.2b. we can calculate the so-called apparent activation energy of a WLF curve at T,

$$d \ln \omega / dT^{-1} \sim T^2(T_0 - T_\infty)/(T - T_\infty)^2 \sim T^2 N_a(T). \qquad (6.20)$$

This conception was already criticized in Sec. 5.2. There is no simple relation to the number of relevant local barriers ($\sim N_a$) as long as the fragility is not high enough to neutralize the T^2 factor in Eq. 6.20. At a given T, e.g. $T_0 = T_g$, we find, of course, that $d \ln \omega / T_g dT^{-1} \sim F$, and there is some correlation with $N_a(T_g)$ inside a substance class.

6.5 Glass transition multiplicity

In Fig. 37a the dynamic shear modulus of oligomer (small molecule) and high-molecular weight polystyrene is compared (Ref. 155). The glass transition temperature for the polymer is much higher than for the oligomer. Starting from the high modulus in the glass zone we see that the glass transition zone of the the oligomer corresponds to the high-frequency edge of the polymer main transition. In polymers the glass transition is not a property specific to chains. The cooperativity of about $N_a \approx 10^2$ monomeric units in a volume of e.g. $15\,\text{nm}^3$ cannot be covered by a single chain (coil diameter $R_0 \approx 7\,\text{nm}$). Rather the 10^2 monomeric units belong to about 10 interpenetrating chains.

The main transition in polymers has a fine structure (see Fig. 29d). Moreover, after a plateau zone, it follows a further dispersion zone at lower frequencies, called the terminal zone or flow transition. Both transitions show steep curvatures in an Arrhenius diagram ($\ln \omega$ vs. T^{-1}) and obey (different) WLF equations (Ref. 108, 109, see Fig. 37b and also Figs. 6b, 7b, 13c, d, 23a, c, f, 27c). Therefore, we can expect common aspects for both transitions. The situation is called, in reference to a spectroscopic term, *multiplicity* of glass transition in polymers (Ref. 157). This concept also includes the weak relaxation between the dispersion zones and the "degenerated" Arrhenius mechanisms of local modes and other secondary relaxation. The terminology is summarized in Fig. 37d.

In the sense of ideal glass transition, WLF scaling and glass transition are synonyms. Therefore, the WLF equation for the terminal translation provokes us to consider it as a special, separate glass transition. According to the general scaling principle we expect larger units for this slower motion. Assuming comparable cooperativity ($N_a \approx 10^2 \ldots 10^3$) the larger units may

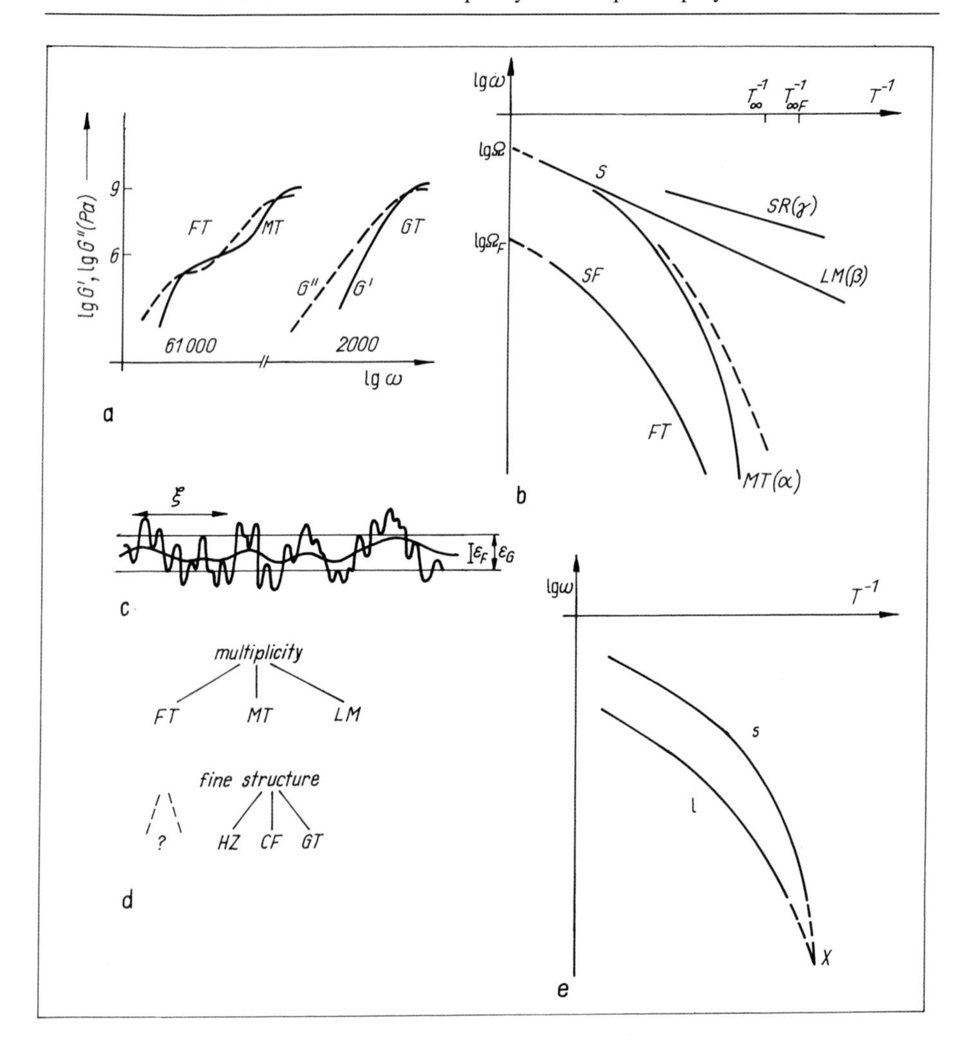
lg G', lg G''(Pa)
9
6
FT
MT
GT
G''
G'
61 000
2000
lg ω
a
lg ω
T_∞^{-1}
$T_{\infty F}^{-1}$
T^{-1}
lg Ω
lg $Ω_F$
S
SR(γ)
LM(β)
SF
FT
MT(α)
b
ξ
$ε_F$
$ε_G$
c
multiplicity
FT
MT
LM
fine structure
?
HZ
CF
GT
d
lg ω
T^{-1}
s
l
X
e

be the tube units of Sec. 5.6. Therefore we have two glass transitions in a polymer, laying "one upon another" and contain molecular motions of the same segments.

The accessible energy landscape of the larger-scale glass transition is coarsened by a preaveraging in the time and length scale of the smaller-scale glass transition. This implies a lower value of mean roughness for the larger-scale landscape ($\overline{\varepsilon_F} < \overline{\varepsilon_G}$, see Fig. 37c, although the details of averaging are not known as yet). According to Eq. 6.19, the Vogel temperature for the flow transition ($T_{\infty F}$) is *lower* than for the glass transition ($T_{\infty G}$),

$$T_{\infty F} < T_{\infty G}. \tag{6.21}$$

It is a consequence of Eq. 6.21 that both dispersion curves must intersect one another at a low frequency (point X in Fig. 37e). This means that the large-scale transition, for high temperatures at smaller times because of the general scaling principle, is not generally triggered by the small-scale transition since $\log(\omega_l/\omega_s) \rightarrow 0$ for $T \rightarrow T_X$.

The scheme of Fig. 37e seems to be generally applicable in sufficiently complex systems. The following terms will be used: *long glass transition* for the large-scale, and *short glass transition* for the small-scale transition. Thus the general picture of Eq. 37e is not restricted to polymer melts. It is also observed in polymer solutions (such as cis polyisoprene in toluene, Ref. 158), in a side-chain smectic polymer (Ref. 160), and in a small-molecule nematic liquid-crystalline glass former (Ref. 161). In liquid-crystalline materials the long transition is usually symbolized by $\parallel$, and the short transition by $\perp$,

Fig. 37. Multiplicity of glass transition.
a. Comparison of shear modulus in polystyrene of low (2000 g/mol) and high (61 000 g/mol) molecular weight. FT flow transition, MT main transition, GT glass transition. b. Arrhenius diagram of a typical amorphous polymer of high molecular weight ($M > M_c$). SR secondary relaxation, LM local mode, S, SF breakdown of scaling (see Sec. 6.8, below), α, β, γ, . . . conventional labels for amorphous polymers (decreasing relaxation temperatures for a given frequency). c. Explanation of Eq. 6.21, $T_{\infty F} < T_{\infty G}$. Preaveraging in the glass (or main) transition scale ξ diminishes the roughness of the remaining energy landscape ($\bar{\varepsilon}_G \rightarrow \bar{\varepsilon}_F$). d. Summary of multiplicity terms for amorphous polymers. GT glass transition in a narrow sense, CF confined flow, HZ hindering zone (see Secs. 5.4 and 6.6, below). e. General scheme of two glass transitions lying one upon another in the same substance. s small scale GT = *short* GT, l larger scale GT = *long* GT, X intersection point (usually at hypothetically low frequencies).

indicating the larger space requirements if umklapp mechanisms of the rods are involved.

The transition multiplicity is further enlarged when different phases in the material have their own (multiple) transitions. This will be discussed in Sec. 8.5.

6.6 Dispersion zones in amorphous polymers

In this section the dispersion zones in amorphous polymers are described in general terms. We refer to shear curves because of their outstanding sensitivity to all zones, and we precede with increasing times, i.e. with increasing mode lengths.

Fig. 38a demonstrates how the different zones can be scanned in one apparatus with a limited frequency interval (e.g. four logarithmic decades) by changing the temperature. Scaled curves, by application of temperature-time superposition, are usually called *master curves*. One has to pay attention to the fact that the shift factors a_T for different dispersion zones are governed by different WLF parameters, i.e. $a_T(MT) \neq a_T(FT)$, with Eq. 6.21; for Arrhenius mechanisms we have $T_\infty = 0$. The common master curves can then cover many decades, e.g. 15 or more. The loss parts are schematically shown in Fig. 38b (details in Refs. 108, 109, 156).

1. Secondary relaxations

Secondary relaxations are not restricted to polymers, they are also observed in other substance classes such as inorganic glasses (Ref. 162) or plastic crystals (Ref. 163). The typical length scale is about 0.5 nm. Polymer relaxations with no relation to the main transition (γ, δ in Fig. 37b) are often interpreted by relatively isolated motions of side groups or isolated foreign molecules (e.g. water in PMMA). The relaxations connected with the main transition at point S in Fig. 37b are usually considered as local modes of the backbone chain (as for PVC), see Ref. 164.

In the 0.5 nm scale mechanical models make sense (Refs. 37, 165–168). For instance, one can classify the values of activation energies or, equivalently, the relaxation temperatures at $\nu = 1$ Hz (Ref. 169). So $T \approx 90$ K is considered as a typical value for a methyl end group ($-CH_2-CH_3$). But there are striking counter-examples where a mechanical model taken for granted

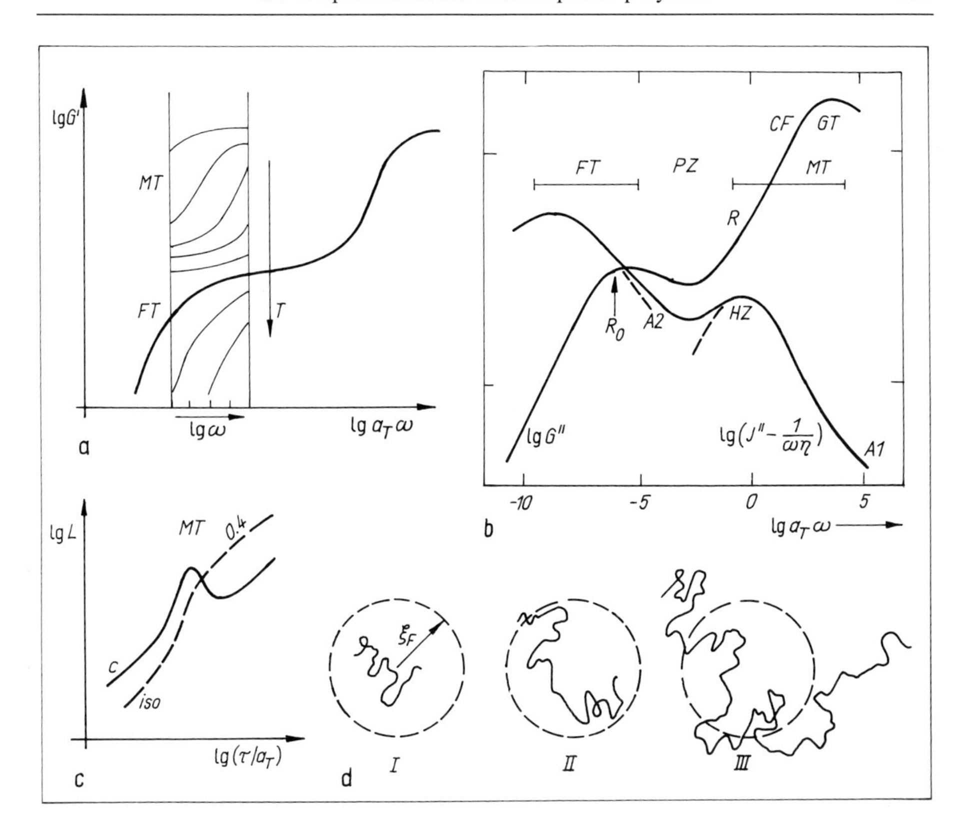

Fig. 38. Dispersion zones in amorphous polymers (schematically).
a. Shear modulus G' from experiments in an apparatus of finite frequency range (e.g. 4 decades) for different temperatures. MT main transition, FT flow transition. b. Log–log master curves of shear loss parts (G'', J''; $a_T \omega$ reduced frequency) for MT *and* FT (simplified, from Ref. 108 for PVAC) A1 Andrade law for soft glassy zone, A2 ditto for plateau zone PZ; GT glass transition zone in narrow sense, HZ hindering zone, R Rouse like modes, R_0 location of chain end-to-end distance scale (uncertain). c. Shear retardation spectrum L for isotactic (iso) and commercial (c) PMMA of high molecular weight. The value 0.4 is the retardation exponent $d \lg L/d \lg \tau$. d. Three regimes of flow transition for increasing chain length. I confined scaling, fluctuating tubes, II crossover, III de Gennes scaling.

before could not be proved by modern NMR methods that can now discriminate many details in this length scale (see e.g. Ref. 169a). The complaisance of the local environments, local packing effects, spreading-out along the chain, and diffusion modes are often important to understand local processes.

The spectrum of a secondary relaxation is usually rather broad, e.g. up to four logarithmic decades. This can at the most be partially explained by broadening due to this complaisance. A large spatial variation of local environments in thermokinetic or amorphous structure must be taken into account, see Fig. 36d, implying a large spatial variation of activation energies. Koppelmann (Ref. 88) concluded from a wide variation of temperature and pressure effects on secondary relaxation in PVC that a widening of intermolecular distances in the range of few per cent would result in spectral narrowing to spectral widths comparable to Lorentz curves. The few per cent are enough to switch off the steep intermolecular repulsion potential of the environment and, therefore, to damp the spatial variation.

2. General remarks to main and flow transition and their sequential predecessors

The two glass transitions of Fig. 37b, main transition MT and flow transition FT, show some similarities that will be described for an increasing time scale (decreasing frequency, increasing length i.e. from the right to the left in Fig. 38b).

(i) The maximum of G'' is before the J'' (or $J'' - 1/\omega\eta$ for FT) maximum. The difference is usually several logarithmic decades. Since the J'' or $J'' - 1/\omega\eta$ maximum indicates the flow onset, the corresponding flow has larger mode lengths than e.g. the entropy fluctuation lying, as expected, near the G'' maximum. This latter maximum indicates the onset of cooperative mobility, a prerequisite for flow.

(ii) Before flow onset in J'' we find universal, broad predecessors with an exponent near 1/3 over several decades of time. (Andrade law, Refs. 170, 108). The predecessor's loss level is considerably higher than expected from an extrapolation of the peaks. There is no relaxation desert here but a special relaxation with a certain independence of the mechanism. The unique exponent and the large time interval indicate, in a space-time picture, that for the preparation of flow a self-similar *sequential process* is necessary that "grows" in time (Ref. 171) and, according to the general scaling principle, also in space. One can perhaps think about a percolation ansatz including time as a coordinate (Ref. 172, hard-softening percolation). This would give an exponent near 0.4 instead of 1/3 (Ref. 173).

(iii) The end of the Andrade zones aproximately coincides with the frequency of the corresponding G'' maxima. Obviously, the onset of cooperativity and therefore of flow ability in polymers coincides with a spatial percolation of the preceding sequential combination of simpler units.

To avoid misunderstanding we should again underline that the interdependence between J'' and G'' by Eqs. 2.13 or 2.35, and between real and imaginary parts of J^* or G^* by the dispersion relation Eq. 2.37, is rather loose if the shear exponent $d \lg G'/d \lg \omega$ tends to or goes beyond one in a larger frequency interval as observed near and at the both glass transitions considered. Small changes in G^* at these flanks (where $d \lg G''/d \lg \omega \approx 1$) result in large shifts of the corresponding J'' or $J'' - 1/\omega\eta$ maxima there (*flank sensitivity*). Therefore we can, to a certain degree, discuss J'' and G'' rather independently.

The dependence on the polymer chain length ($\sim N$) is not shown in Fig. 37b or 38b. The N sensitivity increases with increasing time and length scale. The glass transition is hardly influenced, except that T_g depends on the concentration of free chain ends in the cooperativity region. But the development of fine structure, the separation of main and flow transition for $N > N_c$, and the development of different regimes in the flow transition is increasingly influenced by the chain length. Particularities of the flow transition are, therefore, heavily influenced by the polydispersity (distribution of molecular weights) of the polymer.

The repetition of the glass transition at larger scale, the flow transition, is possible because in polymers the entanglement spacing d_E can define a new unit (tube unit) that can serve as a "particle" for the long glass transition.

3. Main transition

The fine structure of the main transition was described in the discussion of the Rouse modes (see Sec. 5.4 and Figs. 29c, d).

We are now interested in the connection between shear modulus G and compliance J, see Fig. 38b again. The maximum of G'' is linked with a particularity on the high frequency flank of the J' peak. This is probably connected with small shear activities of the density and entropy fluctuation typical for the ordinary glass transition in the 2 nm scale. The confined flow with its large shear exponent $d \lg G'/d \lg \omega$ corresponds to sharp increasing

J'' values towards the J'' peak. The modified Rouse modes R in Figs. 29d and 38b approximately correspond to th J'' peak maximum. They hinder the confined flow (hindering zone). The main transition zone ends at large scales when the entanglements stop Rouse-like chain motion ($d_E \approx 7\,\mathrm{nm}$).

Briefly, the fine structure of the main transition is defined by the G'' maximum (near glass transition zone GT) and the J'' maximum (hindering zone HZ or "entanglement transition"). The between is the strange confined flow CF that is rather sensitive to molecular structure (Ref. 174, 175). The length scale of CF is probably between the characteristic length $\xi_a \approx 2\,\mathrm{nm}$ of the glass transition and the shortest reasonable Rouse-like mode. A finer analysis (Ref. 109) showed that the WLF parameters of GT and HZ are slightly different with

$$T_{\infty H} < T_{\infty G} \tag{6.22}$$

corresponding to Eq. 6.21. This indicates some independence of the two fine structure components.

4. Flow transition (or terminal zone)

The flow transition is separated from the main transition by the so-called rubbery *plateau zone* PZ. This "gap" is proportional to $M^{3.4}$ as qualitatively explained by the reptation model in Secs. 5.5 and 5.6. The shear level G_N^0 of the plateau is scaled with the Kuhn segment as described in Sec. 3.9. As mentioned above the relatively high losses G', J'' show that there is no relaxation desert, the Andrade law indicates a scaled sequential process successively preparing the cooperative flow transition. It is interesting that there is an exception from the usually low Andrade loss (see Fig. 38c, Ref. 176): isostatic PMMA does not have a well developed plateau zone. Obviously the double helix structure of this polymer allows a rather effective, percolative (?) sequential process ($J(t) \sim t^a$ with $a \approx 0.4$) preceding (or being part of) the flow transition.

Conventionally, the functional subsystem of the flow transition corresponds to the function: tangling and loosening of entanglements. There is again a fine structure with broad and separated G'' and J'' peaks. Only a few recent experiments allow to label the spatial scale of the coil radius R_0 in this transition (see Sec. 6.9), R_0 is probably near the G'' peak.

Interpreting the flow transition as a glass transition with tube units as elements, then a new length ξ_F is defined for $T < T_{SF}$ by the characteristic length for cooperativity N_a of this glass transition:

$$\xi_F \approx N_a^{1/3}(T)\, d_E. \tag{6.23}$$

T_{SF} is the temperature of cooperativity onset for the flow transition, see Fig. 37b. If $N_a \approx 300$ is also a typical value for this transition, then we can estimate $\xi_F \approx 50\,\mathrm{nm}$. If d_E does not depend on the temperature T then

$$\xi_F \sim (T - T_{\infty F})^{-2/3}. \tag{6.24}$$

The new length corresponds to a characteristic molecular weight of the chains, the corresponding degree of polymerization can be estimated from $\xi_F = aZ_F^{1/2}$ as

$$Z_F \approx (\xi_F/a)^2 \sim (T - T_{\infty F})^{-4/3} \tag{6.25}$$

if again $d_E \approx$ const. A typical value for PS would be $Z_F \approx (50/0.7)^2 \approx 5000$ with $M_F \approx 5 \cdot 10^5\,\mathrm{g/mol}$ (from $a \approx 0.7\,\mathrm{nm}$).

The existence of this new length would imply three regimes for $T < T_{SF}$, see Fig. 38d. (I) $M_c < M < M_F$. The tube fluctuation corresponds to a motion of units typical for glass transition, but modified because the units are tube units. This model is analyzed in Sec. 6.10 and it will be shown that the new degree of freedom (transverse tube motion) decouples shear and diffusion exponent ($D \sim N^{-2}$, $\eta \sim N^{3.5}$), (II) $M \approx M_F$. Crossover. (III) $M > M_F$. de Gennes scaling. If the tube fluctuation is restricted to the ξ_F (or N_F) scale, then we have obstacles that are fixed when observed from the very large scale $R_0 > \varepsilon_F$. This gives, of course, the reptation coupling ($D \sim N^{-2}$, $\eta \sim N^3$), and has very large relaxation times because of the very high molecular weights of order $10^6\,\mathrm{g/mol}$ and larger.

6.7 Deviations from WLF scaling

As mentioned above the small (when compared to critical phenomena) cooperativity of $N_a \approx 10^2$ implies that intermolecular particularities shine through the universal aspects of glass transitions. Since the cooperativity decreases with higher temperatures increasing deviations from WLF scaling are expected to occur there.

A further possible source of deviations can be expected from the glass transition multiplicity. Even if the short and the long glass transitions are completely independent the spectra are superposed and overlaps cannot be

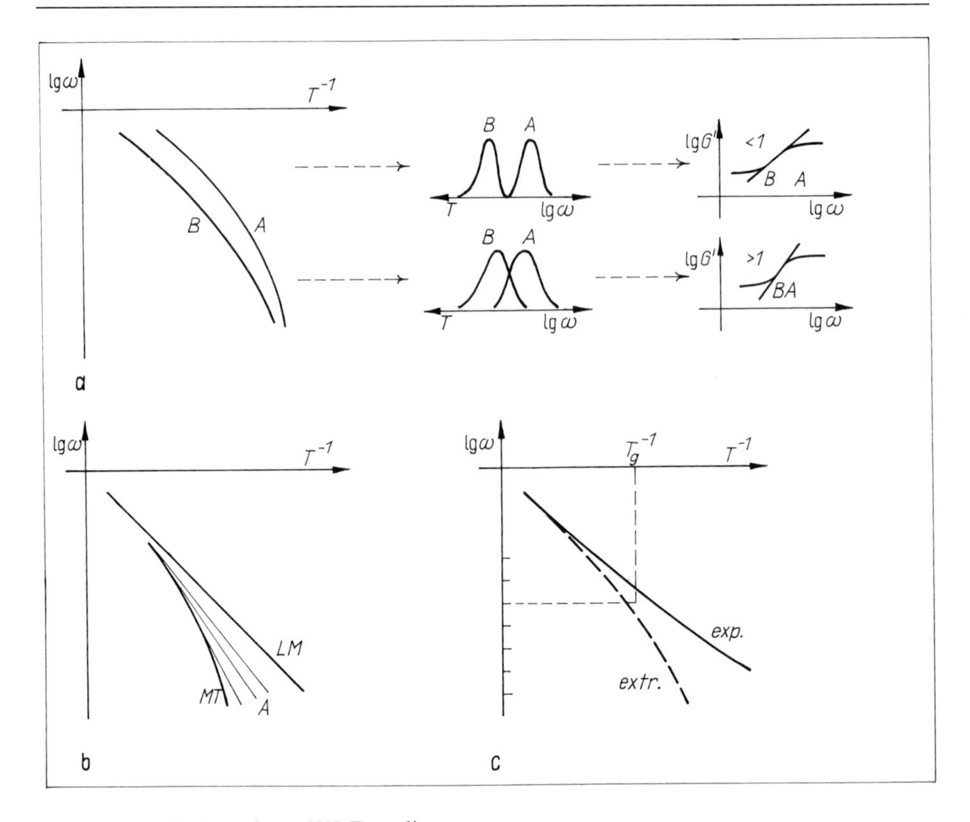

Fig. 39. Deviations from WLF scaling.
a. Superposition of two components (A, B) of a fine structure with some independence of components (Eq. 6.22). b. Proposal for the relaxation traces of an Andrade law A between main transition MT and local modes LM. c. Effect of freezing-in near the thermal glass temperature T_g. exp experimental curve, extr. extrapolated equilibrium WLF curve (obtained after freezing-in correction).

separated by a thermodynamic experiment, see Fig. 23b [Except the two zones can, in a series of experiments, be differently parameterized.] The situation of an overlapping fine structure A, B is schematically shown in Fig. 39a.

As the WLF curves for both components approximate (see Eq. 6.22 and Fig. 37e) the total spectrum narrows for decreasing, sufficiently low temperatures. This tendency is opposite to the widening of the spectrum by WLF scaling for each component (Eq. 6.2a, Fig. 25b). So, e.g. the average shear exponent $d \lg G'/d \lg \omega$ can increase for decreasing temperatures, if the level of the neighboring plateaus is constant. This may explain some features of confined flow.

The two opposite tendencies of Eqs. 6.2a and 6.22 can, in a certain temperature interval, generate a total spectrum with nearly constant width, characterized e.g. by a constant KWW β parameter. Perhaps this was the reason to develop the conception of "thermorheological simplicity" (constant form of spectra) which, in the opinion of this book, actually describes the rather complex situation of Fig. 39a. This is a good example for the dilemma to define what is simple in a complex situation with no accepted theory.

Growing cooperativity of a dispersion zone and growing complexity of neighboring Andrade zones can also change the physical situation at the crossover from one to the other zone. This concerns the flanks of the disperson zone and is very sensitive to the β value. The influence of this effect is difficult to be estimated, especially because there are reasons that the apparent activation energy of the Andrade zone mediates between the neighboring dispersion zones. A proposal is depicted in Fig. 39b.

The freezing-in of fluctuation at the glass temperature T_g (Chap. 7) also changes the slope of the curves in an Arrhenius plot, see Fig. 39c. Because of the broadness and multiplicity of the spectrum the main transition does not freeze-in at once, but successively, starting far above T_g (e.g. $T_g + 10\,\text{K}$) with the slow and long modes and ending far below T_g with quick and short modes leading to the Andrade zone (no desert).

This partial freezing-in leads to partial nonequilibrium that implies driving forces to restore equilibrium. Driving force means "acceleration" of kinetics which means shortening of relaxaton times, i.e. the WLF curve is successively lifted in an Arrhenius plot near T_g. The WLF curvature is therefore decreased, the curves can be straightened or even bended, like as sketched in Fig. 39c.

As a rule, precisely this effect is meant when it is said that WLF is true only from 5 or 10 K above T_g to higher temperatures. This effect should be carefully corrected for if one tries to extrapolate WLF in the direction of the Vogel temperature T_∞ and if one cannot do without the points near T_g. Specialists (e.g. Ref. 177) assure to have always found WLF behavior after freezing-in corrections in amorphous polymers.

6.8 Minimal cooperativity, cage effects, and excess cluster scattering

In this section several high-frequency effects and their consequences are described.

1. Minimal cooperativity and splitting point

Confined scaling of dynamic glass transition as described in Sec. 6.4 requires a minimal size of cooperativity. Values of $N_a = 30 \ldots 300$, depending on the fragility, were found at T_g (Ref. 146); N_a decreases with increasing temperature according to Eq. 6.18a. Consequently there must be a temperature (region near) T_s where a *minimal cooperativity* is reached. For $T > T_s$ WLF scaling breaks down. This breakdown is indicated by S in the Arrhenius diagram Fig. 40a (*splitting point* for glass transition and, if present, local mode).

The size of minimal cooperativity can be estimated from $N_a(T_g)$, Eq. 6.18a, and the onset of curvature in the Arrhenius diagram: The result is

$$N_a(S) \approx 10 \ldots 30. \qquad (6.26)$$

These are rather small numbers for universality: Intermolecular particularities of the substances must influence the frequency and temperature of S to

Fig. 40. High frequency phenomena.
a. Splitting "point" S. Minimal cooperativity at breakdown of scaling. b. Decay of density correlation for high temperatures $T > T_s$. [The dotted curve is a prolongation for the two-step decay (β, α) at $T < T_s$, symbolically.] The arrow indicates increasing q values, q is of order $Å^{-1}$. c. So-called non-ergodicity factor f_q' as a function of temperature T and scattering vector q (of order $Å^{-1}$). The region $T > T_s$ is not cleared up. (What about the structure from self-part correlation if there is no cage?) d. Cage model for the nonactivated process. e. Free play in a cage; $\overline{\Delta r^2}$ from a Debye Waller approximation of curves from Fig. 40c. The typical order of $\overline{\Delta r^2}$ is $x_0^2 \approx 1\,Å^2$. A similar picture is obtained for the scattering intensity I. f. L.h.s. part: Scenario for increasing ability to form activation steps (rattling → mutual change of place). R.h.s. part: mixed situation as for PE. g. Two dynamic glass transitions one upon another according to the scheme Fig. 37e. I normal short glass transition, II long glass transition with stabilized cooperativity regions as units (hypothetical). h. Upper part: two kinds of "states" for cages, living (○), dormant (●). ○ and ● are the symbols used in the text. One-particle model. Lower part: many-particle model for small cages corresponding to small $\overline{\Delta r^2} = 0(Å^2) <$ particle diameter, see Eq. 6.30.

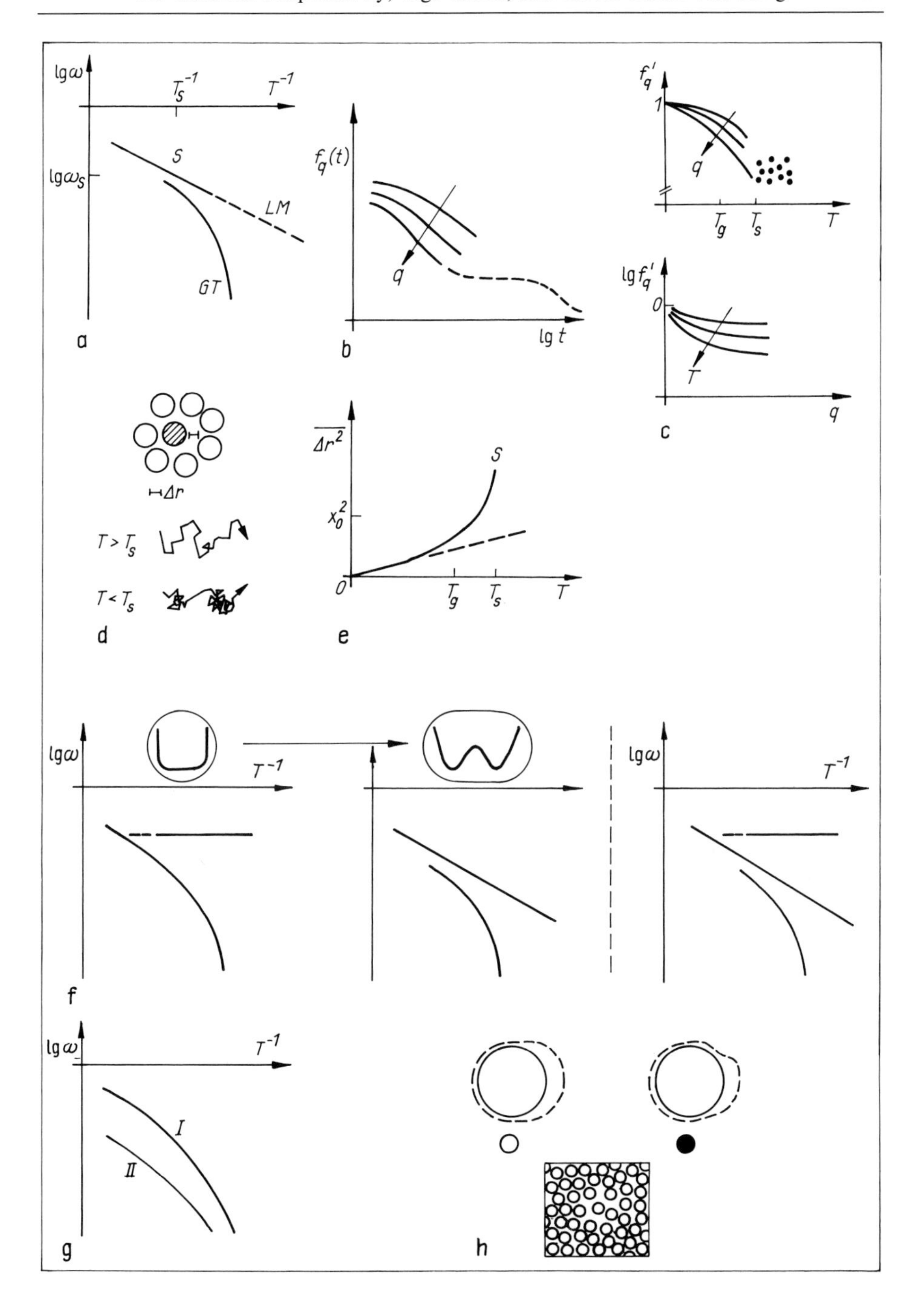
lgω
T_S^{-1}
T^{-1}
lg$ω_S$
S
LM
GT
a
$f_q(t)$
q
lg t
b
f_q'
1
q
T_g
T_s
T
lgf_q'
0
T
q
c
Δr
T > T_s
T < T_s
d
$\overline{Δr^2}$
S
x_0^2
0
T_g
T_s
T
e
lgω
T^{-1}
lgω
T^{-1}
f
lgω
T^{-1}
I
II
g
h

a high degree. If one wants to know which properties of intermolecular interactions promote or hinder the glass transition, or which molecular motions are typical for the "elements" of cooperativity then one should turn to investigations of minimal cooperativity around the splitting point. The usually high frequency ($\nu_s \geqslant$ MHz) also allows the use of dynamic neutron scattering and computer simulation by means of molecular dynamics.

If minimal cooperativity corresponds to a Debye relaxation (KWW parameter $\beta \approx 1$) then we have, from Eq. 6.2a using $\beta \sim (\delta \ln \omega)^{-1}$, for an ideal glass transition

$$\beta \approx (T - T_\infty)/(T_s - T_\infty) \quad \text{for} \quad T < T_s. \tag{6.27}$$

It is typical for polymers (see Ref. 3), but is also found in other substances (see e.g. Ref. 163), that the onset of curvature for a main transition in an Arrhenius diagram is connected with splitting off a local mode (see Fig. 40a again). The local mode has approximately the same activation energy as the original transition at high temperature, $T > T_s$. This suggests to consider the activated local modes with the "compliance" of their environments to be the elements of the cooperative movement at lower temperatures $T < T_s$.

Let us repeat the argument in other words. Consider decreasing temperature approaching T_s. The activated local process (e.g. exchange of places in the first-neighbor's shell accompanied by promoting movements of the other molecules) is confronted with increasing difficulties to overcome local barriers. The role of promoting movements of further particles increases, also the role of competition between different possibilities. Such conceptions can be considered to be precursors of cooperativity in the $\leqslant 1$ nm scale of the ρ level.

Mode coupling theory, proved to be useful for simple liquids (e.g. Ref. 178), can be applied to cooperativity precursors (see e.g. Refs. 179–181). The increasing complexity is monitored by coupling parameters such as λ in equations like Eq. 4.24. This theory describes e.g. the decay of density correlation for $T \gtrsim T_s$ by a von Schweidler law (see Fig. 40b),

$$f_q(t) = \frac{S(q, t)}{S(q, 0)} = 1 - h_q \cdot (t/\tau_0)^b, \quad b < 1, \tag{6.28}$$

separating the q and t dependence, and scaling the time by an exponent b. The time dependence is similar to a KWW law. Temperature enters by the temperature dependence of the coupling $\lambda(T)$ making the reduction time to be a function of T, $\tau_0 = \tau_0(T)$. This describes the continuous increase of viscosity by a factor 10 or 100.

The unexpectedly great success of mode coupling concepts in dynamic critical scaling (see e.g. Ref. 182) induced the application of the mode coupling theory also to the crossover ("point" S) from high temperature behavior to cooperativity. This seems to be an attempt to describe the cooperativity by mathematical structures for collective elements. As mentioned above, decreasing temperature implies increasing coupling, $\lambda = \lambda(T)$. Roughly speaking at a certain $\lambda = \lambda_c$ the kinetic situation of nonlinear mode coupling becomes singular. The result is a sharp kinetic critical temperature T_c widely above T_g, $T_c \approx T_s$. Below T_c, the $f_q(t)$ decay is temporarily stopped for larger time: a so-called ergodic-to-nonergodic transition occurs in the megahertz-and-larger region.

In the case of a cubic basic structure of the equations, e.g. $\rho + \lambda\rho^3$, similar to a van der Waals equation, the nonlinearity generates a sharp T_c via a bifurcation. According to Thom's catastrophe theory it is not so essential to have integro-differential equations for mode coupling instead of simple equations for the van der Waals equation, because analyticity assures a similar germ structure. Of course more complex splitting or parameterizations can be generated by higher nonlinearity, $\rho + \Sigma\lambda_k\rho^k$, $k \geqslant 3$, but then one has to look for reasoning the necessary numbers of λ.

Insisting on (thermo)kinetic causes for glass transitions a sharp T_c and linked topological properties can only be explained by a large cooperativity, $\delta T \sim N_a^{-1/2} \to 0$: A cooperative spectrum cannot be completed in a too small volume. In the framework of ideal glass transition (Sec. 6.4) there is no possibility for a large cooperativity region at higher temperatures, $T > T_g$. Ideal WLF-scaled glass transition with minimal cooperativity and, on the other hand, a sharp critical temperature $T_c > T_\infty$ are mutually excluding conceptions.

One could object that first-order phase transition (e.g. melting) does have a sharp transition temperature without large cooperativity. The answer is that such a transition has global reasons and originates from an intersection of different states that have sharp thermodynamic (not kinetic) state variables because of the statistical independence of subsystems in large phases (large droplets, for instance). Small droplets actually have broader transitions.

A sharp, mean-field like T_c can be produced by a mode coupling because it is no longer sensitive to the spatial scale for $\lambda \to \lambda_c$ (see Ref. 182). [The Mori Zwanzig projection (Sec. 4.1.6) is difficult to handle selectively for a large number of noncollective degrees of freedom.] Therefore the internal restrictions against an uncontrolled increase of the correlation length are canceled. The way out from the dilemma was described by Angell (Ref. 183, see also Ref. 182): The region S in Fig. 40a is a crossover region, as mentioned above, where the coupling of local modes goes over into a new cooperative process with universal features, the dynamic glass transition.

If the dynamic glass transition is ideal, then the new mechanism for $T < T_s$ cannot be expected to be inferred from the mode coupling theory because the strategies of Secs. 4.1.6 (collectivity) and 6.4 (cooperativity) are too different. Seen from the low temperature side the crossover corresponds to the breakdown of WLF scaling. The crossover is degenerated to a point which is best for rough extrapolations in an Arrhenius diagram, when the dispersion zones are represented as lines (see e.g. Fig. 37b). From the size of minimal cooperativity, Eq. 6.26, the temperature uncertainty of the crossover is estimated to be $\delta T(T_s) = 0(10\,\mathrm{K})$. Seen from the high temperature side the cooperativity typical for dynamic glass transition is developed in a temperature interval of this size. It may be characterized by the development of activity splitting (Fig. 27g) on the secondary background. Since the cooperativity region is larger than a local mode we expect the cooperativity onset at a lower frequency, separated from the Arrhenius process by e.g. one logarithmic decade as indicated in Fig. 40a (see e.g. Ref. 183a for PE).

2. Nonactivated process

The bifurcation property of the mode coupling theory drew the attention to weakly or nonactivated dispersions in the high frequency region of $1/\tau \approx 10 \ldots 100\,\mathrm{GHz}$. They were known (e.g. Ref. 184) or simultaneously investigated (e.g. Ref. 185) by several experimental methods.

Fig. 40c shows the decay of the "nonergodicity factor" with increasing temperature,

$$f'_q(T) = \lim_{t \geqslant \tau} S(q, t, T)/S(q, 0, T). \tag{6.29}$$

This factor can be obtained from the incoherent neutron scattering function measuring the self part of the van Hove correlation function, Eq. 1.21. It describes the nondecaying part in the density-density correlation, the "frozen correlations", in other words, the spectral contribution of elastic components in the neutron scattering spectrum. Shortly, this factor describes the "solidity" of the sample. Of course, the term nonergodicity is related to the relaxation time τ in the ns region. [In terms of the mode coupling theory, this factor describes, for decreasing temperatures, how the increasing coupling strength leads to structural arrest of the density fluctuations.]

The low activation energy has led to *cage models* with free play for motions of one or a few particles, or polymer segments, in relatively small spatial or orientational scales. That is, the incoherent neutron scattering is interpreted to measure the self-part correlation of a fast rattling motion (Ref. 186) in local cages built by neighboring particles of thermokinetic structure (see Fig. 40d, cf. also the cage conception from the general scaling principle in Sec. 6.4 Fig. 36d).

The size of free play in such a cage can be estimated by a one-particle Debye-Waller-factor approximation for the curves in Fig. 40c,

$$f'_q = \exp\left(-\tfrac{1}{3}\overline{\Delta r^2}\, q^2\right), \tag{6.30}$$

i.e. looking for a m.s. amplitude from $\overline{\Delta r^2} \cdot q^2 \approx 1$. The result is schematically depicted in Fig. 40e. The typical order is (below) 1 Å. A solid cage would give, according to the Nyquist kT in the FDT, $\overline{\Delta r^2} \sim T$. Fig. 40e shows that the more-than-linear increase starts well below T_g (sequential softening of cages in the Andrade zone A1, Fig. 38b), and that $\overline{\Delta r^2}$ sharply increases when the minimal cooperativity at the crossover S is approached.

The density-corrected scattering intensity I as a function of temperature behaves similar to $\overline{\Delta r^2}$ in Fig. 40e. The excess intensity (beyond the part $\sim kT$) correlates as $I^E \sim (T - T_\theta)^2$ (Ref. 187). Since $(T - T_\infty)^{-2} \sim V_a$ (with V_a the cooperativity volume, see Eq. 6.18a) this could mean that I^E is proportional to the number of cages increasing with V_a^{-1}, provided the scattering intensity per cage is approximately constant.

The behavior at S can be interpreted as the dissolution of cages at the breakdown of their thermokinetic structure. Then the self part of correlations does not longer give a structure information ($f'_q \approx$ const, with considerable scatter of results). Surely there is much free play at $T > T_s$, but there are no thermokinetic cages; the widening of the elastic component of the neutron scattering is perhaps a consequence of missing cooperativity.

Probably, rather specific intermolecular potentials are needed to form cages so that there are probably glass-forming substances with no possibility to form them. Assuming a parameter "increasing ability to form local activation situations instead of (or besides) cages" we can think about the scenario of Fig. 40f. Consider for instance the series of poly (alkyl methacrylates) with increasing C numbers (Me, Et, Pr, . . . , see Fig. 50b below). The splitting-point frequency falls below kHz and the ability mentioned must be considered to be rather high due to the entangling tendency of the α methyl group.

It is interesting that cages are formed in polyethylene PE both in the amorphous phase (rather large cages) and in the crystalline phase (small cages, see Ref. 188). As PE has local modes in the classical sense, we observe in both phases at lower temperatures ($T < T_s$) *three* small-unit dispersions: cage, local mode, and main transition. Therefore no simple bifurcation of a high temperature process is expected to be sufficient for explanation.

In the 100 GHz region a further dispersion is observed, the so-called boson peak (Ref. 192). This is more a collective than a cooperative effect and is connected with dispersions that occur when the wave length of phonons becomes small, comparable with the characteristic length of amorphous thermokinetic structure. Then the phonon lifetime and the period of oscillation is of the same order of magnitude. Of course, the dispersion law is different for collective and cooperative processes.

3. Excess cluster scattering

Many polymers show an excess light scattering that is connected with a large length of order $\xi_D \approx 100$ nm (see e.g. Refs. 189–191). The scattering intensity can by described by a Debye formula

$$I_{ex} \sim \xi_D^3/(1 + q^2\xi_D^2)^2 \qquad (6.31a)$$

corresponding to a simple spatial decay correlation function

$$\varphi(r) \sim \exp(-r/\xi_D). \qquad (6.31b)$$

Two q dependent relaxations are observed in the transport zone of the glass-forming small-molecule mixture of Ref. 191. [The term transport zone will be explained in Sec. 6.10, Fig. 42a. It is an open question whether this doublet can also be observed in pure substances with excess scattering.] Both relaxations can be q extrapolated to WLF dispersion zones I, II (possibly showing linear response signals), see Fig. 40g; the extrapolation uses Eq. 6.32, below. Furthermore, this phenomenon is related to the occurrence of cages, and a particular long-time (days) dynamics is observed with the ξ_D clusters.

These findings allow to consider the following purely hypothetical scenario for the excess clusters. We have to look for relations between three or four length scales that will be considered in the succession: cages (free play of tenths of nm), short glass transition I (2 nm), long glass transition II (order 10 nm), and clusters (ξ_D).

Cages: Because of the steep intermolecular repulsion potentials small deformations of cages are suffient for the induction of an activation or hindering step, see Fig. 40h. Accordingly we have "living" and "dormant" cages (○, ● in Fig. 40h, Ref. 193).

Short (normal) glass transition (I in Fig. 40g): There is, according to Fig. 36d, one cage per cooperativity region of volume V_a with a diameter $2\xi_a$ of order few nanometers. It will be said that this region is stabilized if the cage is dormant. A stabilized region corresponds to a higher structure and should lead to a slight enhancement of T_g. As V_a decreases with increasing temperature we have more cages at higher temperature.

Long glass transition (II in Fig. 40g): If a stabilized cooperativity region can be considered as some new kind of unit (like a tube unit for flow transition) then we have the situation of Fig. 37e with a short and a long glass transition, one upon another. If the cooperativity of both glass transitions is comparable (similar cooperativity values N_a), then the characteristic length of the long glass transition is of order $2N_a^{1/3}\xi_a \approx 10\,\text{nm}$. According to Eq. 6.21 a lower Vogel temperature is expected for the long glass transition. Complications may arise from the "interference" with the scale of concentration fluctuation in mixtures ($\approx 5\,\text{nm}$, better visibility?) or with the entanglement spacing in polymers ($\approx 7\,\text{nm}$).

Clusters: The fraction of dormant cages defines a usual order parameter if there is some different energetic Ψ-level interaction between dormant and living cages. [This means having some kind of spin glass situation.] Then we expect phase separation tendencies if dormant cages (stabilized cooperativity regions) favor a stabilized environment. Clustering is expected if there is some slow equilibrium kinetics ○ $\rightleftharpoons$ ● (Ref. 193). The total phase transition is lodged by awakening dormant cages destabilizing the phases. The cluster size is a certain balance between the different types of kinetics: ○ $\rightleftharpoons$ ●, as well as long and short glass-transition diffusion, and there are several possibilities for long-time effects.

This seems to be the best place to introduce a new general term: The phenomenon, that a sufficiently large structure can have a longer lifetime than their small elements will be called *survival of large structures*. Examples: (i) The critical molecular mass M_c for entanglement onset shows that this concept demands a minimal number of structure elements. Too small structures break down when one element breaks. (ii) Hydrogen bonding kinetics between short chains $M < M_c$ can simulate transient chains long enough ($M > M_c$) to do a transient entanglement (Ref. 194). Then we have temporary tube units that can define a long glass transition. (iii) The reptation paradox of the first paragraph in Sec. 5.6 can probably be solved by using this conception: Sufficiently long chains can diffuse many times their own size (lifetime of elements) without relaxing stress (a property of the surviving large structure of size larger than R_0, see the end of Sec. 6.10).

Possibly, clusters are such a surviving large structure. This conception also allows the construction of large thermokinetic structures, and it seems to be the bases for the organization of a hierarchy of real structures in some polymers. Surely there are also relations to the general scaling principle, to the conception of length-time scaling of statistical independence (Fig. 36, Sec. 6.4) and to the multiplicity of glass transition.

6.9 Activity arrangement across dispersion zones

The guiding set of WLF hyperbolas for a glass transition was obtained from minimal coupling without essential reference to a special activity. [See also Sec. 8.6, below, to this point.] Therefore, the different activities need not have the same maximum position and spectral form. This was really the experimental result of Ferry e.a. (Refs. 52–54) and was in principle described in Sec. 2.7 (Fig. 15) and Sec. 2.8 (Fig. 16e). In Sec. 5.2 (Fig. 27f) the activity splitting

---

Fig. 41. Different activities.
a. Cartoon: Signal splitting distinguishes cooperative molecular motion (I) from local mobilities (II). b. Maximum frequency of different compliance spectra for the main transition in PVAC (J shear, C_p entropy, ε dielectric, NMR T_1 and $T_{1\rho}$ relaxation times). G for shear modulus spectrum. c. Arrangement of different loss functions for PVAC at $T = 318$ K. J'' shear compliance, $J_s'' = C_p''$ entropy compliance, G'' shear modulus. In brackets: length scale in nanometers. d. Dielectric loss in cis polyisoprene (cis PI) of different molecular weight M. FT flow transition, MT main transition. The lower part shows the direction of the monomer dipole moment and the vector sum of their longitudinal components along the chain. e. Comparison of dielectric compliance and shear modulus in the flow transition of cis PI-140 at 273 K. f. WLF curve for a local dielectric process in a low-molecular nematic liquid crystal. g. Map of the defect model for glass transition from Fig. 36d into the lg q–lg ω plane (dispersion law) and into the lg ω–T plane (WLF). q scattering vector. The broken line corresponds to $\xi_a \sim (T - T_\infty)^{-2/3}$. S breakdown of cooperativity; only local processes remain at high temperature.

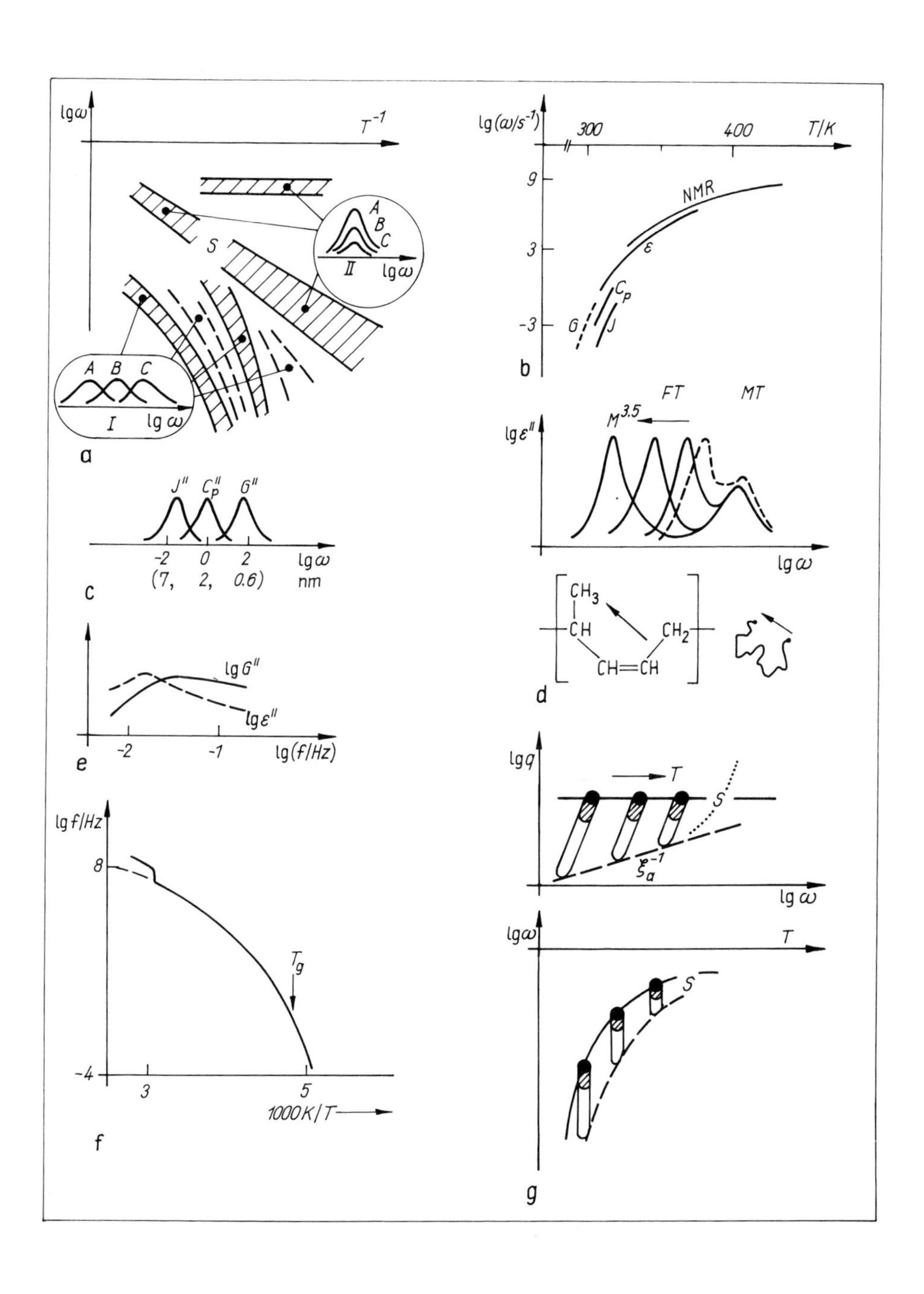
lgω
T^{-1}
S
A B C
I lg ω
II lg ω
a
lg(ω/s⁻¹)
300
400
T/K
9
3
-3
NMR
ε
C_p
G
J
b
FT
MT
$M^{3.5}$
lgε″
lg ω
CH₃
CH
CH₂
CH=CH
d
J″ C_p'' G″
-2 0 2
(7, 2, 0.6)
lgω
nm
c
lg G″
lg ε″
-2
-1
lg(f/Hz)
e
lg f/Hz
8
-4
3
5
T_g
1000K/T
f
lgq
T
S
ξ_a^{-1}
lg ω
lgω
T
S
g

was declared to be an essential characteristic of cooperativity. The general scenario is indicated in Fig. 41a. Three examples will be given.

(i) The maximum frequencies of spectra for an assortment of classical compliances at the main transition of PVAC are schematically depicted in the lg ω–T diagram of Fig. 41b (Ref. 195). Including the shear loss modulus G'', the activities from G'' to the shear loss compliance J'' fan out a stripe up to about three decades in lg ω. For a decreasing frequency across the transition zone (corresponding to an increasing length scale) the succession is: NMR relaxation times T_1 and $T_{1\rho}$, G'', dielectric ε'', $C_p''(\omega)$ = entropy compliance J_s'', J''; the same is found in PS, PMMA and PVC. The arrangement of ε'' shows that in these polymers the dielectric response does not originate from isolated dipoles but from the intercorrelation of many of them (Ref. 196). Fig. 41c shows the arrangement of the entropy J'' peak between the two shear peaks, G'' and J''. According to the general scaling principle this arrangement confirms the length scale of $\xi_a \approx 2$ nm from Eq. 6.16, because the J_s'' position corresponds to the onset of confined flow well located between the onset of mobility (G'' maximum) and hindering by entanglements (J'' maximum, see also Sec. 6.4).

(ii) Fig. 41d shows the dielectric loss $\varepsilon''(\omega)$ in a series of cis polyisoprenes PI with increasing molecular weight $M > M_c$ (Refs. 158, 159). The main (glass) transition peak is little influenced by M, but the l.h.s. peak, corresponding to the vector sum of chain-longitudinal dipol moment components (Fig. 41d), is systematically shifted to lower frequencies. This shift has an exponent of 3.5 corresponding to the increase of the plateau zone broadness. As the dipole sum corresponds to the chain end-to-end distance we have a clear relation of flow transition to this distance.

Recent experiments show, by means of the ω identity, where this label must be attached to shear curves, see Fig. 41e, Ref. 99. For cis PI, M = 140 000 g/mol, the G'' and ε'' curves have maxima only distinct by about 0.3 decades in lg ω. This label is marked by an arrow "R_0" in Fig. 38b. If this picture is universal we can see, that R_0 indicates the *onset* of flow transition since the $J''(\omega)$ maximum has a frequency considerably *lower* than the G'' maximum.

Boyer (Refs. 197, 198) reviewed different activities at freezing-in of the flow transition. This phenomenon was called T_{ll} transition. It corresponds to the activities in the mHz–Hz region of flow transition (see, e.g. Ref. 199). From the review one can see that the flow transition has many activities at rather different positions, also including bulk compliance (density fluctuations). Ultrasonic experiments showed that in the MHz range the maximum of the bulk loss modulus K'' is located near the G'' maximum (Ref. 200). So-called transport activities can be included in such a picture for the flow transition.

(iii) Fig. 41f shows the || dielectric dispersion zone of a small-molecule nematic liquid crystal in an Arrhenius plot (Ref. 201). A WLF curve is observed in the whole frequency region between 10^{-4} Hz (well below conventional T_g) and 10^8 Hz. The dielectric compliance is always a Debye relaxation. This property, and no influence of freezing-in near T_g, indicates that there is an ε active local process at the high frequency end of the dispersion zone that is nevertheless guided by WLF. This corresponds to the defect model of Fig. 36d for glass transition. The particularly active dipol is located near the high free-volume concentration in the cooperativity region.

The situation of Fig. 36d is qualitatively transferred to a dispersion law (log–log plot of scattering vector vs. frequency, Sec. 2.8) and to WLF scaling in Fig. 41g. The latter diagram demonstrates how the short modes (also NMR active) are included in the guiding by the universal aspects of the dynamic glass transition.

6.10 η activity and D activity

This section is to describe how the transport coefficients η (viscosity) and D ((self) diffusion) are related to the generating dispersion zone.

1. Dispersion zone and transport (or flow) zone

Consider a mixture. The concentration fluctuations $\Delta\phi^2$ of the dynamic glass transition form a dispersion zone in an Arrhenius diagram (l.h.s. of Fig. 42a) that can be investigated in principle by linear response methods describing the relaxation of the corresponding susceptibility, Eq. 6.17c, e.g. by means of a maximum in the corresponding loss part. The length scale $\xi_a(\phi)$ is estimated to be about 3 . . . 5 nm for glass transitions in small-molecule mixtures.

Now apply light scattering to the same mixture. Photon correlation spectroscopy of VV light scattering (see e.g. Ref. 191) marks a relaxation step at a typical time τ depending on the scattering vector $\boldsymbol{q}$. This corresponds to diffusion fluctuation, a transport phenomenon described in Sec. 2.8 with $\tau(q)$ according to Eq. 2.67, viz.

$$\tau \;\approx\; 1/\omega \;\approx\; q^2/D. \qquad (6.32a)$$

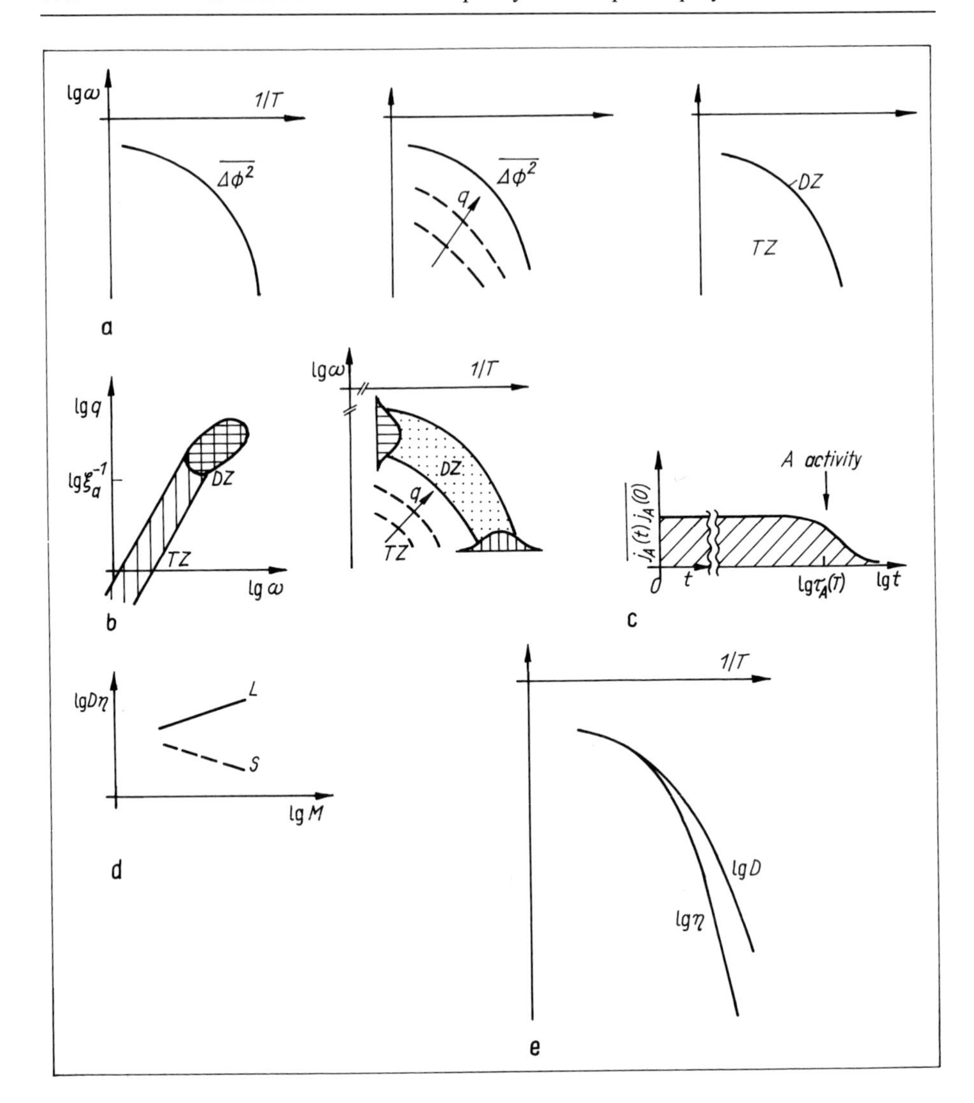
lgω
1/T
$\overline{\Delta\phi^2}$
$\overline{\Delta\phi^2}$
q
DZ
TZ
a
lgq
$lg\xi_a^{-1}$
DZ
TZ
lg ω
b
lgω
1/T
DZ
q
TZ
A activity
$\overline{j_A(t)j_A(0)}$
0
t
$lg\tau_A(T)$
lgt
c
lgDη
L
S
lgM
d
1/T
lgD
lgη
e

Since D is a function of T, by varying the temperature T we obtain $\tau(T, q)$. This gives a set of WLF curves $\tau(T)$ in the Arrhenius diagram that can be parametrized by q (see middle part of Fig. 42a).

Extrapolating this curve set to $q \to 1/\xi_a(\phi)$ we arrive, of course, at the dispersion zone DZ characterized by the loss maxima of linear response. The q-parameterized zone below this dispersion zone is the flow zone of linear response where the shear moduli tend to zero according to $G' \sim \omega^2$ and $G'' \sim \omega$, see Eq. 2.18, or Fig. 23f for polymers. To recall Eq. 6.32a the flow zone will now be called *transport zone* TZ, see the r.h.s. of Fig. 42a. The relaxation in this zone stems from a collective effect ("hydrodynamic modes") governed by diffusion equations with the dispersion law $\tau \sim q^2$, Eq. 6.32a. The transport coefficients themselves (D, η, . . .) are generated in the DZ by a cooperative mechanism. [The term "hydrodynamic limit" is sometimes used for describing the transition from TZ to DZ by transport coefficients.]

The relationship between the scattering experiment and the thermodynamic experiment (linear response) is illustrated in Fig. 42b. The zone in the lg$q$–lgω plane (dispersion law, see l.h.s. of Fig. 42b) where the scattering function is detectable is vertically hatched. For small q it is given by Eq. 6.32a (plus a certain broadness, see Eq. 2.66 and Fig. 16c). For large q, in the order of $\xi_a^{-1}(\phi)$, we observe modifications that can be described by $D \to D(q, \omega)$, see Eq. 4.23; D becomes dependent on q and ω due to the cooperative motion in the dispersion zone DZ. This $q\omega$ dependence of the diffusion coefficient D "generates" the maximum in the corresponding loss part in linear response. The dispersion law ends at a maximal q of order $(0.5\,\text{nm})^{-1}$

The zone that is sensitive to linear response (e.g. maxima of μ modulus

Fig. 42. Transport activities.
a. Dispersion zone DZ and transport zone TZ. $\overline{\Delta\phi^2}$ dispersion zone for volume fraction fluctuation in a mixture, the parameter q denotes increasing values of scattering vector in the transport zone. b. L.h.s. Dispersion law, [vertical hatching] as investigated by scattering, [horizontal hatching] as investigated by thermodynamics (linear response). R.h.s. Linear response spectra for the dispersion zone. (Only spurios signals such as flanks $\sim\omega$ or ω^2 remain in the transport zone.) c. The Kubo integral (hatched) with linear time measure over current correlation $j_A^2(t) = \overline{j_A(t)j_A(0)}$ gives the transport coefficient A. In the logarithmic scale, the main contribution is confined into about one decade of time around the arrow (lg τ_A). d. Stokes Einstein plot for linear chains (L) and 3-arm stars (S). D self diffusion, η viscosity, $M(> M_c)$ molecular weight. e. Comparison of diffusion D and viscosity η for ortho terphenyl OTP (arbitrarily matched at high temperature).

$G''_\mu(\omega)$ or ϕ compliance $J''_\Phi(\omega)$) is horizontally hatched. It is restricted to $q \geqslant \xi_a^{-1}(\phi)$. This length scale (or the corresponding volume $V_a \sim \xi_a^3$) is given by the spatial sensitivity of the thermodynamic experiment (see Secs. 3.6 and 3.7): V_a is the volume of the natural functional subsystem for ϕ fluctuation in the glass transition zone. [Basically, linear response is defined by activities depending only on the frequency ω. The relation to the space is given by the mode-length conception for cooperative motion. (More generally, the thermodynamic sensitivity is given by correlations in the Nyquist quantum cloud for a self experiment.) Larger length scales do not, or less, contribute because of statistical independence between distant natural subsystems.]

The relative position of linear response at a given frequency (or a given temperature) in the Arrhenius plot is schematicaly shown in the r.h.s. of Fig. 42b. For larger values its length scale ξ_a can be estimated from scattering experiments in the transport zone as follows. Determine the dispersion law in the transport zone (like Eq. 6.32a) from scattering experiments at different $q : q(\tau)$. Then extrapolate $q(\tau)$ (or $q(\omega)$) up to τ_R (or ω_R) of linear response and calculate ξ from $q^{-1}(\tau_R)$ or $q^{-1}(\omega_R)$.

Analogous considerations are valid for other activities. The shear properties in the flow zone ($G' \sim \omega^2$, $G'' = \eta\omega$) correspond to a diffusion equation for the transverse momentum density (details see Ref. 49),

$$(\partial/\partial t - \nu\nabla^2)\boldsymbol{g}_t(\boldsymbol{r}, t) = 0, \tag{6.32b}$$

where, in the simplest case, ν is the kinematic viscosity, $\nu = \eta/\rho$, with the dimension m^2/s. The corresponding scattering vector could be calculated from $q^2\tau\nu \approx 1$. The energy transport is more complicated, see Ref. 49 again, Briefly, one important parameter is the temperature diffusion, $a = \lambda/\rho C_p$, also of dimension m^2/s. The differences of numerical values for D, ν, and a – usually a few orders of magnitude – point to differences in the "frequency position of relevant activities" – a few logarithmic decades. The precise meaning of this term will be considered in subsection 2.

The general relationship between the linear transport coefficient ($A = \eta$, $D, \ldots$) in the transport zone TZ and the correlations in the next high-frequency dispersion zone DZ is given by the fluctuation dissipation theorem FDT, in particular by the Kubo formulas,

$$A = kT \int_{(V_a(A))} d\boldsymbol{r}dt\, \overline{\boldsymbol{j}_A(\boldsymbol{r}, t)\boldsymbol{j}_A(0, 0)}. \tag{6.33a}$$

The transport coefficient is obtained from an integration over the correlation function of relevant currents $\boldsymbol{j}_A$ in the dispersion zone (see also Sec. 2.8). The situation is depicted in Fig. 42c. The spatial integration can be restricted to the cooperativity region corresponding to the relevant activity ($V_a(A)$) since outside the correlation tends to zero for length $\lambda > \xi_a(A) \approx V_a^{1/3}(A)$. This corresponds to the steep slope of the activities, e.g. $G' \sim \omega^2$. Thus the transport zone itself does not contribute to the transport coefficient. Therefore

$$A = kT \cdot V_a(A) \int d\tau\, j_A^2(\tau) \tag{6.33b}$$

where $j_A^2(\tau)$ is the spatially averaged current correlation function (numerical factors are absorbed in the currents).

In simple glass formers the temperature dependence of correlation time $\tau_A(T)$ dominates the behavior of transport coefficients (τ approximation): If the mean current fluctuation $\overline{j_A^2}$ and $V_a(A)$ is taken to be constant and the influence of the factor kT is neglected, then the temperature dependence of A is only determined by $\tau_A(T)$,

$$A(T) \approx \text{const}_A \cdot \tau_A(T). \tag{6.33c}$$

For viscosity, $A = \eta$, the temperature dependence is usually well described by the VFT equation (Vogel, Fulcher, Tammann, Hesse, see Refs. 125–127),

$$\lg \eta(T) = A^0 + \frac{B}{T - T_\infty}. \tag{6.34}$$

In the τ approximation Eq. 6.33c, i.e. putting

$$\eta \sim \tau \sim 1/\omega, \tag{6.35a}$$

the VFT and WLF equations are equivalent (Ref. 124),

$$\begin{aligned} B &= c_1^0 c_2^0 = (T_0 - T_\infty) \lg (\Omega/\omega_0), \\ A^0 &= \lg \eta(T_0) - B/(T_0 - T_\infty). \end{aligned} \tag{6.35b}$$

where B is a measure of the shift invariant of Fig. 36a (see Eqs. 6.6, 6.1, 4.36 and 4.38). Beyond the τ approximation, the relaxation time $\tau_A(T)$ and the transport coefficient $A(T)$ cannot be simultaneously guided by the WLF equations. Moreover, the differences contain information about the relevant length scales $\xi_a(A)(T)$, Ref. 99.

2. η and D activity

As has been repeatedly mentioned above (Secs. 2.7, 5.2, 6.9) the activity splitting is considered to be the main point for studying the cooperativity of molecular motion. The transport coefficients can be fit into this picture with the advantage that they just contain more or less direct information about the relevant lengths.

In polymers, the dispersion zones are rather broad, and we have several dispersion zones. Therefore, we have to find an answer to the question, which frequency, or better, which time interval on a logarithmic scale (i.e. which mobility) does mainly contribute to the Kubo integrals Eqs. 6.33a, b. According to the general scaling principle, this mobility is also linked with a most responsible mode length.

Consider an entangled polymer melt. Which disperson zone, the main or the flow transition, is responsible for the melt viscosity? (Which dispersion zone "generates" this viscosity? See also Sec. 3.9 for similar thermodynamic questions). According to the linear time measure $d\tau$ of Eq. 6.33b and Fig. 42c it is the slowest dispersion (largest time τ) that determines the transport coefficient if the ratio of relaxation intensities (step heights in the current correlation) does not compensate the large time ratios of the two zones. Therefore, in the τ approximation,

$$A(T) \approx \text{const}_A \cdot \tau_{AF}(T), \tag{6.36}$$

i.e. the temperature dependence of melt viscosity is governed by the WLF curve of the flow transition and not of the main transition. This is confirmed by experiments (Refs. 108, 109). As mentioned in Sec. 3.9, one can define a dynamic shear viscosity $\eta' = G''/\omega$ in the whole ω region of dispersion zones. This $\eta'(\omega)$ depends on ω, see Fig. 23a. As long as η' is nearly constant this viscosity represents a property similar to a usual viscosity. This is the case in the confined flow zone of the main transition. Corresponding to the much shorter time (as compared to the flow transition), this viscosity is much lower than the melt viscosity. That means that the confined flow between the entanglements, restricted to the length scale $\lambda \leqslant d_E \approx 7\,\text{nm}$, has a high fluidity ($1/\eta'$) and corresponds, in a sense, to the comparably low viscosity of simple glass formers, somewhat modified by the chain structures (see Sec. 5.4). The solidity in the plateau zone (G_N^0 of order MPa) is not destroyed by the confined flow because, as mentioned in Sec. 3.9, this flow cannot percolate in the time scale $\tau < \tau_F(T)$, where the index F stands for the flow transition. Similarly, although not so sharply, one can reflect on the Andrade zones.

Now we can ask for the frequency or time interval inside the flow transition that is responsible for the transport properties of the melt. In the sense of different activities (Sec. 6.9) we expect to find different intervals (mobilities) to be responsible for the (self) diffusion D and viscosity η. The corresponding mobilities or times will be called *D activity* and *η activity*, respectively.

In the subsection 3 it will be shown that the reptation time τ_R corresponds to the D activity. The η activity τ^η is larger, $\tau^\eta > \tau_R$. This can be represented in a so-called *Stokes Einstein plot*, where the ratios of "transport times", τ_A/τ_B, as defined e.g. by Eq. 6.33a or c, are plotted vs. a parameter, e.g. temperature, molecular mass or chain length, concentration etc. Large changes indicate the difference of times, mode lengths or even molecular processes linked with the different activities A and B. An example is given in Fig. 42d (Ref. 202), where the product ηD stands for the ratio τ^η/τ^D ($\tau^D = \tau_R$). For chains this ratio increases with the chain length. This means that the viscosity is "generated" at increasingly larger length than the self diffusion, generated at about R_0. The decreasing ratio for stars indicates a decreasing "hydrodynamic" radius, also if compared to self diffusion scaling by star arm retraction.

Historically such considerations stem from *Stokes Einstein Debye* (SED) *relations* for small-molecule liquids,

$$D = kT/\sigma\eta, \tag{6.37a}$$

$$\tau = v\eta/kT + \tau_0, \tag{6.37b}$$

where σ is a length of order the molecule diameter (or less) and v a volume of order the molecule volume σ^3, τ the rotational diffusion correlation time (e.g. from depolarized light scattering) and τ_0 a time constant of order picoseconds (for $\eta \to 0$). The application of SED relations to glass-forming liquids shows a pronounced bend of τ/η or $D\eta$ vs. T plots at the splitting point S and large deviations from constancy below the splitting-point temperature, $T < T_s$. see Ref. 203a. The problem is the transfer of the constant molecular lengths σ, $v^{1/3}$ for noncooperative molecular motion for $T > T_s$ to characteristic lengths for cooperative motions, $\xi_a(A)$, $\xi_a(B)$, . . . , depending on T for $T < T_s$ ($\xi_a \sim (T - T_\infty)^{-2/3}$ for the ideal glass transition). This problem is illustrated by Fig. 42e showing the large differences of viscosity $\eta(T)$ and self diffusion $D(T)$ curves matched at high temperatures (Ref. 203b).

3. Model for flow transition

This subsection is to give an example for the combination of the transport activity conception with the survival conception (Sec. 6.8.3) for the case of flow transition.

The – purely hypothetical – model of fluctuating tubes (Figs. 31f, g) will be used for a simple Brownian treatment according to Eqs. 1.1–1.5. Differences between D and η activities are possible because a new degree of freedom is introduced: the external tube unit motion (index E) besides the internal in-tube motion (index T),

$$\Delta x(t) = \Delta x^{\mathrm{E}} + \Delta x^{\mathrm{T}}, \tag{6.38}$$

where the coordinate x (for x or y or z) describes the spatial position of a given chain segment (see Fig. 31g). Choose the time step so that the Δx's are statistically independent,

$$\overline{\Delta x_i \, \Delta x_j} = 0 \quad \text{for} \quad i \neq j, \tag{6.39a}$$

irrespective of E, T, or no label. Then

$$\overline{(\Sigma \Delta x_i^2)} = \Sigma\overline{(\Delta x_i^{\mathrm{T}})^2} + \Sigma\overline{(\Delta x_i^{\mathrm{E}})^2} + \Sigma\overline{(\Delta x_i^{\mathrm{T}} \Delta x_i^{\mathrm{E}})}. \tag{6.39b}$$

[The strategy used here is linked to basic features of the ideal glass transition: statistical independence for small scales and confined scaling for a back transfer of R_0 and L scale arguments to the smaller Δx scale (back scaling). As explained in Sec. 6.4, both features can be linked by scaling the statistical independence (Fig. 36c). As these features are very soft conditions, the same arguments can also be applied to the high temperature Arrhenius regime of the flow transition, $T > T_{\mathrm{SF}}$.]

The pristine reptation model with fixed tubes is restricted by $\Delta x^{\mathrm{E}} = 0$, both D and η activity are then defined by the same first term of the r.h.s. of Eq. 6.39b alone. Allowing $\Delta x^{\mathrm{E}} \neq 0$, the situation is changed: If the third term is not D active, then the diffusion is now as before determined by the first term. If the third term is η active and characterized by a longer time than the first-term time, then the η activity is given by the third term (provided that the contributions from the different terms of Eq. 6.39 are not too different, see Sec. 6.10.2). These two if's are effective if the second term is neither D nor η active. Now we shall collect arguments for all three if's.

The second term, $\overline{(\Delta x_i^{\mathrm{E}})^2}$, is certainly shear active (see Figs. 31f and 28e) but not transport active because it contains only tube-unit diffusion. Tube renewal, i.e. the tangling and loosening of entanglements, is necessary for transport, and this can only be described by implying in-tube diffusion, Δx^{T}.

The third term, $\overline{\Delta x_i^{\mathrm{E}} \Delta x_i^{\mathrm{T}}}$, is not diffusion active, because it is odd (linear) in Δx_i^{T}. The first factor, Δx^{E}, is not relevant for D because it does not contribute to a change of the contour segment environment, e.g. concerning tube wall shifts. (Remember that tube-unit diffusion is thought to be some fluctuative microflow of tube units, and that constraint release is not important for monodisperse chain melts, see Sec. 5.6.)

Therefore, the D activity is only given by the first term, $\overline{(\Delta x_i^{\mathrm{T}})^2} \equiv D_{\mathrm{T}}$.

To specify the problem we will make three assumptions, I, II, III. With Assumption I, that the in-tube mobility $D_{\mathrm{T}} = \overline{(\Delta x_i^{\mathrm{T}})^2}$ (not Δx_i^{T}) is statistically independent from the E fluctuation, we get the same expressions for D as for fixed tubes (Eqs. 5.46a, b and 5.47), i.e. $D \sim N^{-2}$, and the reptation time τ_{R} is the D activity. The spatial scale for D is therefore the mean end-to-end distance for the chain coils, R_0.

Now we are to find a reason that the cross term in Eq. 6.39, $\overline{\Delta x^{\mathrm{E}} \Delta x^{\mathrm{T}}}$, is η active and has a longer time, $\tau^{\eta} > \tau_{\mathrm{R}}$. Since Δx^{E} is shear active and Δx^{T} is transport active, both factors can contribute to the η activity. This means, that $\overline{\Delta x^{\mathrm{E}} \Delta x^{\mathrm{T}}}$ is not symmetric with respect to D and η activity. If there is really some statistical dependence between Δx^{E} and Δx^{T} in the same time interval (Assumption II) then we have contributions to the η activity. It remains an estimation of τ^{η}. Multiplying with

$$1 = (\Sigma \Delta x_i^{\mathrm{T}})/(\Sigma \Delta x_i^{\mathrm{T}}) \tag{6.40a}$$

we get

$$\tilde{D}_{\eta} \equiv \Sigma \overline{\Delta x_i^{\mathrm{E}} \Delta x_i^{\mathrm{T}}} = \Sigma \overline{(\Delta x_i^{\mathrm{T}})^2 \cdot (\Sigma \Delta x_i^{\mathrm{E}})/(\Sigma \Delta x_i^{\mathrm{T}})}. \tag{6.40b}$$

The Assumption I implies bar breaking in Eq. 6.40b,

$$\tilde{D}_{\eta} = D_{\mathrm{T}} \cdot Q, \quad Q = \overline{(\Sigma \Delta x_i^{\mathrm{E}})/(\Sigma \Delta x_i^{\mathrm{T}})}. \tag{6.41}$$

The quotient Q is some conditional probability: How large is Δx^{E} for a given Δx_i^{T} (> 0, for instance)? To answer this we use scaling arguments for the back scaling coil $\rightarrow \Delta x$. Since Q is a length ratio, the parameter dependence is directly given by

$$Q \approx \lambda(\text{numerator})/\lambda(\text{denominator}). \tag{6.42a}$$

With Assumption III, that the effective (in the reptation time τ_{R}) tube-fluctuation amplitude is scaled by the coil radius R_0 – longer chains have larger fluctuations in the larger times $\tau_R(R_0)$ – we obtain

$$Q \sim R_0/L_T \sim N^{-1/2}. \tag{6.42b}$$

(According to Sec. 6.6.4 we have $R_0 < \xi_F$ for $T < T_{SF}$, see Sec. 6.6.4.)

The Assumption III does not mean that the tube fluctuation is of order R_0. Such large fluctuations can be excluded by computer simulations (Ref. 203c). Instead it is only assumed that the microflow of tube units in the large time $\tau_R(R_0)$ is back scaled by R_0 also for small fluctuation amplitudes Δx^T, independent whether the tube-unit motion is cooperative (for $T < T_{SF}$) or not ($T > T_{SF}$). The η activity is therefore

$$\tau^\eta \approx L_T^2/\tilde{D}_\eta = L_T^2/D_T Q = \tau_R/Q \sim N^{3.5} \tag{6.43}$$

that is proportional to η in the τ approximation. Therefore the model gives $\eta \sim N^{3.5}$ and $\eta D/N \sim N^{1/2}$ in not too bad agreement with experiments.

In summary we have found that the η and D activities are distinct for the model of fluctuating tubes,

$$D\colon \tau_R \sim N^3, \quad \eta\colon \tau^\eta \sim N^{3.5}, \tag{6.44}$$

with $\tau_R < \tau^\eta$ from $Q < 1$.

According to the general scaling principle larger τ^η values correspond to larger-than-R_0 lengths for the η activity. Since R_0 is located at the high-frequency side of flow transition (when the latter is marked by the maximum in the shear loss maximum of $J'' - 1/\omega\eta$, see Fig. 38b) the η activity can probably be associated with this maximum. This is phenomenologically supported by the finding that there is no additional shoulder in the $G'(\omega)$ and $G''(\omega)$ moduli at the $J'' - 1/\omega\eta$ maximum for monodisperse melts.

Now we can verbally resolve the $M^{3.5}$ paradox (quoted in the first paragraph of Sec. 5.6) by means of the survival conception for large structures (introduced at the end of Sec. 6.8.3). Survival is included in the entanglement concept from the very beginning. For a molecular mass above the critical value, $M > M_c$, breaking of one entanglement or renewal of one tube unit [or scaling this argument: breaking of a not too large subset of entanglements (e.g. tube renewal)] does not mean a breakdown of the whole thermokinetic structure of entanglement network since this structure is a cooperative correlation of a dispersion zone when considered from the low frequency, the "melt side" (see Fig. 23d). Any cooperative dispersion zone can have different activities at rather different time and length scales, and even statistical independence in small scales does not exclude cooperativity in larger scales, see e.g. Figs. 28c and 36c.

7. Thermal glass transition

Let us form a glass by cooling the melt with a given cooling rate $\dot{T}_K = -dT/dt$ and consider the accompanying Arrhenius plot Fig. 43a. The cooling rate corresponds to a certain preparation time interval $\tau_{prep} \approx \Delta T/\dot{T}_K$ or to a preparation frequency $\omega_{prep} \approx \tau_{prep}^{-1}$, where ΔT is a temperature interval with relation to the dispersion zone considered. After crossing the dispersion zone the substance can no longer respond with respect to the fluctuation typical for this zone: The time interval is too short for the thermodynamic experiment related to this functional subsystem, see Sec. 3.9.

Consider the broad spectrum of a dispersion zone. Then the loss of response is successively distributed over a certain temperature interval ΔT, the "transformation interval". Of course, ΔT is of order δT, the mean temperature fluctuation in the functional natural subsystem. The process where the modes successively become too slow for the cooling experiment is called *thermal glass transition*, or freezing-in. The corresponding temperature is the *glass temperature* T_g. It depends on the cooling rate $\dot{T}_K$ (or ω_{prep}, see Fig. 43a again); a conventional value could be $\dot{T}_K \approx 10\,\mathrm{K\,min}$, or $\nu_{prep} = \omega_{prep}/2\pi \approx 10\,\mathrm{mHz}$. Glass transition multiplicity leads to several glass temperatures, e.g. T_F for the flow transition (sometimes called T_{ll}, see Refs. 198, 199), T_g for the main transition, and T_β, T_γ for the secondary relaxation(s).

To define a glass temperature for a given dispersion zone we need three conventions: (i) the cooling rate, (ii) the activity referred to, and (iii) which position in the spectrum or the transformation interval (maximum, inflection point, half value . . .) is to be used.

The freezing-in has far-reaching thermodynamic consequences (see Fig. 43b, according to the famous 1933 paper by Simon, Ref. 89). Consider a material frozen-in at T_g (E in Fig. 43b) and being now in a "glassy state" X at $T < T_g$. Waiting long enough we observe how the relevant modes recover their mobility (X → A). The material properties are isothermally changing with time, i.e. the material is not in equilibrium as long as X ≠ A. As freezing-in at E and the final state A correspond to different thermokinetic structures, the process X → A is called *structural relaxation*. The most remarkable property is that its rate is not only determined by the time scale

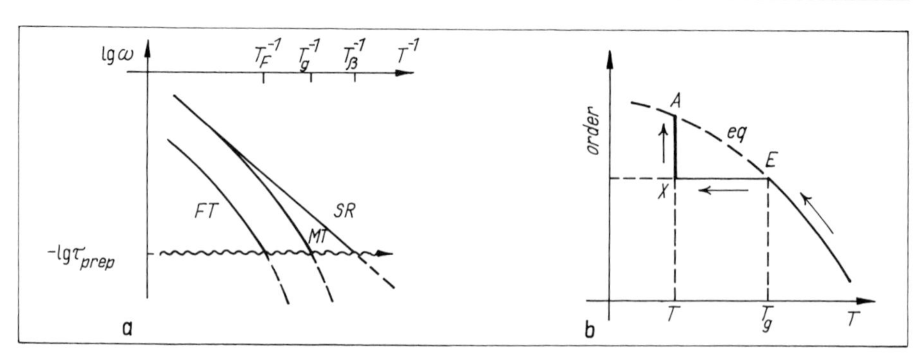

Fig. 43. Thermal glass transition.
a. Definition of glass temperatures. The glass transition multiplicity leads to several glass temperatures: $T_F = T_{11}$ for the flow transition FT, T_g the conventional glass temperature for the main transition MT, and T_β for the secondary relaxation SR. b. Structural relaxation according to Simon (see text). The old term "order" is to be replaced by a parameter describing the (perfection of) "thermokinetic structure".

of the (extrapolated) equilibrium at A but also by the nonequilibrium state X and its prehistory.

According to Simon the structural relaxation is not a process that could be described by an activated process with a macroscopic barrier (see Fig. 27d in Sec. 5.2). One rather has to consider a complicated energy landscape. Let us quote his famous sentence that also regulates the relation to the Third (Nernst) Law (see also Sec. 7.6, below, the insertions [ ] are added by the author): "Der isotherme [reversible] Übergang [X(T_g) → A] in einen Ordnungszustand [A], der dem *inneren* thermischen Gleichgewicht einer anderen Temperatur [$T < T_g$] entspricht und daher vom Zustand kleinster freier Energie nicht durch eine Potentialschwelle getrennt ist, sondern mit ihm durch eine kontinuierliche Folge [the way X → A ~ E . . . A] von stabileren Zuständen zusammenhängt, ist aber offenbar unmöglich." In other words: The relaxation X → A is not too far away from the reversible way E . . . A in the landscape, X → A is without a macroscopic activation barrier since E...A is without, but X → A is irreversible, because the stability is gradually enlarged at constant temperature.

The structural relaxation is a complicated situation because now it is necessary to consider three time scales:

(1) Fluctuation or relaxation time (spectrum) of the system
(2) Frequency or time to probe the system
(3) Time scale (or program) to prepare the glassy system (τ_{prep}).

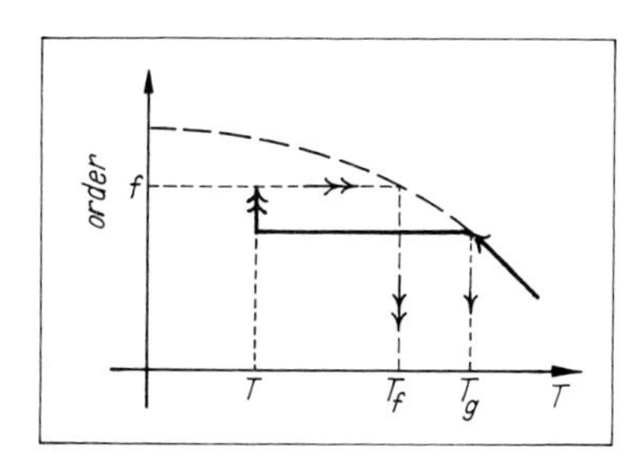

Fig. 44. Definition of fictive temperature T_f.
The order f can be defined by volume, enthalpy, or other observables.

Structural relaxation is often investigated by isobaric enthalpy or volume retardation (related e.g. to the refractive index). The corresponding susceptibilities

$$T(\partial S/\partial T)_P = (\partial H/\partial T)_P = C_P = \overline{\Delta S^2}/k = T\overline{\Delta S^2}/kT \qquad (7.1)$$

and

$$(\partial V/\partial T)_P \equiv V\alpha_P = \overline{\Delta S \Delta V}/kT \qquad (7.2)$$

are connected with the mean entropy fluctuation and the cross fluctuation between entropy and volume, respectively. These are different actitivies, their spectral densities are therefore different, $\Delta S^2(\omega) \neq (\Delta V \Delta S)(\omega)$. The density fluctuation (interesting for scattering experiments) can thermodynamically be characterized by isothermal compressibility,

$$-(\partial V/\partial p)_T \equiv V\kappa_T = \overline{\Delta V^2}/kT \qquad (7.3)$$

corresponding to a third spectral density, $\Delta V^2(\omega)$, and to bulk compliance. Dielectric permittivity corresponds to a fourth spectrum, and so on.

7.1 Fictive temperature. Material time

1. The conceptions

Tool (Refs. 204, 205) introduced the term *fictive temperature* T_f, instead of order, to characterize the stage of structural relaxation. The idea is illustrated in Fig. 44. Any frozen-in order f is mapped via the equilibrium curve to a hypothetical freezing-in temperature T_f.

This concept has three weak points.

(i) It is not sure that the structure ("order") during the structural relaxation at T is the same as the thermokinetic structure ("equilibrium order") at the corresponding T_f, or even at any equilibrium point.

(ii) The fictive temperature is not an intensive variable in the sense of an equivalence class index of the Zeroth Law. Moreover, if the order is represented by different activities ($f = V$, or $f = H$, etc.), then we have different T_f for the same stage of structural relaxation, $T_f^V \neq T_f^H \neq \ldots$, because the different activities have different spectral densities.

(iii) Even for the same density and temperature after preparation, at $T < T_g$, the microscopic order as characterized e.g. by the radial distribution function $g(r)$ depends on the parameters of the preparation program (pressure-temperature-time . . .program, see ;
e.g. Ref. 206). For instance, the short range order (the height of the first peak in $g(r)$) is lower for pressure-densified glasses, whereas volume retardation favors the packing.

As the uncertainties of T_f remain small, usually of order 1 K, the fictive temperature is often a good tool for the correlation of structural relaxation.

It is more critical to characterize the historical aspect of the state at $T < T_g$ by a reduced or *material time*. The idea of this concept (see Ref. 207) is to connect a given stage during structure relaxation with a certain number of successful activation steps in the history. This success rate in the energy landscape depends on the actual temperature and on the structure, characterized e.g. by the fictive temperature. The material time, therefore, counts the beats of a mean success-rate clock.

The weakness of this concept is obvious. The variety of activation steps in the complicated energy landscape can really be collected only in a spectrum. This cannot easily be characterized by one time when different parts of it simultaneously contribute to the structure relaxation.

2. Narayanaswamy scheme

The combination of fictive temperature, material time, and elements of linear response to describe the structural relaxation is called *Narayanaswamy scheme* (see Refs. 208–210). Consider a temperature-time program $T(t')$ starting at $t' = t^0$ in the equilibrium. The material time ζ at actual time t,

$$\zeta(t) = \int_{t^0}^{t} dt' \tilde{a}_T(T(t'), T_f(t')) \equiv \int_{t^0}^{t} dt' \tilde{\tau}(t'), \tag{7.4}$$

is expressed as an integral over a corresponding shift factor $\tilde{a}_T = a_T/\tau_0$ describing how the number of beats $\tilde{\tau} = \tau/\tau_0$ of the material clock depends on the temperature $T(t')$ and the fictive temperature $T_f(t')$; $t' < t$, and τ_0 being a time constant, e.g. 1 second. A material equation with the compliance $J(t - t')$ is used for an extensive observable,

$$x_B(t) = \int_{-\infty}^{t} \dot{f}(t')J(t - t')dt'. \tag{7.5}$$

This equation is adapted for structure relaxation by

$$x_B \rightarrow T_f, \dot{f} \rightarrow \dot{T}, J \rightarrow \Phi, t - t' \rightarrow \zeta - \zeta', \tag{7.6}$$

where the disturbation force f is now given by the temperature-time program, and Φ is a cross compliance $T \rightarrow x_B$. [It is practice to make a Kohlrausch Williams Watt (KWW) ansatz for $1 - \Phi$ with parameter $\beta = \beta_{KWW}$, Eq. 2.48.] We have

$$T_f(t) = \int_{t0}^{t} \dot{T}(t')\Phi(\zeta - \zeta')dt' \tag{7.7}$$

with $\zeta' = \zeta(t')$. The dependence of $\tilde{\tau}$ on both T and T_f is approximated by a mixing formula, e.g.

$$\tilde{\tau}(t'') \sim \exp\left[BT_g - BT_g^2\left(\frac{x}{T(t'')} - \frac{(1 - x)}{T_f(t'')}\right)\right] \tag{7.8a}$$

with a mixing parameter x, $0 < x < 1$. The first term of a Taylor series expansion for small $T - T_g$ is

$$\zeta - \zeta' = \int_{t'}^{t} dt'' \exp\{B[x(T(t'') - T_g) + (1 - x)(T_f(t'') - T_g)]\} \tag{7.8b}$$

where B is the slope of the relevant dispersion zone in a $\ln \omega$ vs. T plot,

$$B = (d \ln a_T/dT)_{T_g} = \Delta H^*/kT_g^2, \tag{7.9}$$

a constant of order 1/K, with ΔH^* the apparent activation energy at T_g.

The pair {Eq. 7.7 and 7.8a or b} defines a nonlinear system of equations that allows the calculation of $T_f(t)$ for any $T(t')$ program.

3. Discussion

The output $T_f(t)$ is monitored by a set of four parameters,

$$\{\tau_g, B, x, \beta\} \tag{7.10}$$

where $\tau_g = \tau(T_g)$. The parameter $\beta = \beta_{KWW}$ stands for the inverse broadness of the Φ decay. Hodge (Ref. 211), attempting to understand the structural relaxation as determined by the same factors that influence equilibrium behavior above T_g, observed that the mixing parameter can be correlated with the fragility (parameter F according to Eq. 6.5a). Substitute Eq. 7.8a by the Adam Gibbs notation Eq. 5.21a,

$$\tau = \tau_0' \exp [Q/RT(1 - T_\infty/T_f)] \qquad (7.11a)$$

where T_f is now for T, T_∞ for T_2, and $\tau_0' \sim 1/\nu_0$, see also Ref. 212. Then from an integral like

$$\int_{T_\infty}^{T_f} C_P \, d \ln T, \qquad (7.11b)$$

and from comparing Eqs. 7.8a and 7.11a we see that

$$x = 1 - T_\infty/T_f \approx 1/F. \qquad (7.12)$$

The four-parameter set is now $\{\tau_g, Q, T_\infty, \beta\}$ or $\{\tau_g, F, T_\infty, \beta\}$. Since T_∞/T_f ($\approx T_\infty/T_g$) is varying in a considerable degree for different substances, general correlations such as Eq. 6.27, $\beta \sim (T - T_\infty)$, and the dependence of parameters on the degree of cooperativity can be tested by correlation with the fragility (Ref. 213).

Regardless of the improvements the basic shortcomings of the material-time concept and of the Narayanaswamy scheme cannot easily be overcome. An internal contradiction emerges from the attempt to use the time conception of one internal clock as an argument for Φ describing a broad time spectrum. In equilibrium the FDT produces the universal time t for $\Phi(t)$ from the many time intervals producing τ in a correlation function $\varphi(\tau)$, but in complicated nonequilibrium histories the material time cannot be universal for different parts of the spectrum.

Nevertheless, for simple $T(t')$ programs the structural relaxation can be successfully described by the Narayanaswamy scheme after parameter adjustments. The parameters obtained differ from those that can be gained from equilibrium properties at $T > T_g$. But there are also experiments that cannot be, in principle, described by the Narayanaswamy scheme (see Sec. 7.3, below).

The arrangement of different activities across the glass transition zone (Figs. 27g, 35e, and 41a,c) is also reflected by the thermal glass transition. The activities with lower frequencies (longer times) freeze first. It was found that, in a broad assortment of polymers, the average dielectric relaxation times are spread over three logarithmic decades at a T_g defined calorimetrically for a given cooling rate (Ref. 214). This indicates that the position of the $\varepsilon''(\omega)$ peak in the main transition zone can be rather different from the peak position of the imaginary part of entropy compliance $J_s''(\omega) = C_p''(\omega)$.

7.2 Freezing-in at thermal glass transition

Consider an isobaric cooling process with constant cooling rate $\dot{T}_K$. In Fig. 45a the freezing-in of the entropy susceptibility at cooling ($C_p = \dot{Q}/\dot{T}$) is compared with the linear response from periodic calorimetry Ref. 139 in the equilibrium ($C_p'(\omega) = J_s'(\omega)$, $C_p''(\omega) = J_s''(\omega)$, real and imaginary part of the entropy compliance). Both responses (for glycerol) are represented over the temperature axis, the first is parametrized by $\dot{T}_K$, the second by the frequency ω.

The step in C_p corresponds to the step in $C_p'(\omega)$ and to the peak in $C_p''(\omega)$, the C_p deficit expresses the inability of thermodynamic experiments to register fluctuations which are too slow. Nonetheless there is a cardinal difference to the linear response plots comparing retardation $J(\tau, T)$ and dynamic compliance $J^*(\omega, T)$ in equilibrium (e.g. Fig. 12a,b). Freezing-in means the transition to non-equilibrium, any C_p step is successively influenced by the fictive temperature T_f.

Fig. 45b is an integral representation of freezing-in, $S = \int C_p d\ln T$. This diagram corresponds to the Simon plots Figs. 43b and 44, parametrized by the cooling rate and specified for entropy S or enthalpy H, a similar plot is obtained for the volume.

Diagrams similar to Fig. 45a, b can be obtained for the refractive index and the dielectric permittivity. For pure cooling, the differences between equilibrium and non-equilibrium response can well be explained by the Narayanaswamy scheme.

Fig. 45a demonstrates a rather parallel dynamics parametrized by $\dot{T}_K$ and ω, respectively. Since the high-temperature flank of the $C_p(T, \dot{T}_K)$ step is less influenced by T_f we can there compare $\dot{T}_K$ and ω rather directly, see Fig. 45c. For glycerol one obtains (Ref. 214a)

$$\lg(\omega \delta T/\dot{T}_K) \approx 0.6 \pm 0.3 \tag{7.13}$$

where $\delta T \approx 4\,\mathrm{K}$ is the dispersion of the $C_p''(\omega, T)$ curves corresponding to the mean temperature fluctuation of the natural functional subsystem.

This comparison is related to the time aspect of subsystem definition (Sec. 3.7). Eq. 7.13 defines, for the entropy temperature pair, the minimum time interval

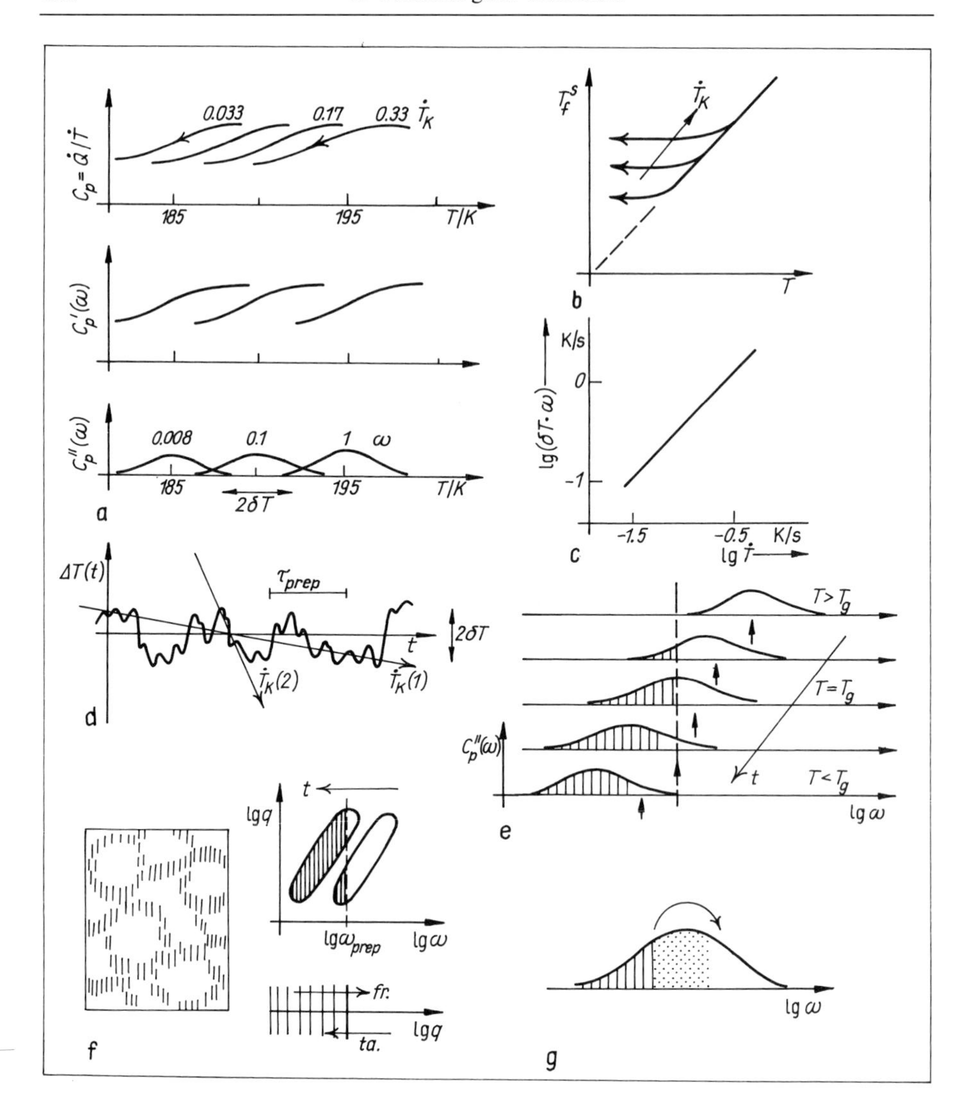
0.033
0.17
0.33
$\dot{T}_K$
$C_p = \dot{Q}/\dot{T}$
185
195
T/K
$C_p'(\omega)$
$C_p''(\omega)$
0.008
0.1
1
ω
2δT
a
T_f^s
b
T
K/s
0
-1
$\lg(\delta T \cdot \omega)$
-1.5
-0.5
c
$\lg \dot{T}$
$\Delta T(t)$
τ_{prep}
t
2δT
$\dot{T}_K(2)$
$\dot{T}_K(1)$
d
T > T_g
T = T_g
T < T_g
e
lg ω
lg q
$\lg \omega_{prep}$
fr.
ta.
f
g

needed in the quasi classical case for a thermodynamic experiment to measure the contribution of a frequency interval $\Delta\omega$ at ω to the entropy susceptibility,

$$\Delta C_P(\Delta\omega) = k^{-1} \int_{(\Delta\omega)} \Delta S^2(\omega)\, d\omega. \qquad (7.14)$$

The situation is illustrated in Fig. 45d. Obviously, the slow cooling rate $\dot{T}_K(1)$ can collect the contributions of fluctuation to ΔC_p, but the fast rate $\dot{T}_K(2)$ cannot. There is a cooling rate $\dot{T}_K$ in between that marks the sensitivity threshold. From Eq. 7.13 one can estimate a corresponding frequency ω and a relevant minimum preparation time, $\tau_{prep} \approx 2\pi/\omega$ necessary for the experiment at ω and T. The numerical value in Eq. 7.13 shows that for glycerol a time interval comparable to the correlation time is needed for the thermodynamic experiment in the quasiclassical case.

Of course, this case also allows a pure classical description. Consider the cooling of a classical glass-former. At T_g the time becomes too short to permit the system to find the optimal configuration of particles. This means that any thermodynamic potential H is larger than the optimum = equilibrium value, $H > H_{eq}$, and the susceptibility is therefore lower, $C_p = \partial H/\partial T < C_p$ (eq.), see Fig. 45b again where T_f stands for H. This process can completely be calculated from Newton's equation of classical mechanics, $C_p = (dH/dt) \cdot (dt/dT)$, if one finds a reasonable method for the simulation of continuous and homogeneous energy loss corresponding to a given cooling rate dT/dt

Fig. 45. Freezing-in.
a. Comparison of entropy compliance for freezing-in in a thermogram ($C_p = \dot{Q}/\dot{T}$ vs. T) with equilibrium linear response ($C_p^*(\omega) = C_p'(\omega) - iC_p''(\omega) = J_s^*(\omega)$) for glycerol. $\dot{T}_K$ cooling rate in K/s, ω frequency in rad/s, δT mean temperature fluctuation in the natural subsystem. Schematically. b. Fictive temperature related to entropy response. c. Comparison of frequency ω and cooling rate $\dot{T}$ at given values of temperature T for the high frequency flank of the spectrum. This map $\dot{T} \to \omega$ defines the preparation frequency. d. Thermodynamic experiment: The fluctuation represented (e.g. $\overline{\Delta T^2}$) can be measured with slow cooling rate, $\dot{T}_K(1)$, but not with too high rate, $\dot{T}_K(2)$. e. Freezing-in of a spectrum, schematically. The frozen-in part of the spectrum is hatched. The small arrow indicates the drift of material clock rate. f. Spatial aspects of freezing-in. l.h.s. The one-connected frozen-in region prevents percolation of flow (solidification). The diameter of a mobility cell is about $2\xi_a$. r.h.s. Upper part: Freezing-in of dispersion law (cf. with Figs. 16d and 41g). Lower part: Motion of front line for freezing-in (→ fr.) and "tawing" (← ta). q scattering vector. g. Spectral effects. [hatched box] frozen-in part, [dotted box] grey region. The arrow indicates spatial correspondence for the mobilization (e.g. during annealing at constant temperature).

(see e.g. Ref. 215). In the following we shall consider the effects of spectral broadness to freezing-in.

The following rough picture can explain that the transformation interval of freezing-in is related to the mean temperature fluctuation δT of the natural subsystems. Such subsystems fluctuate independently of each other so that they have "different temperatures". Therefore, they can freeze-in sooner or later. This links the transformation interval with properties of the natural subsystems, e.g. with its volume V_a, see Eq. 6.16.

Now we use temperature-time equivalence to discuss the glass transformation in the frequency representation. For the time being we use a simplified yes-no picture: No action for frozen-in frequencies $\omega < \omega_{prep}$ (hatched), full action for working frequencies $\omega > \omega_{prep}$ (no hatched). Fig. 45e shows the corresponding freezing-in process with the further approximation that the shape of the spectrum is not changed by freezing. From Fig. 45e we can estimate the evolution of $T_f(t)$ and $\zeta(t)$ as mean values of the nonhatched parts, the better the smaller this part is.

Passing over to a spatial picture we should use the general scaling principle Sec. 2.8 applied to Fig. 36d. Since the slow modes freeze first, it is the long modes that are first to become immobilized. Hatching again the frozen-in parts we obtain a negative of Fig. 36d at T_g, see Fig. 45f. Now the regions with a low concentration of free volume are hatched. They form a continuous, one-connected region enclosing dispersed regions of size ξ_a that are yet working because of a higher concentration of free volume. The corresponding solidification at T_g is essentially the restriction and finally the prohibition of percolation for the island mobility by the hatched network. Computer simulation along the lines of Ref. 133 allows to construct such structures (Ref. 215a).

Attempts to understand the glass transition as a corresponding percolation singularity (Ref. 216) failed (Ref. 216a), because the relevant mobility in the cells of nearly constant size $2\xi_a$ is too low to give the hydrodynamic mobility clusters of size $> 2\xi_a$ a thermodynamic significance. In other words: The percolation singularity would correspond to large length scales larger than $2\xi_a$. If the scale $2\xi_a$ is actually relevant for the generation of entropy, then such a percolation would perhaps be interesting for the flow onset in the transport zone but not for the heat capacity in the dispersion zone. The situation is changed for the smaller scales $< 2\xi_a$ of the glass transition predecessors Sec. 6.6.2, but the singularity cannot be reached, too.

In the $\boldsymbol{q}$ space (scattering vector) the picture is more simple, see the middle and right hand parts of Fig. 45f. Decreasing temperature shifts the borderline gradually to lower ω and higher q values. The topological aspects, however, are hidden by the Fourier transformation from real space to $\boldsymbol{q}$ space.

The yes-no picture is rather crude. We may ask, for instance: what could be the meaning of a hatched frequency that cannot work, i.e. that cannot respond to an external disturbance? What is the effect of structures defined by freezing-in with a considerable uncertainty in q and ω, see Sec. 2.8, Fig. 16d?

To remain in the ω picture one can construct a grey area of about a half or one logarithmic decade in time or frequency corresponding to the $\Delta \lg \omega$ broadness of the dispersion law (diameter of the cigars in Figs. 16d, 41g and 45f). The spatial (topological) relation between frozen-in and full-working frequency regions is characterized by an arrow ↷ in Fig. 45g. Because of the FDT ω identity Secs. 2.6, 3.7 this arrow does not mean a frequency change between emitter (hatched, for tawing) and absorber but the spatial reference "about the same mode length". Furthermore, the transmission includes a modification of modes, e.g. due to a production of additional free volume in the mobile islands because the solid network is not able to follow immediately the equilibration of the islands.

Two deviations from the material time conception are possible. (i) The working part of the spectrum is too broad for describing it by an average time. (ii) During special temperature-time programs it seems possible that a mode ω frozen-in at T would correspond (↷) to another frequency ω' at the new equilibrium T' for the tawing. We would be confronted with a situation that violates the ω identity of the FDT, ω (emitter) would be different from ω' (absorber). Both deviations will be collected in the term *spectral effect*.

7.3 Nonlinearity, overshoots, and Kovacs' expansion gap

The following terms are generally used for a description of structural relaxation during complicated temperature-time (T–t) programs. *Nonlinearity* means that for temperature jumps ΔT larger than the mean temperature fluctuation of the natural subsystem δT the shape of response is considerably influenced by sign and amount of ΔT. *Nonexponentiality* means that the response $\Phi(\tau)$ in Eq. 7.7 is different from (broader than) the exponential decay. This property is often expressed by a smaller-than-one KWWβ exponent Eq. 2.48.

The thermograms ($\dot{Q}/\dot{T}$ vs. T, or integral curves H vs. T, V vs. T) for heating-after-cooling processes depend on whether there was an annealing stage (time t_e, temperature T_e) in the glassy state or in the transformation interval, and on the heating-to-cooling rate ratio, $\dot{T}_{\mathrm{H}}/\dot{T}_{\mathrm{K}}$. Some instructive examples are shown in Figs. 46a–c, see Refs. 217–219. A peak in the heating curve is called *overshoot*. The height and, in less degree, the temperature position of the overshoot increase with increasing annealing time and temperature and increasing $\dot{T}_{\mathrm{H}}/\dot{T}_{\mathrm{K}}$ ratio. This property can be used to enlarge the visibility of glass transitions with a small step height ΔC_p in thermograms.

Curves like Figs. 46a–c can well be modelled by the Narayanaswamy scheme. The parameter set Eq. 7.10, especially x and β_{KWW}, also varies inside a substance class. So, as a rule, PS shows a small peak also for $\dot{T}_{\mathrm{K}} = \dot{T}_{\mathrm{H}}$, and PVAC does not. The overshoot can also be modelled for exponential decays, $\beta_{\mathrm{KWW}} = 1$. The overshoot is, therefore, not an effect of nonexponentiality.

A famous example of nonlinerity in the response to a complicated temperature-time program is shown in Fig. 46d, see Ref. 220. The program has several stages. (1) The sample (PVAC here) was equilibrated at $T + \Delta T$ (upper case) or $T - \Delta T$ (lower case, $T = 35°\mathrm{C}$, $\Delta T = 2.5\,\mathrm{K} \approx \delta T$). (2) Then the sample was quenched to T. (3) Now the response was isothermally measured at $T = \mathrm{const}$ (volume here); it is, of course, a contraction for the $+\Delta T$ history and an expansion for the $-\Delta T$ history (upper and lower curve in Fig. 46d, respectively). The curves are highly asymmetrical.

In the main stage the expansion is somewhat faster than the contraction.

Fig. 46. Overshoots and expansion gap.
a. Enthalpy H or volume V in a simple cooling–heating cycle (——) or with annealing at $T_{\mathrm{e}} < T_{\mathrm{g}}$ (· · · ·). b. Thermogram ($\dot{Q}/\dot{T}$ vs. T or $\dot{V}/\dot{T}$ vs. T, i.e. $C_p(T) = dH/dT$ or $\alpha(T) = dV/dT$) for a simple cooling–heating cycle. c. As before with an annealing stage at T_{e}. The curves in b and c are derivatives from a. d. Nonlinearity. Isothermal change of $V(t)$ at temperature T after quenching from equilibrium at $T + \Delta T$ (+, contraction) and from equilibrium at $T + \Delta T$ (−, expansion), schematically. e. Kovacs expansion gap. The limit ($t \to \infty, \delta \to 0$) value of effective time for contraction is the equilibrium relaxation time at T, $\tau(T)$, but for expansion the limit time is different from $\tau(T)$ and depends on the ΔT in the history; the same isothermal experiment in PVAC for $T = 35°\mathrm{C}$ as in Fig. 46d is shown. $\tau_{\mathrm{eff}}^{-1} = -(d\delta/dt)/\delta$, $\delta(t) = (V(t) - V_\infty)/V_\infty$. The parameter is the equilibrium start temperature $T \pm \Delta T$ before the quenching $T \pm \Delta T \to T$. f. Stages of the Kovacs experiment in spectral representation for $\Delta T \approx 2\delta T$. r.h.s.: contraction, l.h.s. expansion experiment. (0) Reference equilibrium at T. (1) Equilibration at $T \pm \Delta T$. (2) Quenching from $T \pm \Delta T \to T$. (3) Start of relaxation (hatched = frozen-in, arrow see Fig. 45g). (4) Limit time for contraction as determined by the absorber. (5) l.h.s. Limit time for expansion as determined by the emitter. (6) The new equilibrium.

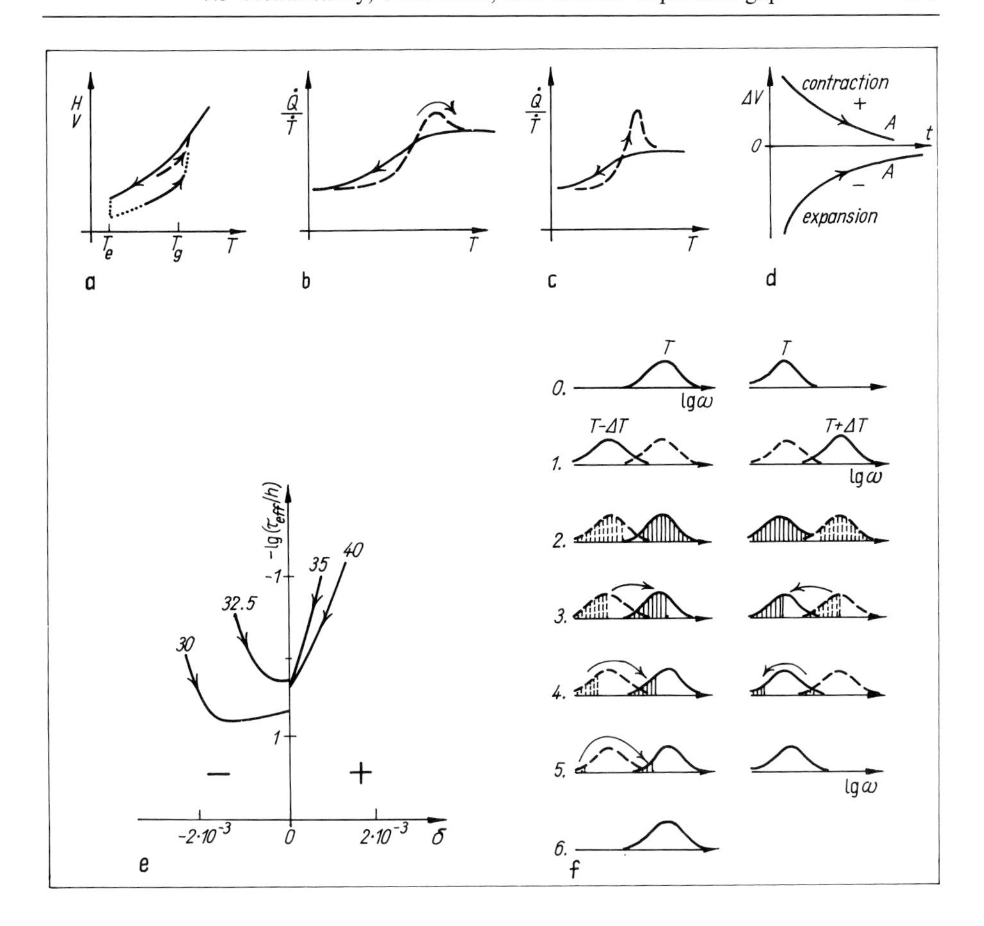
H
V
T_e
T_g
T
a
$\dot{Q}/\dot{T}$
T
b
$\dot{Q}/\dot{T}$
T
c
ΔV
contraction
+
A
0
t
A
−
expansion
d
$-\lg(\tau_{eff}/h)$
-1
1
32.5
30
35
40
−
+
$-2\cdot10^{-3}$
0
$2\cdot10^{-3}$
δ
e
T
T
0.
lgω
T-ΔT
T+ΔT
1.
lgω
2.
3.
4.
5.
lgω
6.
f

This corresponds to the overshoot. But the most interesting stage is the tail for long times, A in Fig. 46d. It can well be discussed in a Kovacs diagram, Fig. 46e. The abscissa is the reduced isothermal response: the relative volume difference to equilibrium

$$\delta(t) = (V(t) - V_\infty)/V_\infty, \tag{7.15a}$$

with

$$\delta(t) \to 0 \quad \text{for} \quad t \to \infty. \tag{7.15b}$$

The ordinate is the logarithm of a local time scale

$$\tau_{\text{eff}}^{-1} = -(d\delta(t)/dt)/\delta(t) \tag{7.16}$$

($-\lg \tau_{\text{eff}} = \lg \tau_{\text{eff}}^{-1}$, the higher the retardation rate the higher $-\lg \tau_{\text{eff}}$). The asymmetry of this diagram is striking, there are crucial differences between contraction and expansion. For contraction at long times, $\delta \to +0$, the effective times for different ΔT converge to the equilibrium value $\tau(T)$,

$$\delta \to +0, \tau_{\text{eff}} \to \tau^+ = \tau. \tag{7.17}$$

The expansion, however, does not seem (for $|\delta| > 10^{-4}$) to converge to the universal time τ,

$$\delta \to -0, \tau_{\text{eff}} \to \tau^-(\Delta T) \neq \tau. \tag{7.18}$$

The larger the ΔT jump in the cooling history, the lower the limit velocity and therefore the larger the τ^- limit: *Kovacs expansion gap*.

The staggered expansion wings and the nearly linear contractions can qualitatively be reproduced by a nonexponential Narayanaswamy scheme. The different limits for $|\delta| \lesssim 10^{-4}$, however, cannot be obtained because τ_{eff} diverges for $t \to \infty$ (artificially, by ignoring other subsystem fluctuation?) due to reflecting the long KWW tails by τ_{eff}.

A qualitative explanation of the gap can be obtained from a discussion in terms of a quasiclassical thermodynamic experiment (see Secs. 2.6 and 3.7). The gap is a spectral effect as explained at the end of Sec. 7.3. The situation is depicted in Fig. 46f drawn for $\Delta T \approx 2\delta T$, the time evolution is from the top to the bottom.

After quenching (stage 2) both spectra are virtual: The frozen-in spectrum at $T + \Delta T$ (or $T - \Delta T$) and the goal spectrum for T. Relaxation occurs by exchange of quanta from the emitter (frozen-in spectrum) to the absorber (goal spectrum). The exchange arrows connect about the same parts of the spectra, e.g. right flank with right flank: tawing a long mode here means mobilizing of a long mode, and so on.

The later states are firstly considered for contraction (r.h.s. of Fig. 46f). Generally, the relaxation rate is dominated by the slowest process. For contraction, this is the absorber spectrum having a lower frequency than the emitter spectrum. Arriving at equilibrium it is the goal spectrum that makes linear response. Thus we obtain Eq. 7.17.

Secondly consider the expansion experiment on the l.h.s. of Fig. 46f. Now the emitter spectrum, due to its lower frequency, is the slowest step. If true this means that τ_{eff} is determined by the success rate of the frozen-in spectrum and not by the absorption ability of the goal spectrum. This difference generates the asymmetry of Fig. 46e. For larger $|\Delta T| > \delta T$, the limit of τ_{eff}, τ^-, is then approximately determined by the time of the structural relaxation at $T - \Delta T$,

$$\tau^- \approx \tau(T - \Delta T) > \tau(T) \text{ for } \Delta T > \delta T. \tag{7.19}$$

The term spectral effect here means that we have to consider two spectra with *different* typical frequencies: the controlling frozen-in emitter spectrum at $T - \Delta T$, low frequency, and the goal (later equilibrium) spectrum at T with higher frequency. The two are to be connected by the Nyquist cloud of Fig. 21e having the ω identity of the FDT. That means that the experiment requires *equal* frequencies. Furthermore, the mode length before freezing-in and after thawing is also the same. But the controlling frequency belongs to the low-temperature freezing-in at $T - \Delta T$ and is therefore lower. This is in disagreement with the dispersion law for equilibrium, also demanding about the same frequencies. Therefore, the expansion is heavily restricted by the frequency difference for two reasons: the fluctuation dissipation theorem and the general scaling principle.

Details of the spectra are needed for the case $|\Delta T| \lesssim \delta T$, e.g. $|\Delta T| = 2.5\,\text{K}$.

7.4 Physical aging

A structural relaxation so far below T_g that it cannot be described by main transition modes alone is called *physical aging*. Modes of main transition predecessors must be used, e.g. the shear activities in the Andrade zone A1 of Fig. 38b. This is a self-similar sequential process in space and time that is kept going by the fast secondary relaxation, e.g. by the local modes β of Fig. 37b.

The basic experiment (Ref. 221) is sketched in Fig. 47a. The sample is cooled to $T_e < T_g$ and then annealed there (annealing, preparing or aging

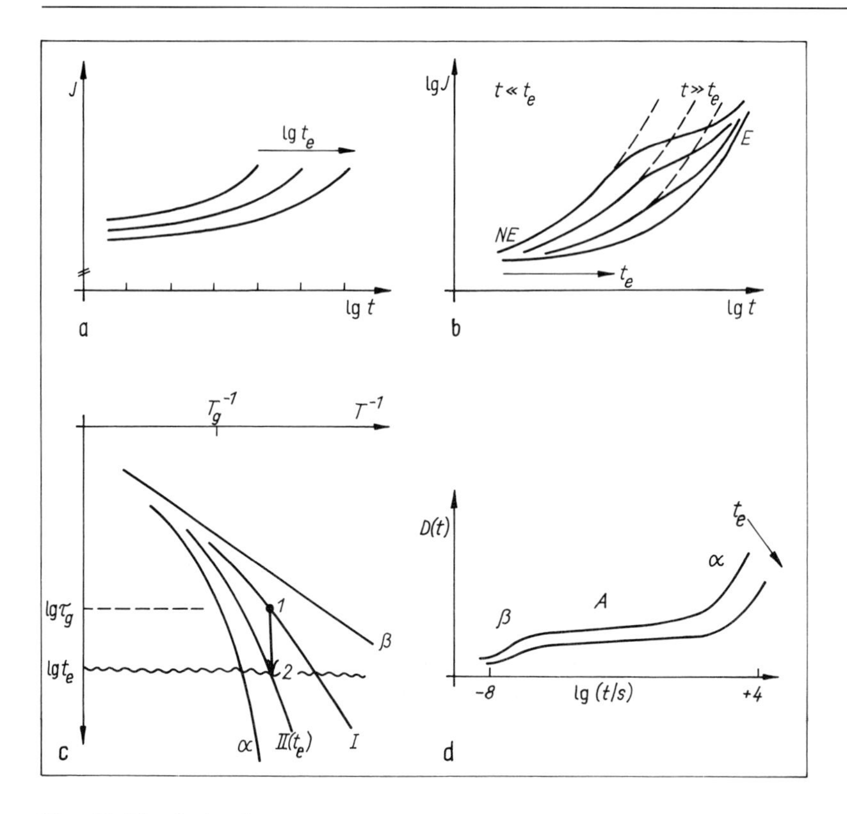

Fig. 47. Physical aging.
a. Shear compliance J at $T_e < T_g$ as a function of probing time interval t after aging (= preparation) time (parameter t_e) for $t \ll t_e$. b. The same experiment continued to $t \gg t_e$ up to equilibrium E, schematically. NE nonequilibrium. c. The same experiment mapped into an Arrhenius diagram (ordinate lg$t$). The annealing time t_e acts near 2 and shifts the Andrade mode II = II (t_e) to the left. The probing time t acts near 1 and goes from "below β" to 2. The case $t \gg t_e$ corresponds, in a way, to $t = t_e$ because then the experiment actually observes the shear response during the shift II (t_e). d. Schematical picture of a creep experiment in PVC near T_g (minute scale) from very short time (near β) to very long times. $D(t)$ creep function Eq. 2.61b. β local mode, A Andrade modes, α main transition.

time t_e at T_e). After annealing the shear creep experiment is performed (time interval t, creep or probing time). The result is the shear compliance $J(t, t_e, T_e)$, usually measured for $t \ll t_e$. The experimental curves are similar to an Andrade law (power law with exponent $\approx 1/3$) and can be mastered by a shift factor a_T. It was found (Struik law) that

$$\mu \equiv -d \ln a_{T_e}/d \ln t_e \approx 1, T > T_\beta \tag{7.20}$$

and that μ decreases for temperatures below T_β where the aging effect is small.

The crossover from short $t \ll t_e$ to long $t \gg t_e$ creep times is indicated in Fig. 47b (Ref. 222). At very long times t of order the equilibrium retardation time at $T = T_e$ the curves lead to the equilibrium compliance.

The Struik law Eq. 7.20 can be interpreted by using the scaling concept to sequential aging Sec. 6.6.2. Two preliminary notes ahead. (i) Consider the Arrhenius diagram of Fig. 47c. I and II are Andrade modes with length scales $\lambda_I < \lambda_{II}$. Choose T_e so that, for a given cooling rate, II is frozen-in and I is at the borderline between frozen-in and till working for arriving T_e at $t_e = 0$. Physical aging ($1 \rightarrow 2$) is the successive tawing (mobilization) of the Andrade spectrum from I to II. Let II be arrived at the aging time t_e, II $=$ II(t_e). Then all modes β...I...II are working, even if modified by the larger frozen-in modes between II and glass transition ($\lambda_{II} \ldots \xi_a$). The creep experiment (creep time interval t starting from t_e) tests successively, from β to e.g. II, the mobilized modes with regard to their shear retardation activity. (ii) Fig. 47a tells us how the structure relaxation (II shifts with increasing t_e towards to the glass transition, $\lambda_{II} \rightarrow \xi_a$) modifies the mobilized modes (β...II): the growing density (decreasing free volume) diminishes the shear compliance.

Eq. 7.20, including the statement that it is valid for the whole time scale t (but $t \ll t_e$), expresses the fact that both t_e (shift or borderline II $=$ II(t_e)) and t (test or probing time for sequential aging) proceed with the same rate. Both are real times, t_e for the shift factor a_{Te} and t for the successive testing of all mobilized Andrade modes (cf. the ω identity: here t identity, of the FDT Sec. 2.6). Because of self-similarity both processes make "the same", it is only the length scale (smaller for t than for t_e) that is different. Deviations from the Struik law Eq. 7.20 can arise from a new typical or characteristic length (ρ level for β, ξ_a for glass transition α), or if the free volume annihilation during aging would define a new length.

Let us repeat this complicated matter in other words. The frozen-in structure beyond the borderline II(t_e) defines a length $\lambda_{II}(t_e)$ characterizing the frozen-in state like as in Fig. 45f, l.h.s., now for smaller scales because $\lambda_{II} < \xi_a$. According to the scaling conception of Andrade modes there is no additional typical length between 0.5 nm and $2 \cdot \lambda_{II}$. Therefore the smaller free volume for larger t_e does not alter the power law, and the process remains self-similar for $t \ll t_e$ as shown in Fig. 47a. It is only for $t \gtrsim t_e$ that the length $\lambda_{II}(t_e)$ is involved in the response, consequently we observe a deviation from the power law (see Fig. 47b again). The driving force to equilibrium acts directly for $t \gtrsim t_e$, the lower compliances for the larger modes are due to this additional acceleration towards the equilibrium compliance.

When the aging experiment, not too far below T_g, is continued to very large times, e.g. over 12 logarithmic decades, then the time scale comes up with the glass transition zone α and the shear compliance again increases, see Fig. 47d (Ref. 223).

The physical aging damps not only the shorter Andrade modes but also the β relaxation and even the low-temperature processes (Ref. 224). The accompanied densification lowers the relaxation strength of the β relaxation and the C_p values at $T < 1$ K. The possibilities of short modes are generally restricted because they are influenced by the loss of free volume in the environment.

Nonlinear aging experiments (large stress values beyond the linear response range) can be described by the competition between the forced generation of free volume due to the stress and annihilation of free volume due to structural relaxation (Ref. 225). A "rejuvenation" by shear, however, has still not been observed yet although it was looked for e.g. in Ref. 225. Obviously the physical aging, being a fine, sequential structure relaxation with smoothly proceeding time and length scale, cannot be inverted by a "global" shear action to the subsystems. A better candidate may be (Ref. 226) the pressure, especially negative Δp, because it has a more direct relation to free volume.

7.5 Conventional thermodynamics of thermal glass transition

Usually the thermal glass transition is *thermoreversible*. This term means that a cooling-heating cycle starting and ending in equilibrium $T > T_g$ at the same state T, p, ϕ has the same state variables U, H, S, . . . as before irrespective of temporary freezing-in and structural relaxation below T_g.

It is quite another question whether such a cooling-heating cycle is a reversible cycle at all. This is not the case: Since each cycle contains structural relaxation in its parts below T_g, and since structural relaxation is irreversible, any cycle with freezing-in parts is an *irreversible cycle*. The degree of irreversibility for a cooling-heating cycle without annealing below T_g can simply be estimated (see Refs. 227, 228 and Fig. 48a). From thermoreversibility and from the First Law it follows that the two hatched areas are of equal size.

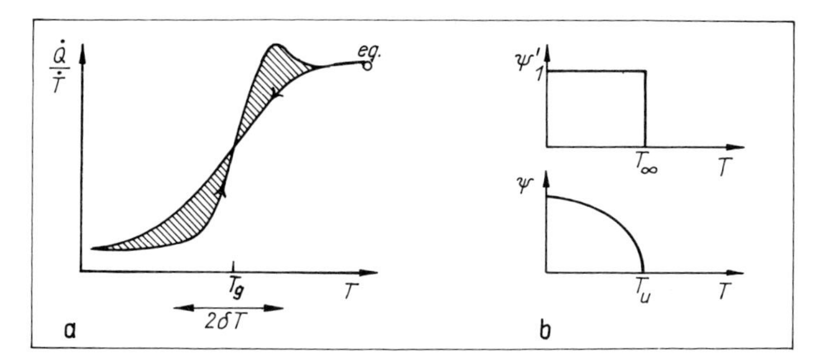

Fig. 48. Thermodynamics of thermal glass transition.
a. This cycle is irreversible. b. Comparison of nonergodicity parameter ψ' at the Vogel temperature with order parameter $\psi(T)$ at a phase transition.

Therefore, the irreversibility is of second order in δT. We have, using $C_p \equiv \dot{Q}/\dot{T}$, the First Law

$$\Delta H = \oint C_P dT = 0. \tag{7.21}$$

The Second Law gives the entropy production per cycle

$$\Delta S_i = \oint \frac{C_P}{T} dT \approx \Delta C_P(\delta T/T)^2, \tag{7.22}$$

although, of course, S(initial) $=$ S(final) for the cycle. The main parameter is δT, the mean temperature fluctuation of the natural functional subsystem.

Considering the reversibility as an essential criterion of a phase transition, the glass transition is not a phase transition as long as $\delta T > 0$. The concept of an "underlying phase transition" (Refs. 92, 134, 154, 299) means, therefore, the requirement that $\delta T \to 0$ for infinite time, $t \to \infty$. This depends on then art of extrapolation. If we use WLF scaling (i.e. the ideal glass transition of Sec. 6.4) for extrapolation then we come to the Vogel temperature T_∞. According to Eq. 6.2a $\delta T \to 0$ there, as for a critical state, and the transition would be reversible.

A further criterion for a phase transition is the appearance of an order parameter. One could think (Refs. 230, 231) to take a so-called non-ergodicity parameter. Such a parameter, if finite > 0, is defined as the limit of a correlation function for large time, $\psi' = \varphi_\infty = \varphi(\tau \to \infty)$ with $\infty < \tau$ (equilibrium). Obviously, $\varphi_\infty > 0$ does exist in a frozen-in state, because φ_∞ is a characteristic for any structural arrest. But, trivially, $\varphi_\infty \to 0$ when the time scale τ grows up to the (extrapolated) equilibrium WLF relaxation time

τ (equilibrium). For an ideal glass transition there is always such a time as long as $T > T_\infty$: The glass transition remains a thermokinetic phenomenon with no true order parameter, because all the hyperbolas in the lg ω–T diagram have a finite slope for $T > T_\infty$. There is no ergodicity violation for ideal glass transition since the whole spectrum is guided by the hyperbolas.

The situation is abruptly changed at the Vogel temperature, $T = T_\infty$. Ignoring Simon's warning (Ref. 89) against a discussion of two diverging time scales we find a switch on of the ψ' order parameter at $T = T_\infty$ for $t \to \infty$ (truly infinite), see Fig. 48b. But this is a rather pathological case when compared, for instance, to a second-order phase transition: For ideal glass transition, we have $\psi' \equiv 1$ for $T < T_\infty$ where the largest ($\xi_a \to \infty$) thermokinetic structure remains arrested for all times. For a phase transition, however, the behavior of the order parameter $\psi(T)$ in the ordered phase is of central interest. The critical exponent for $\psi(T)$, β, is imbedded in the singular environment by scaling laws.

Thus, the ideal glass transition is a thermokinetic phenomenon that has nothing to do with a phase transition except on condition that one is interested in a discussion of the extrapolated pathological case $\beta = 0$ at $T = T_\infty$.

It is of more practical interest whether a glass can be in mutual equilibrium with its environment. This is essentially a question of contacts. Exchange of heat makes thermal equilibrium (equal temperature), exchange of work makes mechanical (equal pressure), and exchange of particles makes phase or "chemical" equilibrium (equal chemical potential). Even the ideal frozen-in glass at $T < T_\infty$ does not forbid the atoms to oscillate which can ensure thermal and mechanical equilibrium.

The equilibrium situation is more complicated in partially frozen-in states for multiple glass transitions, e.g. if local modes work below T_g. On the one hand, the partial or successive freezing-in can lead to internal stress distributions that can even result in self-destruction of the material (Ref. 232). On the other hand the still working functional subsystems can allow a particle exchange with the environment (ions, small solvent molecules etc.) so that phase equilibria of glasses with respect to this exchange are practically possible (Ref. 233).

7.6 Low temperature behavior

Low temperature is the domain of the Third Law (Nernst heat theorem,

NHT). Simon found a formulation that does not exclude glasses from the discussion: The entropy change associated with any isothermal reversible process of a condensed system approaches zero as the temperature approaches zero (see e.g. Ref. 234).

We shall consider two aspects of the NHT. (i) Consider two samples of the same substance that are cooled from equilibrium to glass via different ways, e.g. different temperature-time-pressure histories. Then we have two glasses that cannot, for $T < T_g$ (or $T < T_\infty$), mutually be transformed by an isothermal reversible process. This means that any thermodynamic comparison of the two glasses with regard to the NHT is excluded below T_g; we have, in a sense, two "different substances" (Ref. 235). (ii) Consider one of the two glasses. Then we can, using the remaining molecular movements, do normal thermodynamics as long as T_g is not crossed, including isothermal reversible processes and the normal implications of the NHT (e.g. $C_p \to 0$ for $T \to 0$).

Firstly we consider aspect (ii). Let the process under consideration be parametrized by pressure. Then we have the general fluctuation formula for the entropy susceptibility,

$$-\left(\frac{\partial S}{\partial p}\right)_T = \frac{\overline{\Delta S \Delta V}}{kT}. \tag{7.23}$$

The NHT can therefore be substantiated by the fact that quantum mechanics generally decouples entropy and volume fluctuation in condensed matter at low tempertures. [The S fluctuation includes the overcoming of gaps between discrete energy levels – the eigenvalues of the Schrödinger equation – whereas the V fluctuation rests on the continuous change of these levels with the boundary conditions of this equation. If the system feels the discreteness at low temperature then the cross fluctuation in Eq. 7.23 decouples because of the difference between the two underlying topologies: discrete and continuous.] This is valid for each glass, and since comparison between the different glasses is excluded, we may put $S = 0$ for $T = 0$ for each of them. As mentioned before,

$$\text{NHT} \Rightarrow C_p \to 0 \text{ for } T \to 0. \tag{7.24}$$

From Eq. 7.23 we see, using the Schwarz inequality, Eqs. 7.1–7.3, and $(\partial p/\partial V)_T \neq 0$ for $T \to 0$, that

$$C_p/T \to 0 \Rightarrow (\partial S/\partial p)_T \to 0 \Leftrightarrow \text{NHT}. \tag{7.25}$$

Thus although $C_p \to 0$ (e.g. $C_p \sim T^\alpha$, $\alpha > 0$) is an implication of the NHT,

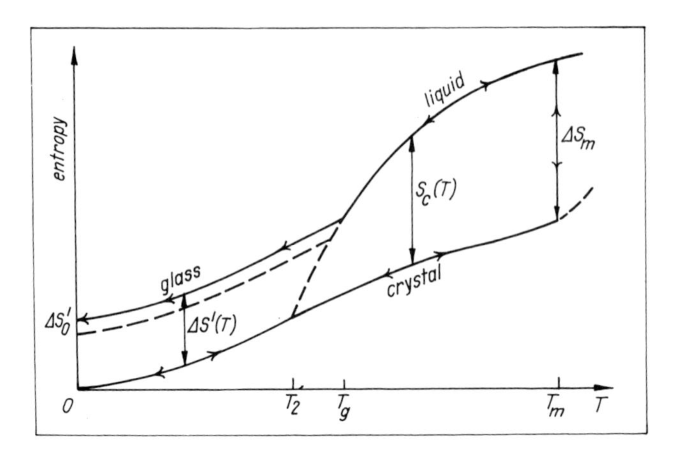

Fig. 49. Ersatz entropy of a glass.
$\Delta S'(T)$ excess entropy of a glass as compared to the crystal if the glass entropy is connected to equilibrium (↔) at T_g. The broken lines are for a lower cooling rate. T_2 Kauzmann temperature, $S_c(T)$ configurational entropy.

one must calculate $\alpha > 1$ for explaining the NHT from the properties of state density at low temperature ($(\partial p/\partial V)_T \neq 0$ supposed).

Obviously, serious problems arise for $\alpha \to 1$. In all glasses (Refs. 236, 237) a C_p exponent near 1 (e.g. $\alpha \approx 1.1$) and some universality in the prefactor was observed. This means that in all glasses (inclusive polymers, of course) there are molecular processes that do work below 1 K and that dodge the discreteness of Debye's phonon spectrum ($C_p \sim T^3$). Tunneling two-level systems promoted by low energy barriers are suggested for explanation (Refs. 238, 239), perhaps associated by soft modes localized to about 100 ($\approx N_a(T_g)$) participating particles (see e.g. Ref. 240).

As mentioned in Sec. 6.8 confined scaling with a dispersion law includes one local concentration of free volume (defect, cage) per cooperativity region of volume V_a for any thermokinetic structure. From the $T < 1$ K properties we learn that the spectrum of associated energy barriers universally goes down to zero, and that, therefore, some white cores of the disperse structure in the l.h.s. of Fig. 45f can also work at low temperatures $T < 1$ K, far below the Vogel temperature.

Now we shall consider aspect (i). Since, according to Eq. 7.22, the irreversibility of a cooling process at T_g is of second order in $(\delta T/T_g)$, a glass excess entropy $\Delta S'(T)$ (ersatz entropy) can be measured – and, if necessary, corrected in second order – that would be obtained by an extrapolation from a liquid equilibrium state above T_g, see Fig. 49. The entropy of the liquid state can be determined from the equilibrium heat capacities of the crystal and the liquid and the heat of melting. Values of order 10 J/mol · K are obtained for the ersatz entropy at $T = 0$, $\Delta S'_0 = \Delta S'(T = 0)$, and $\Delta S'(T)$ increases with temperature, up to 10% at T_g (Refs. 241, 242).

The discussion of $\Delta S'(T) > 0$ has little to do with the Third Law. The change of $\Delta S'(T)$ compares harmonic and inharmonic oscillations and, in some cases, other local movements between crystal and glass, i.e. between two thermodynamically "different substances". The increase of $\Delta S'(T)$ indicates that the entropy generation in the glass is larger than in the crystal: a glass is more alive than a crystal. It is this property that opens a wide field for application of glasses that is even enlarged for polymers because of the glass transition multiplicity.

The value of $\Delta S'$ at the glass temperature T_g of course depends on the cooling rate, because T_g depends on it. $\Delta S'(T_g) = S_c(T_g)$ is a measure for that part of additional entropy generation in the liquid (as compared to the crystal) that is "lost" between T_g and the Kauzmann temperature T_2 by freezing-in at T_g. [The symbol $S_c(T)$ means the configurational entropy explained in Eq. 5.18 and Fig. 27f.] The ratio to the melting entropy, $\Delta S'(T_g)/\Delta S_m$ ($\Delta S_m = S_c(T_m)$ or $\Delta S'(T_m)$ with T_m melting temperature) is of order few 10%, decreasing with increasing fragility.

8. Examples

This chapter presents illustrative examples for the systematic treatment of dynamic and thermal glass transition in Chaps. 6 and 7.

8.1 Relaxation cards for amorphous polymers

The two basic types of Arrhenius diagrams for amorphous polymers are shown in Fig. 50a. The details Sec. 6.8 near the splitting S and the high frequency properties $>$ GHz are dropped.

Typical relaxation temperatures where the dispersion zones are at a frequency of about 1 Hz (e.g. the maximum of $\tan \delta$ for shear) are listed for some conventional samples in Tab. IV.

The extrapolation of the united $\alpha\beta$ relaxation to $T \to \infty$ gives about $\lg \nu_0/\mathrm{Hz} \approx 13$, see Eq. 5.12. The temperatures of the flow transitions for $M > M_c$, sometimes denoted by T_{ll}, heavily depend on the molecular weight and are, therefore, not listed, see Secs. 5.5, 6.6.4, and 6.10. The T_α temperatures for the main transition nearly correspond to the calorimetric glass temperatures T_g. For a more precise statement of this relation an analysis of the activity order across the transition and of the relation between frequency and cooling rate is necessary. The β relaxations Sec. 6.6.1 are local modes for PVC, for the other polymers side group motions are shared in a certain degree. The γ relaxations are usually connected with side group motions (phenyl group for PS, α methyl group for PMMA), and the δ relaxation is usually considered as a special low temperature phenomenon: tunneling. The activation energies for the β and γ relaxation can be estimated from the listed temperatures using Heijboer's formula Eq. 5.11.

A comprehensive discussion of the relaxation cards can be found in the famous book by McCrum, Read, and Williams, Ref. 3. Each polymer has peculiarities, e.g. tacticity (especially important for PMMA) or partial crystallinity. The latter amounts to about 5 per cent for conventional PVC and is zero for the conventional types of the other three. For PVC we observe two effects: (i) The lower the polymerization temperature the higher is the syndio

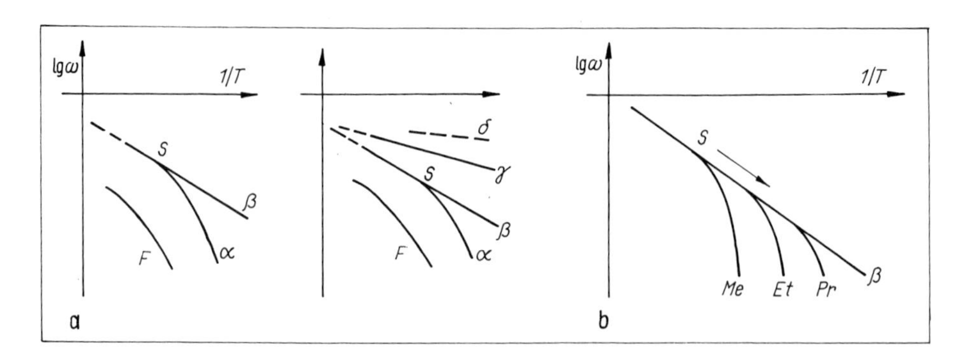

Fig. 50. Relaxation cards for amorphous polymers.
a. For PVAC, PVC (l.h.s.) and PS, PMMA (r.h.s.). Numerical values can be found in Table IV. No γ relaxation is usually observed without side groups or guest molecules. b. Simplified representation of splitting-point shift for alcyl methacrylates due to a fixed local mode β. Me methyl, Et ethyl, Pr propyl.

tacticity, the higher the degree of crystallization, the higher the material order, and the higher, therefore, T_g. (ii) Because of the crystallite structure in the 5 nm scale the extension of thermokinetic structures $\xi_a(T)$ is hindered at low temperatures and, therefore, deviations from the WLF equation are observed below the mHz range, also in equilibrium (Ref. 243). Further peculiarities are the molecular mass distributions, chain branching, and impurities (e.g. water in PMMA, additives, etc.). The figures of T_α are, so to speak, approximate values for widespread, i.e. conventional types.

The splitting-point S coordinates are also listed in Tab. IV. It is interesting to compare these coordinates in the homologue series of alkyl methacrylate polymers, see Fig. 50b. Increasing the size of the alkyl group (Me, Et, Pr ...) does not or little influence the β relaxation "arrested" by an "entanglement" of chains due to the α methyl groups. But internal softening due to the alkyl

Table IV. Typical relaxation temperatures for some amorphous polymers (in °C)

polymer	T_s	$\lg \nu_s/\mathrm{Hz}$	T_∞	T_α	T_β	T_γ	T_δ
PVAC	160	≈9	−26	35	−115	–	–
PVC	150	7.3	37	90	− 50	–	–
PMMA	≈145	≈4	30	105	10	−170	–
PS	160	7	37	100	30	−120	(30 K)

Table V. Glass temperatures T_g or some frequently investigated glass formers

	T_g/K		T_g/K
crown glass	≈ 730	PS	373
boron oxide B_2O_3	540	ortho terphenyl	242
nitrate salt*	340	glycerol	190

*40 $Ca(NO_3)_2$ 60 KNO_3

groups lowers T_g and T_α so that the splitting point is shifted to low frequencies in the Hz range.

Relaxation cards are important tools for the discussion of relaxation experiments of all kinds. Since often only a small frequency region can be investigated, sometimes restricted to one or a few activities, the primary information is, as a rule, rather meager. Drawing, however, the points in a relaxation card usually allows the attachment to a definite relaxation (e.g. flow, α, β, ...). This opens the way to use the information from other experiments in the literature (other temperature or frequency region, other activities, comparison with "neighbored" substances, ...).

8.2 Control of T_g

The glass temperature T_g is controlled by many parameters. The classical paper for polymers is Ref. 244, tables for T_g values are in Refs. 245–247, see also Refs. 154, 248. A small selection of T_g values for polymers is presented in Table I, page 22, the T_g values of a few typical glass formers are given in Table V.

Tammann (Ref. 249) pointed to the universality of glass forming: In 1933 he stated that practically all substances could at least partially be transformed into a glassy state if they are cooled down, or quenched, with sufficiently high rate. Actually, glass transitions are found in some quenched liquid metals (Ref. 250), in plastic (Ref. 94) and liquid crystals (Refs. 251–253), in water ($T_g = 136$ K) and ice ($T_g = 100$ K), see Ref. 254, and many other materials, e.g. spin glasses. Even a Peierls transition in the 1 K region of charge density waves in special crystal structures with highly anisotropic band structure shows features of a glass transition (Ref. 255), also a jam of motor cars on a Friday evening and so on.

Generally spoken, T_g (and T_∞, see Eq. 6.5b) is controlled by two properties: order and intermolecular potentials. T_g is the higher the larger the molecular order and the larger the energy of molecular interaction are. The influence of intermolecular parameters was discussed in Sec. 6.4, Eq. 6.19, the influence of order on the mean roughness of the energy landscape is explained in Fig. 51a (cf. with Figs. 36e and 37c). For given potential parameters the roughness of the energy landscape (or more precisely, of its accessible part that is influenced by stochastics) is enlarged by order, which implies an increase of T_∞ Eq. 6.19 and T_g Eq. 6.5b.

The same conclusion can be obtained from a thermodynamic analysis. Describing T_g as an enthalpy-to-entropy ratio,

$$T_g = \Delta H/\Delta S, \tag{8.1}$$

then ΔH and ΔS mean the deviations of ersatz enthalpy and ersatz entropy, respectively, from the corresponding equilibrium properties of the liquid at a given temperature below T_g, see Fig. 51b. The increase of T_g with increasing energy follows immediately from the numerator. The influence of order comes from the denominator. The larger the order of the material the smaller the disorder that can freeze-in, the smaller, therefore, ΔS, and the larger T_g.

What follows is a short discussion of six control parameters.

(i) **Pressure**. Roughly speaking, pressure enlarges energy and order, so that T_g increases with increasing pressure p. Typical dp/dT_g values for polymers are of order a few MPa/K, the values are usually somewhat larger than the isochoric tension coefficient in equilibrium, $\gamma_V = (\partial p/\partial T)_V = (\partial S/\partial V)_T$.

An analysis with the aid of ideal-glass-transition methods points to some universality. When the temperature in Sec. 6.4 is substituted by the intensive pressure variable then, for T = const, one can expect to find WLF hyperbolas in a p–lg ω diagram. The set of hyperbolas defines a Vogel pressure, $p_\infty(T)$, where the cooperativity goes to infinity. It is the same to say that the Vogel temperature depends on pressure,

$$T_\infty = T_\infty(p). \tag{8.2}$$

Having WLF sets of hyperbolas in a lg ω–T diagram *and* in a lg ω–p diagram implies (Ref. 256) that the isochronals (ω = const) in a pT diagram are either parallel straight lines or hyperbolas with common asymptotes for different frequencies: $p \cong \Pi < 0$, Π is called the *pressure asymptote*, and $T \cong \Theta > T_g$. Since, obviously, too large negative pressures exclude any glass transition (too much free volume) we can forget the straight lines, and we arrive at the situation sketched in Fig. 51c: $d^2p/dT_g^2 > 0$ as usually observed (Ref. 257).

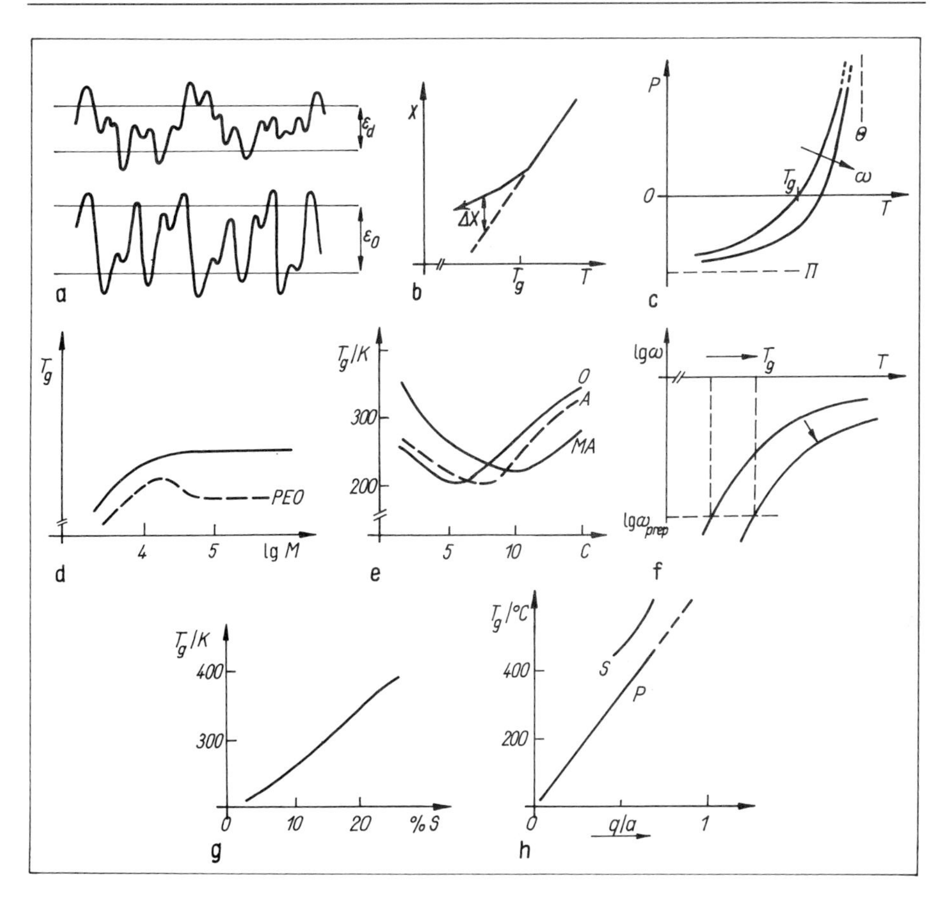

Fig. 51. Control of T_g.
a. Additional order in the material (lower part) enlarges the mean roughness of the energy landscape. Remark: The stochastics of the accessible part is essential for T_g. b. Definition for $\Delta X = \Delta H, \Delta S$ in Eq. 8.1. c. (Dynamic) glass transition in a pressure-temperature diagram. $\Pi < 0$ pressure asymptote, Θ temperature asymptote. d. Dependence of T_g on polymer chain length. The figure 4 means 10^4 g/mol. e. Internal softening of polyolefins (0), polyacrylates (A), and polymethacrylates (M) by linear alkyl side chains with length C ($C = 1$ Me, $C = 2$ Et, . . .). f. Lowering of mobility lg ω enlarges T_g. g. Increase of T_g in natural rubber due to vulcanisation with sulfur. %S per cent bounded sulfur. h. Polyelectrolytes. Increasing T_g due to additional Coulomb energy in phosphates (P) and silicates (S).

Extrapolations for amorphous polymers yield values of order several 10^8 Pa for the pressure asymptote Π, and Θ values of about 100 ... 200 K above T_g. As mentioned above, the physical meaning of Π is that for large negative pressures (large intermolecular distances) the necessity for cooperative movements decreases and gets lost for too large values of $|\Pi|$. A similar argument should be valid for the Θ asymptote, but the situation is more complicated because the cooperativity generally decreases at large temperatures and the intermolecular potentials are heavily modified at large positive pressures.

The existence of a pressure asymptote implies that for low temperatures relatively small negative pressures are sufficient to induce flow processes. Negative normal stress can be generated by impacts which can, therefore, lead to material failure due to flow processes (crazes, shear bands in polymers) also at very low temperatures.

(ii) **Chain length**. Chain ends in polymer materials disturb the ϱ level order in their environment which means more free volume there. Therefore, as a rule, T_g decreases with decreasing molecular weight M, see Fig. 51d. The plateau at higher chain lengths is a consequence of the characteristic length for glass transition: The effect of M on T_g is expected to be small if we find, on the average, much less than one chain end per cooperativity region. From $N_a \approx 100$ units and $M_o \approx 100$ g/mol it follows that the plateau onset is at molecular weights of order 10^4 g/mol. This is accidentally the same order as M_c for chain entanglements. The T_g decrease at low M shows some details (Ref. 258). The number of units per cooperativity region remains of order $N_a \approx 100$, but the intramolecular contacts (e.g. barriers of conformation change) are systematically substituted by (lower) intermolecular contacts so that there are also energetic aspects in the T_g slope.

A maximum of T_g in the M crossover region is observed in the amorphous phase of PEO (Fig. 51d, Ref. 259). Obviously the crystallization ability of the amorphous material can enlarge the order as long as the structure is loosened up by the chain end concentration.

(iii) **Softening**. The dependence of T_g on the length of linear side chains for differently alkylized poly acrylates, methacrylates and olefines is shown in Fig. 51e (Ref. 260). T_g decreases for short lengths because the material tends to a structure like poly ethylene ($T_g \approx 200$ K for the amorphous phase) with no foreign sidegroups (that would enlarge the energy e.g. the acrylate). This decrease of T_g is called *internal softening*. (External softening means lowering of T_g by addition of a proper small-molecule solute.) For larger length of side

Table VI. T_g variations with tacticity (T_g in °C)

	poly acrylates		poly methacrylates	
	iso	meanly syndio	iso	syndio
Me	10	8	43	160
Et	−25	−24	8	120

chains, however, T_g increases. Obviously the larger alkyl side groups can organize structures inside each cooperativity region that enlarge the order and therefore T_g.

(iv) **Networks** lower the segment mobility. Since $d \ln \omega/dT > 0$ (Fig. 51f) networks increase T_g. This is shown in Fig. 51g for a natural rubber vulcanized by sulfur. In terms of energy and order we can say that the sulfur enlarges the intermolecular energy, and the chain order in the vicinity of crosslinks can also contribute.

Another well-known example is the increase of T_g at curing of epoxide resins.

(v) **Polyelectrolytes**. Polyelectrolytes are a good example for the effect of additional energy on T_g. Fig. 51h (Ref. 261) shows a linear dependence of T_g on the Coulomb term q/a, where q is the average cation charge (in elementary charge units) and a is the distance (in Å) between 0^{--} and the cation, estimated from values of van der Waals radii. The straight lines are for phosphates XPO_3 with X = H, Na, Ca, K, Li inclusive their mixtures, and for the system $Na_2SiO_3 + CaSiO_3$, respectively.

(vi) **Chemical configuration**. Changing the chemical configuration of simple organic polymers can lead to a considerable variation of T_g. For instance, the tacticity of polyacrylates has only a little effect on T_g, but the tacticity of poly methacrylates has a large one, see Table VI.

Obviously the ability of the α methyl groups to "entangle", which was mentioned above, generates some order in the ϱ level (1 nm scale) where, despite of the amorphous state, the different tacticity gains a considerable influence (cf. also the remark on helicity in Sec. 6.6.4).

A further well-known example for the influence of chemical configuration are the isomers of poly butadiene: 1.2 $T_g = 269$ K, 1.4 trans $T_g = 255$ K, 1.4 cis $T_g = 165$ K.

8.3 Relaxation in polymer solutions

The diversity of structure in polymer solutions and its change with polymer concentration (Sec. 1.4, Fig. 4) corresponds to a diversity of relaxation phenomena. This subsection brings only a few illustrative examples. A review of polymer rheology in solutions and mixtures – outside the scope of this book – can be found in Ref. 9.

The composition dependence of dielectric relaxation temperatures at a frequency of 1 kHz in the system PVC + tetrahydro furane THF is shown in Fig. 52a (Ref. 262). The α and β' temperatures increase with the polymer concentration since the energy steps and barriers of polymer configurations have a higher energy, and because the polymer has a higher order than the solvent. The proper glass transition is α. The continuous increase of T_α with polymer concentration can, in a certain degree, be understood as a continuous substituting of solvent units in a cooperativity region by polymer units with higher energy and order. A splitting point S′ comes up at about 70% wt PVC. In a way, the increase of polymer concentration in a solution corresponds to a decrease of temperature in homopolymers: A minimal polymer concentration is necessary to generate the conditions for a local mode β typical for polymers. A possible discontinuity (or bend) in the α curve near S′ indicates a structural change connected with the occurrence of β.

In the semi region of the solution an additional local γ and a more cooperative β' mode are observed that have no direct correspondence to relaxation in the pure components. They are possibly based on (nano-) heterogeneity in the polymer solution Fig. 4b. Thus the PVC + THF system is an instructive example for changing multiplicity of polymer relaxation.

The example cis poly isoprene (cis PI, $M \approx 24\,000$ g/mol) + toluene is

---

Fig. 52. Relaxation in polymer solutions.
a. Dielectric relaxation temperatures in the THF (tetrahydrofurane) + PVC system for $\nu = 1$ kHz. b. Ditto in toluene + cis polyisoprene (cis PI), $M = 24\,000$ g/mol. α_1 flow transition. The marginal symbols are the conventional symbols for the pure components. c. Dielectric relaxation for this toluene + cis PI system in an Arrhenius diagram. — 30% wt polymer, --- 50%. d. Glass temperature in the TCP (tri cresyl phosphate) + PS system. e. Transition from Rouse–Zimm (RZ) behavior to flow transition (FT) by enlarging the polymer concentration c in the Araclor + PS system. — lg G', --- lg $(G'' - \omega\eta_s)$, $G^* = G' + iG''$ dynamic shear modulus, parameter: polymer concentration in g per cm^3. PZ plateau zone. Ordinates suitably displaced.

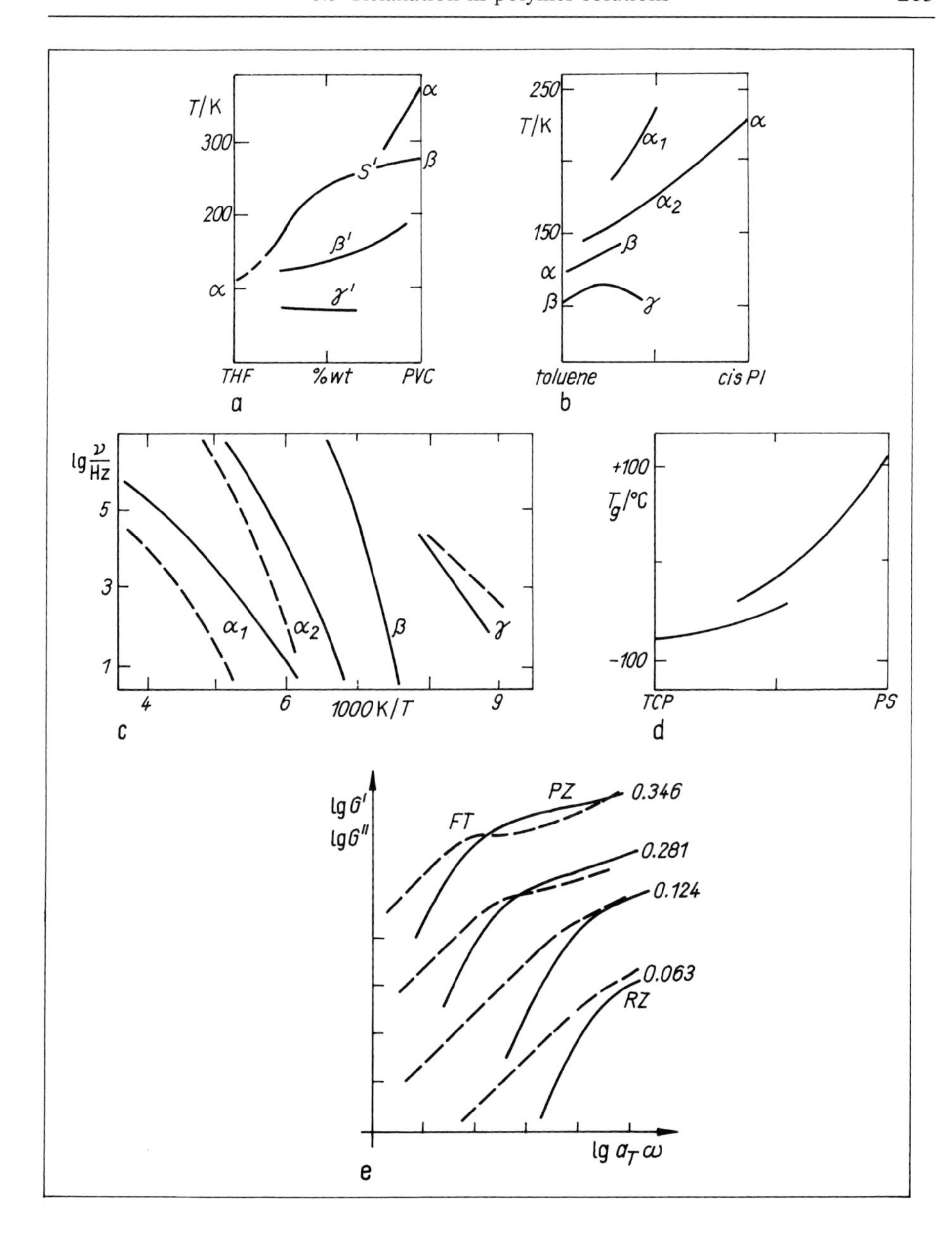
T/K
300
200
α
THF
%wt
PVC
a
α
β
S′
β′
γ′
250
T/K
150
α
β
toluene
cis PI
b
α
α1
α2
β
γ
lg ν/Hz
5
3
1
4
6
1000 K/T
9
c
α1
α2
β
γ
+100
Tg/°C
-100
TCP
PS
d
lg G′
lg G″
FT
PZ
0.346
0.281
0.124
0.063
RZ
lg aTω
e

Table VII. Vogel temperatures for the glass transitions in the cis PI + toluene system (T_∞ in K)

polymer conc., wt.%	α_1	α_2	β
30	95	118	120
50	95	135	–

very instructive because the flow transition is ε active here, i.e. it can be seen in dielectric investigations, Fig. 41d (Ref. 158). The concentration dependence of dielectric relaxation temperatures is shown in Fig. 52b, the Arrhenius diagram for two concentrations in Fig. 52c. β is the glass transition in toluene and γ is a local process in toluene. The increase of the temperatures shows that both small-molecule motions are influenced by sharing or the presence of the polymer.

The glass transition β of toluene does not continuously pass over to the glass transition of the polymer, α, although the latter is also heavily modified by the other component. This effect can also be observed in other systems, e.g. TCP (tri cresyl phosphate) + PS, Fig. 52d (Ref. 263). The nanoheterogeneity at median concentration Fig. 4b is so large that even two glass transitions can simultaneously be observed: one of the solvent (modified by the polymer), the other of the polymer (modified by the solvent). This means that the nanoheterogeneity is large enough to allow cooperativity regions of "diameter" $2\xi_a \approx 4$ nm with rather different average polymer concentration.

The polymer glass transition α wins at higher polymer concentration because the solvent regions get too seldom or too small, for higher concentrations also too small for the local γ relaxation of toluene.

For the given molecular weight a minimal polymer concentration is also needed for the onset of flow transition α_1, cf. Fig. 4d. The spatial scale indicated by α_2 is the polymer coil radius. The flow–transition character of α_2 is proved (Ref. 158) by the molecular–weight dependence of the lg ω difference between α_1 and α_2 (exponent 3.4, ω frequency). In the Arrhenius diagram Fig. 52c both α curves (and, of course, β) are of WLF type. Table VII reports the Vogel temperatures.

The values for α_1 and α_2 show that $T_{\infty 1} < T_{\infty 2}$ for both concentrations. This confirms Eq. 6.21, $T_{\infty F} < T_{\infty G}$: the long glass transition α_1 has the smaller Vogel temperature, as explained in Fig. 37c.

In both cases PVC + THF and cis PI + toluene the solvent glass temperature increases with polymer concentration. A famous counter example is the PEO + water system. It is sometimes used by the fire brigade. Adding PEO lowers the order in liquid water, the glass temperature therefore decreases with the consequence of higher mobility, i.e. lower viscosity: more liters per second through the fire hose.

The influence of the polymer concentration c on the typical cooperative movement of solvent molecules can be observed in different ways, e.g. by light scattering (Ref. 264). The polymer chain changes the thermokinetic structure of the solvent in the environment. This effect is also the core of the so-called η'_∞ problem, where η'_∞ is the high frequency limit of the real part of complex viscosity, high frequency, of course, as compared to the dispersion zone considered. The point is that $\eta'_\infty \neq \eta_s$ for $c \neq 0$, where η_s is the pure solvent viscosity (Ref. 264a). The ratio η'_∞/η_s sharply increases with c like

$$\eta'_\infty(c, T) = \eta_s(T) \exp(c[\eta'_\infty]) \tag{8.3}$$

and can reach values of order ten. The effect is described by the parameter $[\eta'_\infty]$ where the brackets indicate a so-called intrinsic viscosity, $[\eta'] = \lim (\eta' - \eta_s)/\eta_s c$ for $c \to 0$. It was confirmed by a broad variation of conditions that $[\eta'_\infty] \neq 0$ is mainly due to the modified solvent friction, whereas internal viscosity, chain stiffness and local constraints are not so important (Refs. 265, 266). The relation $[\eta'_\infty] > 0$ is in correspondence with an increase of T_g in dilute polymer solutions or nanoheterogeneity regions with low local polymer concentration, see curve α in Fig. 52a and β in Fig. 52b, where the thermokinetic structure has a higher order than in the pure solvent.

Consider Fig. 4d for a system with high M, i.e. with long enough chains. Enlarging the polymer concentration c can lead to entanglement. The entanglement onset can be demonstrated in the series of shear G^* curves Fig. 52e (Ref. 267) for the PS (M = 267 kmol/g) + chlorinated diphenyl (Araclor 1232) system. We see the formation of the flow transition and the plateau zone from the Rouse Zimm behavior in dilute solution. The corresponding formation of the flow transition in melts by increasing the molecular weight was presented in Fig. 6b. Fig. 52e shows, in a way, the transition from Figs. 28f or 29b (low c) to Fig. 29c (high c). Further examples are discussed in Ref. 200 including the relation to corresponding bulk activities.

Crossing the thermal glass transition isothermally by a variation of concentration gives interesting effects on the diffusion (Ref. 268). The nonsteady-state diffusion of a solvent into a glass enlarges the solvent concentration and lowers the glass temperature, $T_g(c)$. When $T_g(c)$ gets to the experiment temperature T we observe a coincidence between relaxation time and diffusion time in the diffusion layer. This situation is called *viscoelastic diffusion*, see Sec. 10.5.

8.4 Mixtures and statistical copolymers

The two examples of Fig. 53a (Ref. 269) show that thermokinetic averaging over the cooperative movement of glass transition gives smoother curves than the thermodynamic averaging for getting chemical potentials. The latter is rather selective in the crystalline phase. The huge liquidus amplitudes near the compounds As_2Se and P_4Se_{10} are only reflected by moderate peaks in the T_g function, and the variation of the latter curve is generally smaller. The left eutecticum in P_xSe_{1-x} is below T_g. This is one of the few examples for a conventional glass transition without metastability (see also Sec. 6.1. Of course, dynamic glass transitions at high frequency are more often observed in stable systems.)

As a rule of thumb

$$T_g \approx \tfrac{2}{3} T_c \tag{8.4}$$

where T_c is the "crystallization temperature". But there are many exceptions, not only in mixtures. As the crystal growth rate tends to zero at T_c the glass forming ability increases near a eutecticum, also for $T_E > T_g$.

In mixtures of organic polymers without mixing gap, or outside the gap, and in "forced mixtures" such as statistical copolymers (AABABBAB...), as a rule, a continuous, smooth $T_g(\phi)$ function is observed that connects the T_g's of pure components at $\phi = 0$ and $\phi = 1$ (T_{gA}, T_{gB}). Peculiarities like those as generated by nanoheterogeneities in polymer solutions are not expected to occur far from the spinodal (no blobs and heaps) and, for the copolymers, for not too long sequences of one component. The ρ-level cooperativity region always contains about a hundred monomeric units. Therefore, in the scale of fluctuations, increasing ϕ means continuous substitution of one component by the other resulting in the smooth $T_g(\phi)$ function. Usually, it can well be approximated by the so-called Gordon Taylor formula (Refs. 270, 271)

$$A\phi_A(T_g(\phi) - T_{gA}) + B\phi_B(T_g(\phi) - T_{gB}) = 0 \tag{8.5}$$

with A, B the constants, and ϕ_A, ϕ_B the volume fractions with $\phi_A + \phi_B = 1$,

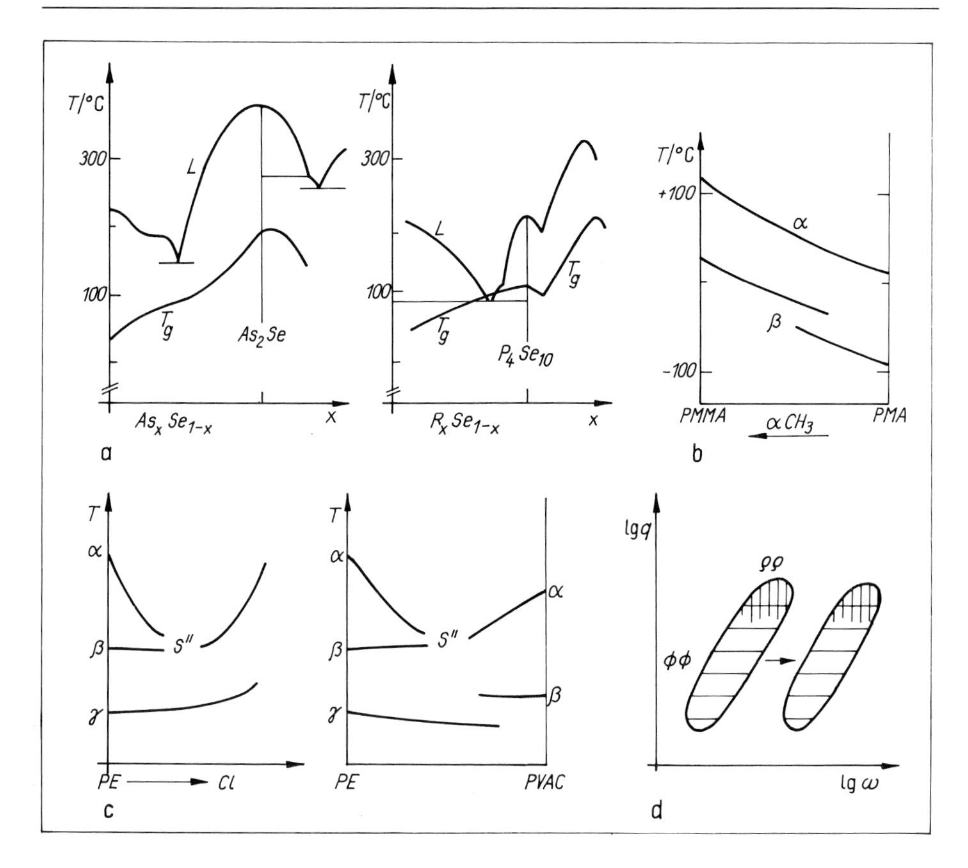

Fig. 53. Relaxation temperatures in mixtures and statistical copolymers.
a. Glass transition T_g and liquidus line L in two inorganic polymer-melt systems. b. Glass transition α and local modes β in the PMMA + PMA statistical copolymer system. c. Relaxation temperatures in statistically chlorinated polyethylene and in the PE + PVAC ("EVA") copolymer system. d. Dispersion law for mixtures. $\rho\rho$ density fluctuation, $\phi\phi$ concentration fluctuation, q scattering vector, ω frequency. The shift corresponds to a change of average concentration.

of course. As a rule, the ratio A/B usually observed makes the curves to be below the straight line between the T_g's of pure components.

This rule can be substantiated by Eq. 8.1. Compatibility is characterized by not too large values of Gibbs free mixing energy

$$\Delta G_m = \Delta H_m - T\Delta S_m. \tag{8.6}$$

For $\Delta H_m > 0$ this means a tendency to small ΔH_m and large ΔS_m. If ΔH_m

and ΔS_m are parallel to the ersatz values of ΔH and ΔS in Eq. 8.1, then smaller ΔH and larger ΔS mean lower $T_g = \Delta H/\Delta S$ for the mixture.

There are systems that do not follow Eq. 8.6 because of good conditions for specific interactions or specific order in certain ϕ regions. Positive deviations from T_g additivity are also observed, see Ref. 9.

The main transition α and the secondary relaxations β in the PMMA + PMA statistical copolymer system are shown in Fig. 53b (Ref. 164). Here ϕ could be replaced by the fraction of statistically distributed α methyl groups. The continuous $T_\alpha(\phi)$ function corresponds to the continuous deposition of α methyl groups in the cooperativity region. The local modes β, however, show a discontinuity. It is important whether a αCH_3 group with local entanglement tendency participates (higher activation energy corresponds to higher T_β, see Eq. 5.11) or not. Continuous substitution is not a conception for local modes with participation of only a small number of monomeric units. For mean αCH_3 concentrations even two local modes are observed: There are places where the αCH_3 groups participate, and there are other places where they do not.

Since the volume fraction ϕ is a ψ-level order parameter we expect larger characteristic lengths for concentration fluctuation than for density fluctuation of the ϱ level, see Fig. 53d. Larger length corresponds to a smaller mean time and to a broader spectrum. Exploring the ϕ fluctuation spectrum by a ϱ level method, e.g. by the specific heat C_p or the dielectric permittivity ε, then this smaller sond experiences some relaxation inhomogeneity. In this way the relatively broad transformation interval of C_p or ε curves in mixtures reflects the mean ϕ fluctuation $\delta\phi$ according to Eq. 6.17c. The Eq. 6.16 for V_a from δT and ΔC_p is not applicable without correction of this ϕ inhomogeneity. A relation between ϱ and ϕ fluctuation necessitates the knowledge whether the inhomogeneity can be described by a local relaxation time $\tilde{\tau}$ depending on local $\tilde{\phi}$, $\tilde{\tau}(\tilde{\phi})$, or not, see Sec. 8.8.

Statistical copolymerization of polyethylene PE leads to the demolition of the semi crystalline structure if, as usual, the new groups cannot or not easily be implemented in the lamellas. This loss of order results in decreasing T_g for the crystalline phase, see Fig. 53c (Refs. 272, 273). The disappearance of the last crystallites is indicated by a splitting point S″ where the glass transitions of the crystallites and the interfacial merge. After that the main transition temperature increases because of the higher energy due to the chlorine atom, T_g (PVC) > T_g (PE), or to the higher T_g of PVAC. The local process in the

PE + PVAC system shows a discontinuity that can be explained along the lines used for the PMA + PMMA system.

8.5 Relaxation in semicrystalline polymers

The relaxation cards of polyethylene PE and polyethylene oxide PEO are shown in Figs. 54a, b. The Arrhenius diagrams are roofed by a general local mode γ_{III} from which a larger number of cooperative processes branches off [see Figs. 50a, b for a comparison to amorphous polymers. Usually, the relaxations are labelled by α, β, γ, ... in direction of decreasing relaxation temperatures irrespectively of the physical nature of the underlying process. Therefore, the local (chain) mode is called β for the amorphous polymers and γ for the semicrystalline polymers.] Typical relaxation temperatures in the Hz range are listed in Table VIII.

The following discussion is restricted to PE, general issues are discussed in Ref. 275. It should be stated before that there are many types of PE differing not only by the molecular weight and its distribution but also in short and long chain branching (number, kind and distribution along the chain). Further crystallization parameters arise from the temperature–pressure–solvent–time program of the sample (Chap. 11). We have, therefore, a broad assortment of PE states which differ not only in the fractions of three phases according to Fig. 8d but also in their size, shape, perfection and spatial arrangement. Most of the relaxations (temperature, shape and intensity of response), especially α_c and $\beta(U)$, are rather sensitive to the crystallization parameters. This is indicated in the second row of Table VIII.

Based on a broad variation of crystallization parameters (see e.g. Refs. 276–279) arguments can be collected for the following assignment of the steep relaxations in Fig. 54a to the phases in PE: α_c is the glass transition in the layer crystallites (as is a glass transition in a plastic crystal, see Sec. 1.7), $\beta(U)$, the so-called upper glass transition, is the glass transition in the interfacial layer, and $\beta(L)$, the lower, is the main transition in the amorphous phase (hard to observe in PE, see Ref. 280). The more local processes (γ) can also be modified by the phase and, probably, by the phase boundaries. The considerable graduation of glass transitions in PE,

$$T_\beta(L) < T_\beta(U) < T_{\alpha c}, \tag{8.7}$$

can well be explained by Eq. 8.1: The main variation in PE is an increasing order in the corresponding phases, i.e. Eq. 8.7 is a consequence of the

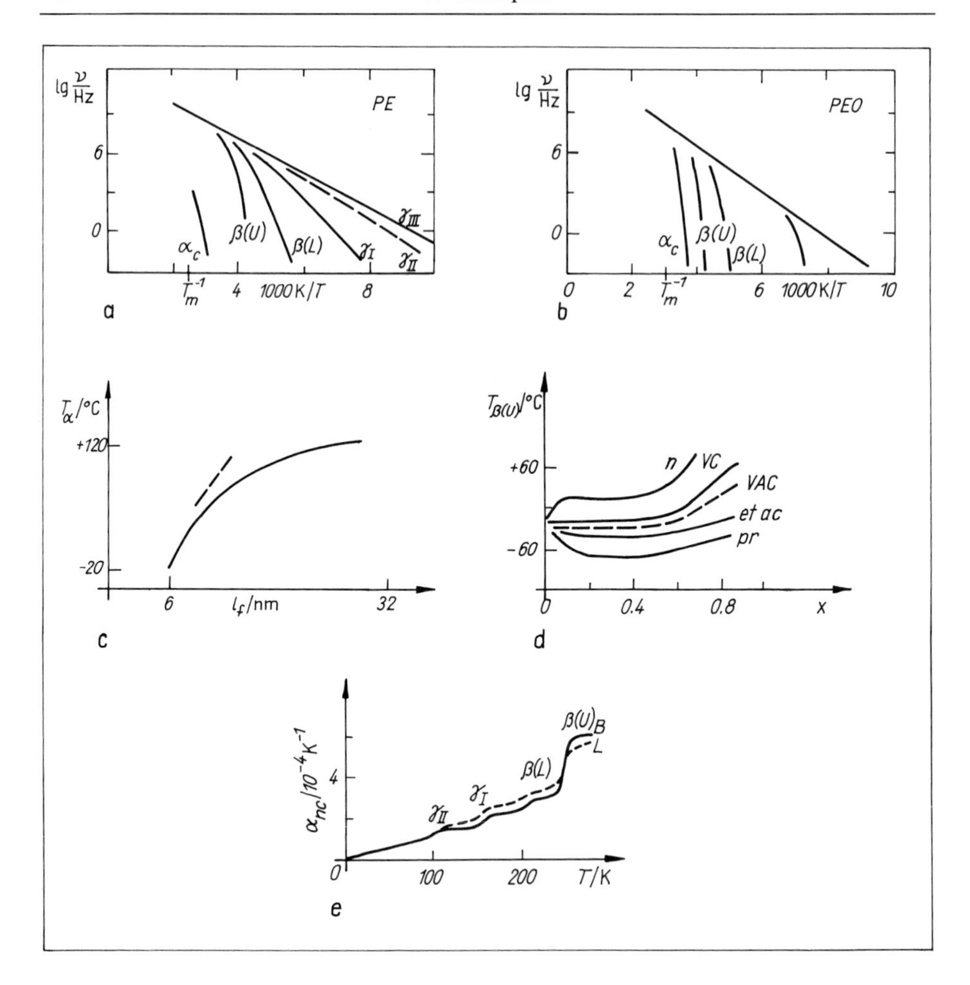

Fig. 54. Relaxation in semicrystalline polymers.
a. Arrhenius diagram in PE. b. Ditto in PEO. The flow transition is dropped in both relaxation cards. c. Correlation of glass temperature T_α in the crystallites with lamellar thickness l_f for PE. The exceptions are for LLDPE (Linear Low Density Poly Ethylene). d. Dependence of glass temperatures $T_{\beta(U)}$ in the interphase as a function of co-unit mole fraction: n norbornene, VC vinyl chloride, VAC vinyl acetate, etac ethylacetate, pr propylene. e. Thermal expansion coefficient for two ethylene types as a function of temperature. B branched, L linear.

Table VIII. Typical relaxation temperatures for two semicrystalline polymers (in °C)

polymer	α_c	β(U)	β(L)	γ_{I}	γ_{II}	γ_{III}
PE	70	−20	−80	−120	−150	−170
	−20 ... +120	−50 ... +20				
PEO*	30	−5	−60			−130

*from Ref. 274 where further relaxations are indicated also for PEO.

different ΔS in Eq. 8.1. The same order is expected for the Vogel temperatures in the phases. Eq. 8.7 does not describe short, median, and long glass transitions in the sense of Sec. 6.5, Fig. 37e.

The α_c temperature strongly depends on the lamella thickness l_f, see Fig. 54c, Ref. 278 (although there are significant deviations from this correlation for so-called LLDPE types, Ref. 281). The larger l_f gets the higher is $T_{\alpha c}$. Referring to Eq. 8.1 again, obviously, the order in the lamellas is the higher the thicker they are. This is supported by the observation (Ref. 33, cf. also Fig. 8g in Sec. 1.7) that the unit cell widening as a measure of disorder does also depend on the lamella thickness.

An alternative conception for the α_c transition is described in Ref. 282. It is assigned to the interfacial phase but triggered by rotational (screw) jumps of chains in the crystalline lamellas, like a marionette. But this is improbable because usually long chain parts do not play an essential role for glass transitions, and because the glass transition is a more cooperative than collective phenomenon.

The concentration dependence of $\beta(U)$ temperatures in the interphase, parameterized by the kind of the counit added, is shown in Fig. 54d (Ref. 283). Broad plateaus are observed in a large concentration interval. Since the $\beta(U)$ transition is a cooperative phenomenon obeying Eq. 8.1, the plateau means that the interphase is loosened up only in a different volume fraction but without increasing disorder. This view is supported by increasing relaxation strengths.

Usually, the thermal glass transitions from γ_{II} to $\beta(U)$ are well indicated by steps in the thermal expansion coefficient, often better than in the heat capacity (see also Ref. 274 for PEO). According to Eq. 7.2 these steps are freezing-in cross fluctuations between entropy and volume. Therefore, the freezing-in "order" fluctuation ΔC_p is accompanied by a strong density fluctuation that should well be indicated by compressibility steps.

Fig. 54e shows the thermal expansion $\alpha(T)$ for two PE types as measured by integral small angle X ray scattering (Ref. 284). Although the relaxation intensities of the same relaxation can be rather different in different types, the reduced sum $\Sigma(\Delta\alpha_i/\alpha_i)$, $i = 1, 2, 3, 4$, over the four non-crystalline relaxations is rather universal (Ref. 285). There is probably a fixed total free play for the order–density variation in the non-crystalline phases of PE that can, in dependence of crystallization parameters, differently be distributed to the less (γ_{II}, γ_I) and more ($\beta(L)$, $\beta(U)$) cooperative transitions.

8.6 Free volume and configurational entropy

The aim of the *free volume* conception is to connect the large variation of mobility lg ω (due to temperature or pressure change) with a variation of extensive variables such as volume or entropy. The basic idea is: the lower the density is the higher are the free play and the mobility.

According to Kovacs (Ref. 220) there are two aspects in the constructive part of the conception. (i) *volumes libres de relaxation*. Starting point is a well known ansatz for the liquid viscosity η (Refs. 286, 287) that represents $\eta(V,T)$ as a function of one variable f alone, $\eta(f(V,T))$,

$$\eta \sim \tau \sim \exp(B'/f) \tag{8.8}$$

(Doolittle equation) with f the *fractional free volume*, $[f] = 1$. This is the relative part of the total volume V that controls the mobility. Obviously, $f \ll 1$, and the large part $1 - f$ is occupied by effective molecular cores depending somewhat on the temperature T, due to the high frequency vibrations, and by the temporarily not accessible packing gores; $f = 0$ means $\tau = \infty$.

(ii) *volumes libres d'extrapolation* corresponds, in a sense, to the conception of fictive temperature. A linear ansatz is often used,

$$f = \alpha_f(T - T_\infty), \tag{8.9}$$

with α_f the thermal expansion coefficient for f; $f = 0$ means $T = T_\infty$, the Vogel temperature.

Combination of the two Eqs. 8.8 and 8.9 gives a surprisingly simple reason for the VFT and therefore for the WLF equation (Ref. 124), with

$$c_1^0 = B'/2.3f_0, \; c_2^0 = f_0/\alpha_f, \tag{8.10}$$

where $f_0 = f(T_0)$. Typical values with reference to the glass temperature $T_g = T_0$ are $B' \lesssim 1$ (e.g. $B' \approx 0.4$), $f_0/B' \approx 0.02 \ldots 0.04$, and $\alpha_f/B' \approx$ a few 10^{-4} K^{-1}. The numerical values show that very small changes, $\Delta f \ll 1$, a few per cent, have large effects in controlling the mobility.

Bounds of the free volume conception are set by the fact that the linkage between f and measurable density changes is rather uncertain. Neither $\alpha_f = \Delta\alpha$ is generally valid, where $\Delta\alpha$ is the relaxation intensity of thermal expansion at T_g, nor is WLF excluded for isochoric conditions (see J. Koppelmann in Ref. 164, p. 229). The free volume is a useful but not an ab initio concept. The core of the concept is cooperativity: The free volume can easily be redistributed, there is no or merely a minor energetic aspect, and the molecular nature of the energy landscape is referred to by a minimal void size of a few angstroms assumed to be necessary for elementary change-of-place processes (Ref. 288). The free volume is also the root of defect diffusion models and of the minimal coupling concept for the ideal glass transition (Sec. 6.4). Moreover, the free volume is used for the correlation of thermodynamic excess variables, see Sec. 9.4.

Many models and computer simulations provide the mobility necessary for thermokinetics and thermodynamics by the introduction of *voids* V, a very special variant of free volume. Sometimes the voids are vested with the component status as if they had an identity. A consequence of this investment is that they cannot have a chemical potential, rather

$$\mu_v \equiv 0 \tag{8.11}$$

because their number cannot arbitrarily be given but is an implication of the equilibrium condition, $\partial G/\partial N_V = 0$, like as for phonons.

The free volume and the configurational entropy S_c are compared in Figs. 55a and b, see also Sec. 5.4. Operationally they are equivalent, especially after linearization, $S_c \sim (T - T_\infty)$, similar to Eq. 8.9. Their utility is determined by the criterion whether thermokinetics (neither exactly determined by the one or the other) can be better correlated by S_c (Adam & Gibbs) or by V_c (Fig. 55a) or f. [V_c could also be called configurational volume.]

It is from experience that the configurational entropy is the better variant. A general reason is as follows. The essential quantity is the mobility $\ln \omega$. Because of the multifarious details of the energy landscape there is no crucial difference between probabilities W of energy steps and transition probabilities ω for energy barriers, $k \ln W \sim k \ln \omega$ (symmetry property of ideal glass transition). Therefore, $\ln \omega$ could also be named "thermokinetic entropy" (Ref. 141).

In terms of the ideal glass transition concept we can argue that the block

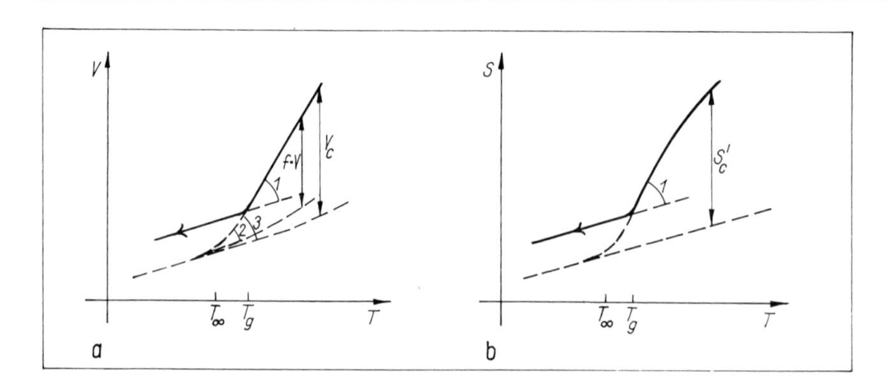

Fig. 55. Free volume and configurational entropy.
a. Free volume of extrapolation. The angles can be described as follows. 1 $\Delta\alpha_p(T - T_g)$, 2 $f = \alpha_f(T - T_\infty)$, 3 $\Delta\alpha_\infty(T - T_\infty)$. All the three angles are different, especially $\angle 1 \neq \angle 2$. V_c "configurational volume" defined by analogy to S_c'. b. Configurational entropy S_c' related to an equilibrium glass. $S_c' \approx S_c$ of Figs. 49 and 27f where S_c is related to the crystalline state. 1 $\Delta C_p(T - T_g)$. T_∞ Vogel temperature.

volume fluctuation ΔV_i in Eq. 6.10 can be substituted by a corresponding entropy fluctuation ΔS_i,

$$\Delta\omega_i = \Delta\omega_i(\Delta S_i). \tag{8.12a}$$

Then, since the entropy is an extensive variable as well, and since the fluctuations are reversible, we have, instead of Eq. 6.12,

$$\Delta S_a = \Delta S_1 + \Delta S_2 + \ldots + \Delta S_n. \tag{8.12b}$$

Eqs. 8.12a and 8.12b are an entropy variant of the minimal coupling.

The ideal glass transition is invariant against $\Delta V_i \leftrightarrow \Delta S_i$ (or $\Delta E, \ldots$), but the efforts to correlate the mobility $\ln \omega_i$ with thermodynamic variables are not.

The application of the Keesom Ehrenfest relations to the glass transition, the so-called Prigogine Defay ratio, can also be discussed with respect to this issue. These relations are implications of the fact that the variables of the First Law fundamental form are continuous at a second order phase transition u. We have a caloric relation

$$\left(\frac{dp}{dT}\right)_u = \frac{\Delta C_p}{TV\Delta\alpha_p} = \frac{\Delta(\overline{\Delta S^2})}{\Delta(\overline{\Delta S \Delta V})} \tag{8.13a}$$

and a thermal relation

$$\left(\frac{dT}{dp}\right)_u = \frac{\Delta\kappa_T}{\Delta\alpha_p} = \frac{\Delta(\overline{\Delta V^2})}{\Delta(\overline{\Delta S \Delta V})}. \tag{8.13b}$$

The Δ's before the mechanical coefficients (C_p, κ_T, α_p) are differences in the susceptibilities of the two phases, the Δ before the mean fluctuations are the differences between the mean fluctuations of the two phases, each being in dynamic equilibrium [with overlapping fluctuations in the minimal subsystems]; see also Eqs. 7.1 to 7.3 for the terminology.

Applied to a thermal glass transition ΔC_p means the relaxation intensity of the relevant functional subsystem. Then $\Delta(\overline{\Delta S^2})$ means the corresponding step in the spectral density

$$\Delta(\overline{\Delta S^2}) \rightarrow \overline{\Delta S^2} = \int_{(\Delta)} \Delta S^2(\omega)\, d\omega. \tag{8.14}$$

For these steps Eqs. 8.13a and b are not valid because the low temperature side ($T < T_g$) cannot be in mutual equilibrium with the high temperature side ($T > T_g$). Small deviations were found for the caloric, and large for the thermal relation (Ref. 289),

$$(dp/dT)_g \approx \Delta C_p / TV\Delta\alpha_p, \tag{8.15a}$$

$$(dp/dT)_g \approx (2 \dots 4)\Delta\alpha_p/\Delta\kappa_T. \tag{8.15b}$$

This means that the attempt to correlate relaxation intensities with thermodynamic variables is expected to be of lower success for the volume than for the entropy.

The Prigogine Defay ratio P is the quotient of Eq. 8.15a over Eq. 8.15b,

$$P = \frac{\Delta\kappa_T \Delta C_p}{(\Delta\alpha_p)^2 TV} = \frac{\overline{\Delta S^2}\,\overline{\Delta V^2}}{(\overline{\Delta S \Delta V})^2} \geqslant 1. \tag{8.16}$$

$P \geqslant 1$ follows from the Schwarz inequality of mathematics. From Eqs. 8.15a and 8.15b we see that $P \approx 2 \dots 4$. $P > 1$ indicates a certain decoupling of entropy and volume fluctuations in the relevant functional subsystem at T_g. This effect is on line with the tendency of the Third Law although the glass transition is not a quantum effect in the usual sense (cf. Sec. 3.7 to this point). $P > 1$ is rather a dispersion effect. Modelling the dispersion by a set of "internal parameters λ" clearly more than one λ is necessary to find $P > 1$ (Ref. 290).

Generally speaking, $P > 1$ implies that configurational entropy and configurational volume cannot simultaneously serve as a correlation basis for the glass transition.

8.7 Ionic conductivity

This section gives an example for an extreme violation of the Stokes Einstein relation which must be explained by an additional mechanism for the mobility of ions. To be precise we shall confine ourselves to the ionic conductivity in inorganic glasses. To understand the electric conductivity in polymeric insulators further conceptions must be included: Injection of charge carriers, field deformation by space and surface charge, and others.

The typical dc conductivity behavior of a glass with median conductivity is shown in Fig. 56a. The typical time for conductivity is derived from pictures of charges in an "electric landscape". The more frequent a jump of a charge is the shorter is the time, and the larger the conductivity σ, i.e. $\sigma \sim \tau^{-1}$. Conductivity and viscosity times are arbitrarily matched at high temperatures assuming a definite relation at high temperatures. The Stokes–Einstein relation ($\tau_\eta/\tau_\sigma \approx$ const) is violated at low temperatures, this ratio can reach large values (several logarithmic decades).

Some essential experimental findings are presented in a slight generalization: The dependence on substance (conductor ... insulator) is shown in Fig. 56b. For a given substance, Figs. 56c, d illustrate the dependence on temperature and frequency.

At high temperatures, the temperature dependence of σ, and, possibly, the mechanism of charge transport is largely determined by the molecular motion in the main relaxation (dynamic glass transition zone). At low temperatures, $T \lesssim T_g$, where the relaxation becomes very slow, the conductivity is comparatively high so that one has to think about additional, faster

---

Fig. 56. Ionic conductivity of glasses.
a. Comparison of characteristic times for conductivity σ and viscosity η, cf. Fig. 42e. b. Comparison of good, median and poor conductors. c. ac conductivity σ parameterized by frequency ω. d. Frequency dependence of conductivity. Parameter: dc conductivity (plateau $\sigma_0(T)$) in regime (i), increasing with temperature T. (ii) Jonscher regime, (iii) loss free region with exponent 1. e. Electric loss modulus M'' peak in the Jonscher region. f. Representation in the complex permittivity plot. Upper part: ohmic non-hindered dc conductivity. Lower part: partially free hopping also in regime (i). g. Electric to shear analogy.

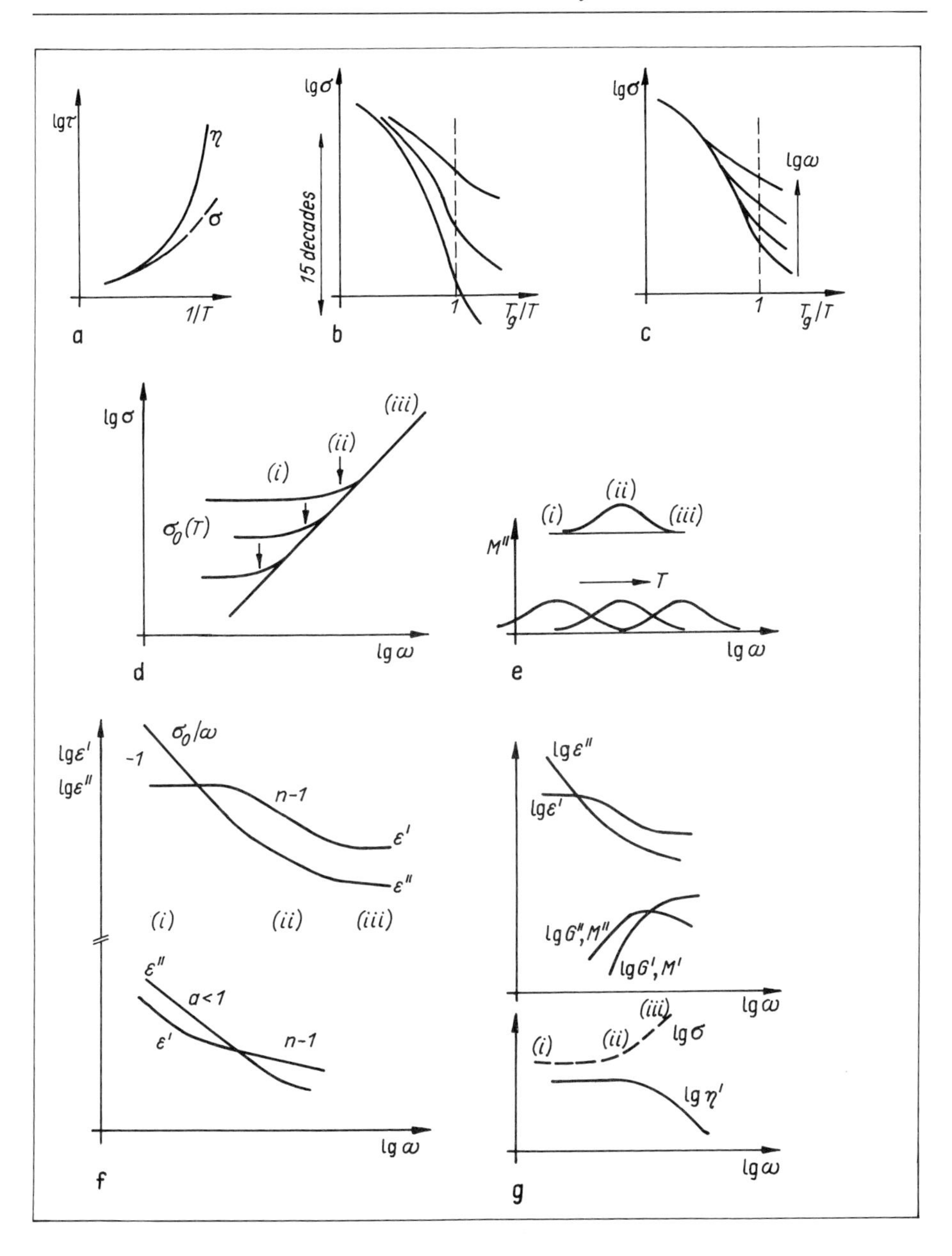
lgτ
η
σ
1/T
a
lgσ
15 decades
1
T_g/T
b
lgσ
lgω
1
T_g/T
c
lg σ
(iii)
(ii)
(i)
$\sigma_0(T)$
lg ω
d
(ii)
(i)
(iii)
M''
T
lg ω
e
lgε'
lgε''
σ_0/ω
-1
n-1
ε'
ε''
(i)
(ii)
(iii)
ε''
a<1
ε'
n-1
lg ω
f
lg ε''
lgε'
lg G'',M''
lgG',M'
lgω
(iii)
(i)
(ii)
lg σ
lg η'
lgω
g

mechanisms for charge transport: Thermokinetics is too slow to explain the conductivity experiments. The main conception is the electric landscape, i.e. a picture of conductivity bands in a thermokinetic structure.

As mentioned above, the conductivity also depends on frequency, the dynamic (ac) conductivity increases with higher frequency. The dispersion zone can be described by the middle one of three regimes, see Figs. 56d, e, f. They will be characterized by the real part σ of the complex conductivity

$$\sigma^* = i\omega\varepsilon_0\varepsilon^* \tag{8.17}$$

where ε^* is the dynamic electrical permittivity, a dielectric compliance.

(i) dc conductivity,

$$\sigma = \sigma_0(T) \sim \omega^0. \tag{8.18}$$

In the first regime the conductivity does not depend on the frequency ω, the temperature dependence is (or can be) described by an activation energy E_A^*, usually starred though real, of course. The complex permittivity in this regime is described by $\varepsilon^* = \varepsilon' + i(\varepsilon'' + \sigma_0/\omega\varepsilon_0)$.

(ii) Jonscher regime (Ref. 291),

$$\sigma \sim \omega^n, n < 1. \tag{8.19}$$

The frequency dependence is approximately described by an exponent n, the temperature dependence is usually described by an activation energy $E_A < E_A^*$ (see Fig. 56c), the ratio $E_A/E_A^* \approx 1 - n < 1$. This regime is characterized by a peak in the electric loss modulus M'', the imaginary part of $M^* = 1/\varepsilon^*$, see Fig. 56e. [As has been often mentioned above the modulus displays the restrictions against mobility, i.e. the resistance. The flow ability, i.e. the conductivity, however, is better displayed by the compliances, σ^* or ε^*, see Fig. 56f. Their interrelation by $M^*\varepsilon^* = 1$ is weak for the case of large changes across the dispersion zones so that they can be considered with a certain independence].

(iii) (So-called) loss free regime,

$$\sigma \approx \varepsilon''\omega \sim \omega, E_A < E_A^*. \tag{8.20}$$

There is no generally accepted picture about this behavior, the issue is still discussed controversially (see e.g. Ref. 292). Therefore, the situation will only be described by means of an analogy with the shear phenomena at the flow transition FT of polymers,

$$\text{(i)} - \text{(ii)} - \text{(iii)} \Leftrightarrow \text{FZ} - \text{FT} - \text{PZ}, \tag{8.21}$$

[although the spatial scale would be better represented by the glass transition in small-molecule glass formers, FZ – GT – GZ. The acronyms are FZ flow zone (viscosity η'), PZ plateau zone $\sim$ GZ glassy zone. The question of cooperativity remains open in the case of charge mobility.] Eq. 8.21 will be called electric-shear analogy of ionic conductivity (Fig. 56g). This analogy is not a discussion of the activity arrangement across the dispersion zone (the case $T < T_g$ is included for conductivity) but a comparison of viscosity for the thermokinetic energy landscape with conductivity for the electric landscape.

(i) The dc conductivity (σ = const) corresponds to the constant viscosity η' in the flow zone. The transport coefficients are determined in the dispersion (Jonscher) zone. This means that σ is determined by the transitions between more or less localized charge states.

(ii) The Jonscher regime corresponds to the flow transition. The loss modulus G'' maximum is, of course, at a higher frequency than the corresponding peculiarities in the compliances (maximum in $J'' - 1/\omega\eta$ for shear). The dispersion corresponds to the mobility generation by disentanglement if one relates entanglement points to localized charge states. The low exponents in ε' or σ ($n < 1$) are often associated with the picture of a partially (or hindered) "free hopping" of charges.

(iii) The loss free (or low grade loss) regime corresponds to the decrease of $\eta' = G''/\omega$ in the plateau zone (cf. Fig. 23a). Viscosity is here related to the modulus, $\eta^* = G^*/i\omega$, whereas conductivity is related to a complaince, see Eq. 8.17. According to the general scaling principle, a higher frequency means a shorter mode. Nevertheless, in the *mechanical* case the behavior in this zone is better described by the relevant (Andrade) compliance; the term quasi elastic is only used with regard to the rubbery plateau in the storage modulus G'. As compared to the transition itself the Andrade process likewise has a lower activation energy, see e.g. Fig. 41a for T = const. In the *electric* case the situation is described by the term quasi elastic motion "inside" the localized charge states. The current is a displacement current of restricted charge motions. [Special models are exponential conductivity and ion-pair dipole relaxation (Ref. 293), or thermally activated motions of ions between nearly equivalent potential wells (Ref. 294). Being a compliance, the dielectric activity of the short modes gives a large contribution to the ac conductivity. This picture is supported by an activation analysis from NMR measurements (Ref. 295). The short modes show the same activation energy in the relevant T–$\ln\omega$ regions as the corresponding dc or ac activities, respectively.]

The electric shear analogy helps to understand how particle (charged or not) motion is transformed into modulus and compliance activities. To explain a conductivity we have to make models for compliances which are much more inconvenient than models for moduli because of the larger mode lengths responsible for the former.

8.8 Glass structure

In this section we ask if there are principles for the glass structure that follow from the thermokinetics of the ideal dynamic glass transition. According to this approach the glass structure at $T < T_g$ is the frozen-in thermokinetic structure of the liquid in equilibrium near T_g (see Fig. 36d, and the negative Fig. 45f at T_g) modified by local relaxations remaining for $T < T_g$ and by internal strains due to different local thermal expansion coefficients of the parts with different structure. Three well known conceptions (Ref. 296a) can be substantiated by this approach.

1. Microcrystallinity. Inside of any cooperativity region of size $2\xi_a \approx 3$ nm (larger for higher fragility) there is, caused by the thermokinetic scaling, a place of high mobility due to low local density (defect, cage; see Secs. 6.4, 6.8, 7.2 and 7.6). This is the disperse structure of the Figs. 36d and 45f. The loosened structure there allows a more or better local order organized at cooling in the liquid above T_g. That is we find "microcrystalline clusters" in an amorphous matrix ("tissue") of higher local density, see Fig. 57a. In the clusters, the groups of coordination complexes (e.g. tetraeder) can have an order ("symmetry") that would not be allowed for long reaching periodicity, but a certain similarity to possible crystalline structures is usually assumed. As the length scale is 1 nm for the cluster as well as for the matrix significant differences in the second shell are expected to be observable so that the second peak of the radial distribution function $g(r)$ should show some fine structure, e.g. Fig. 57b. For $T > T_g$, the thermokinetic structure (see Fig. 23d) is characterized by the characteristic length ξ_α that depends on temperature according to Eq. 6.18. The number of clusters, therefore, decreases for lower temperatures according to $N_a^{-1} \sim (T - T_\infty)^2$. Quenched glasses have therefore more clusters and shorter distances.

2. Medium range order. Many glasses of interest are mixtures. There are

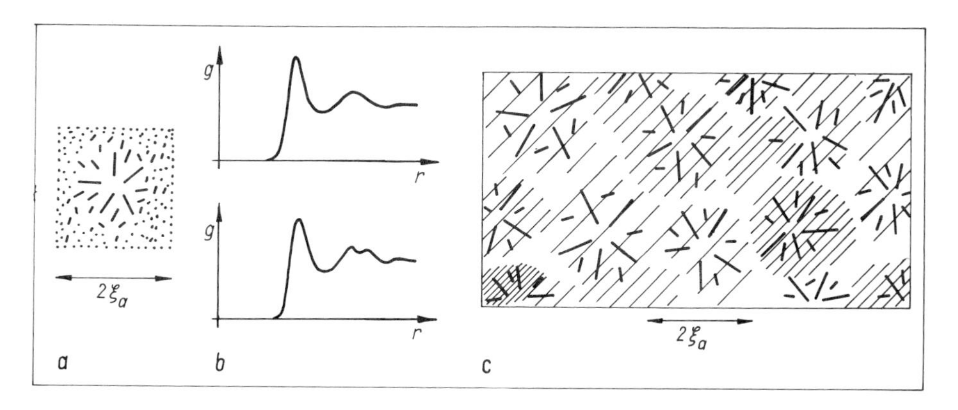

Fig. 57. Glass structure.
a. Structure of one cooperativity region. Inside, near the defect or cage, there is more structure (cluster) than in the denser matrix (tissue). b. The second-shell peak of the radial distribution function $g(r)$ can have a fine structure if the volume fractions of cluster and tissue are comparable. c. Larger-scale composition fluctuation leads to composition differences between neighbored ρ-level cooperativity regions.

two consequences. (i) The cluster can have a composition different from that of the matrix. Since the number of clusters changes during the cooling of the liquid a considerable local component (mass) transfer is expected to occur. This will be called *thermokinetic diffusion*. Its origin is thermokinetic scaling, and an explanation by chemical potentials seems to be questionable because of the small length scale ξ_a of the natural subsystem. (ii) As mentioned in Sec. 8.4 the spatial scale of composition fluctuation is larger than for density fluctuation, because the volume fraction ϕ is an order parameter. The clusters

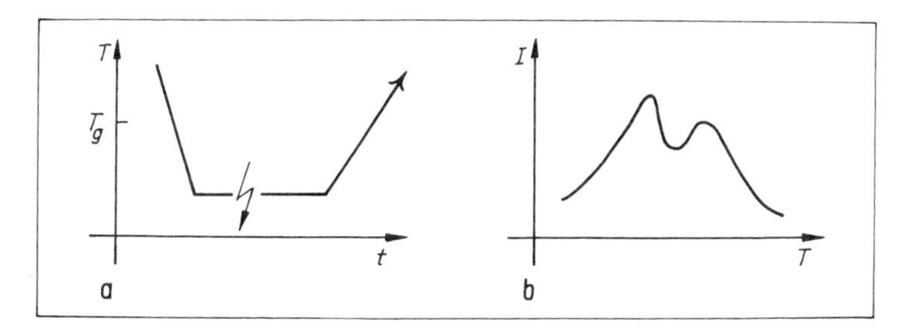

Fig. 57A. Thermoluminescence.
a. Temperature-time program. b. Thermostimulated light intensity I as a function of temperature.

and their environments, i.e. the ρ-level cooperativity regions, have therefore a different average composition with a ψ-level correlation length of about 5–7 nanometers, see Fig. 57c.

The spatial aspect of composition fluctuation is thus described by two parameters, one for the relation between the cluster and its neighbored matrix (thermokinetic diffusion), and the other for the ϕ correlation between the different cooperativity regions of density fluctuation. This prevents a simple interpretation of the relaxation spectra in mixtures. In general, they cannot be described by a local relaxation time $\tilde{\tau}$ depending on one local concentration $\tilde{\phi}$, $\tilde{\tau}(\tilde{\phi})$, see Sec. 8.4.

3. Anomalies. Consider a glass series with varying composition, e.g. the alkali substitution $LiO_2 \rightarrow NaO_2$ in silicate glasses (mixed alkali effect). With regard to the length scale of 1 nm the clusters consist of about 10 coordination complexes. If they form a definite structure then the change of it, due to the substitution, can be a discrete event in a small-number situation with possibly important consequences. Structure-sensitive properties (e.g. electric conductivity) may therefore have dramatic changes in a small concentration interval (which is somewhat enlarged by the compliance of the matrix and by the composition fluctuation of the cooperativity regions Fig. 57c).

Summing up: The glass structure is mainly organized in the liquid state. Some often discussed properties of glass structure can be considered as direct consequences of the thermokinetic scaling Eq. 6.13 (including a dispersion law, the general scaling principle) behind the WLF scaling of the ideal dynamic glass transition above T_g.

The same principles can also be applied to polymer structures as prepared by the short glass transition. Modifications are expected from two properties of polymers: the cooperativity regions, as a rule, are larger than in inorganic glasses ($2\xi_a \approx 4 \ldots 5$ nm for polymers), and the possibility to form definite cluster structures is reduced by the chain nature and by low tacticity. Therefore, the clusters ("nodules" in polymers) are expected to be somewhat larger, and the microcrystallinity and the anomalies are (much) weaker. A transition between micro (thermokinetic) and thermodynamic crystallinity may be represented by conventional polyvinylchloride.

8.9 Thermostimulation

The stepwise tawing of relaxations was presented in Fig. 54a for the example of thermal expansion. Such experiments are called *thermostimulation* if additional, external parameters are frozen-in in the history. Examples: (i) Thermostimulated depolarization (TSD). The sample is frozen-in under an electric field. The field is switched off at low temperatures, and depolarization steps are observed during heating. (ii) Thermoluminescence (TL, see Fig. 57Aa, b). A sample containing dyes is cooled down to e.g. 78 K. The activation (charge separation) is by X rays, for example, and the light intensity (from recombination of charge carriers after transport) is registered for the heating stage.

Care is necessary for the interpretation of thermostimulation in terms of relaxation. TSD, for example, can almost completely be reduced to the complex permmittivity, i.e. we learn something about the ε activity. But TL (or thermostimulated current, TSC, Ref. 295a) is a more complex phenomenon. Apart from molecular aspects (e.g. spectral distribution of light), not only the influence of tawing ("activated") relaxations on the charge transport (see Ref. 296) but also other aspects (electric landscape, trap distribution) should be considered. Furthermore, dyes can develop the host environment. The local induction of free volume can generate relaxations that are more characteristic for the polymer host than for a special dye guest although they are missing without the latter. Highly cooperative relaxations (e.g. the main or the flow transitions) are not active if all charges are already removed by local or coupled local processes at lower temperatures.

III Thermodynamics

As mentioned in the Introduction the analysis of thermodynamic data is less systematic than of relaxation data. The additional spectral parameter, the frequency or the time, is missing. Therefore the analysis of length scales, possible from linear response due to the general scaling principle, is nearly impossible from the thermodynamic data alone. They can report, by superposition, from different length scales. The compressibility equation Eq. 1.30 displays the central role of the direct correlation function for generating the thermodynamic variables. This function corresponds to the ρ level of one or a few nenometers. The energy equation Eq. 1.31 confirms this length: The longer-reaching radial distribution function is damped by the short range of the intermolecular potential.

The general situation is modified in polymers by the coil structure of chains, by entanglements, and for long chains, perhaps by the far reaching cooperativity of tube units. It is expected that the influence of the ψ level on thermodynamic properties is decreasing with larger lengths. We find an overall term, the combinatoric entropy of Sec. 3.3 and, of course, modifications from networks, order, crystalline phases etc.

The calculation of thermodynamic variables is confronted with, so to speak, strong interaction of several particles. The first shell contains about 10 particles, and the thermokinetic structure of the environment cannot always be embraced by a mean field correction. The extremely nonlinear and chaotic molecular situation is always linearly explored by usual thermodynamics being a limit case for the linear response. A quasistatic ersatz process always ensures that for natural subsystems the disturbance is smaller than the fluctuation. This fact is expressed by linear forms (named fundamental) for the potentials, e.g.

$$dG = -SdT + Vdp + \sum_B \mu_B \, dN_B. \tag{9.1}$$

The Gibbs free energy G is a function of the natural variables temperature, pressure, and particle (or mol) number of components B in mixtures (polymer blends), $G = G(T, p, N_B)$. The equations of state are rather complicated functions because the susceptibilities (second derivatives of G) are functions of p, T, and N_B (or ϕ_B, the volume fractions). The Gibbs map of a complicated molecular situation onto the thermodynamic functions is difficult to survey, and the inverse problem is not solvable at all. (One should think about the high dimensionality of the molecular situation as compared to the small number of thermodynamic variables.)

This lack of systematics is reflected by the state of the art: The thermodynamics of mixtures (in German: Mischphasenthermodynamik) is usually accompanied by some philosophy and, nowadays, by large computers. Accepting the situation the author has decided to bring examples. They are selected from thermodynamics of mixtures, compatibility, and partial crystallinity; kinetic aspects are increasingly inserted. The literature is terrific in extent, also because of the technical interest. The intention to represent the progress is leading to encyclopedic features (see Ref. 9, e.g.), especially when no colleague should be ignored. The reader is referred to this book (and the other reviews, see below) for the work not considered here.

9. Thermodynamics of mixtures

An essential point for polymer mixtures is the stability against phase decomposition. The local material or *diffusion stability* condition is given by the monotony of the chemical potential as a function of volume fraction,

$$(\partial \mu_B / \partial \phi_B)_{T,p} > 0. \tag{9.2}$$

Violation of this inequality leads to phase decomposition. The matter is complicated by metastability; similar to the situation of the van der Waals equation (there with respect to mechanical stability, $(\partial p / \partial V)_T < 0$) a phase decomposition can also be achieved for local stability Eq. 9.2, if we have a global equilibrium situation with a smaller thermodynamic potential G; the difference to the global equilibrium is called δG.

The mixture situation is shown in Figs. 58a–d, see also the review Ref. 297 used as a reference for several paragraphs of this chapter. It is useful to suppress terms linear in ϕ_B, $B = 1$ or 2, by definition of a ΔG_{m}, shortly ΔG,

$$\Delta G(\phi_2; T, p) \quad = \quad G(\phi_2, T, p) - \phi_1 G_{01}(p, T) - \phi_2 G_{02}(p, T), \tag{9.3}$$

where ϕ_1 in G is substituted by $\phi_1 = 1 - \phi_2$ and $G_{0\mathrm{B}}$ are the Gibbs free energies of the pure components B; $\Delta G = 0$ for $\phi_1 = 0$ and for $\phi_2 = 0$ (cf. Eq. 3.17a).

The stable case for isothermal-isobaric conditions is presented in Fig. 58a. The chemical potentials related to ΔG are denoted by $\Delta \mu_B$, the partial molar quantities of ΔG, which can be obtained by Roozeboom's tangent method as indicated in Fig. 58a. There is no composition where a δG could be gained from any phase decomposition.

Suppose a given pressure p. If the $\Delta G(\phi_2, T)$ surface has a plait (Fig. 58b), then a δG can be gained there by phase decomposition, see Fig. 58c. (A plait can be obtained from a model that can also represent unstable states). The global equilibrium is obtained from the *double tangent construction* of Fig. 58c, because the equilibrium condition between the two phases I and II, $\mu_B^{\mathrm{I}} = \mu_B^{\mathrm{II}}$, defines a common tangent at the $\Delta G(\phi_2)$ curve for given T. The

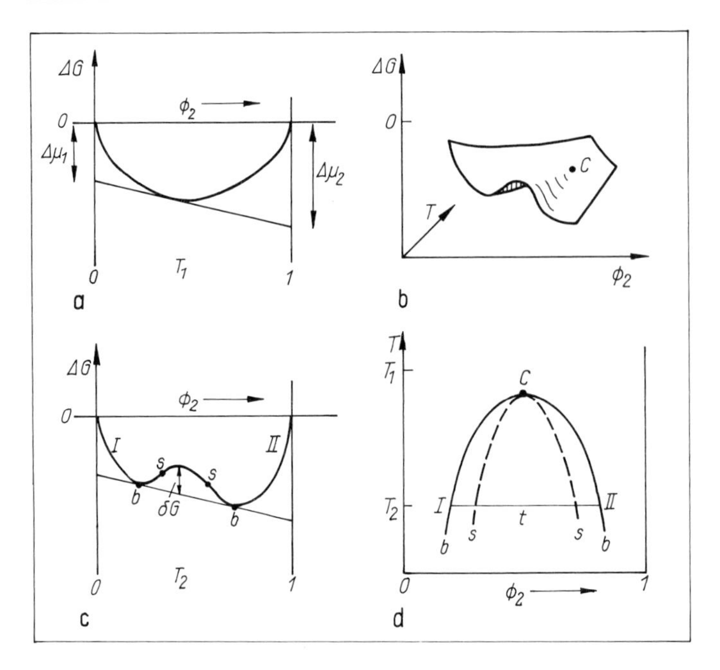

Fig. 58. State diagrams near phase decomposition in mixtures.
a. Gibbs free energy of mixing ΔG in a stable phase. $\Delta\mu_B$, $B = 1$ or 2, chemical potentials, ϕ_2 volume fraction of component 2. b. Plait in the ΔG surface over the $\phi_2 T$ plane, T temperature. c. Double tangent construction for determination of binodal (b) composition in phase I and II. δG gain of potential for phase decomposition, s spinodal (limit of local stability). d. Phase diagram of UCST type. C critical point, b binodal, s spinodal, t tie line for equilibrium at temperature $T = T_2$.

double tangent defines the two equilibrium phase compositions that are called binodal. The local stability limit

$$\partial\mu_B/\partial\phi_B = 0 \tag{9.4}$$

is represented by the inflection points of the $\Delta G(\phi_2)$ curve. The corresponding compositions are called *spinodal.* The projection of binodal and spinodal compositions in the $T\phi$ plane is shown in Fig. 58d. For any plait, of course, the spinodal is "inside" the binodal.

The two phases become identical at the critical point C where the spinodal and binodal have the same extremum (common horizontal tangent in the $T\phi$ or $p\phi$ plane). As mentioned in Sec. 1.4, for the case of a maximum in the $T\phi$ plane we have a UCST, upper critical solution temperature, for a minimum in this plane we have a LCST, lower c.s.t. The horizontal straight lines in a $T\phi$ or $p\phi$ diagram that connect equilibrium compositions on the binodals are called con(n)ods or tie lines.

The different stability concepts can also be explained by Figs. 58c and d. Outside b we have global and local stability, no δG, and Eq. 9.2 is valid. Between binodal b and spinodal s we have metastability: the local Eq. 9.2 is valid, but globally we find a δG. Inside the spinodals the mixture is locally and globally unstable.

On the plait and its vicinity the following tendencies are generally observed for the Gibbs free energy (since $\Delta G = 0$ for $\phi_1 = 0$ and $\phi_2 = 0$):

$$\begin{array}{l}\text{Increasing } \Delta G \text{ means increasing instability,}\\ \text{Decreasing } \Delta G \text{ means increasing stability.}\end{array} \tag{9.5}$$

At the spinodal the ϕ fluctuations become very large (turbidity, opalescence). This leads to modifications of the $\Delta G(\phi_2, T, p)$ surface near the spinodal. For polymers particularities arise from the fact that different length scales are differently sensitive to instability.

9.1 Mixing and excess variables

There are two basic conceptions in the thermodynamics of polymer mixtures: A transfer of methods developed for small-molecule mixtures (Sec. 9.1 and 9.2) and a special Flory-Huggins philosophy (Sec. 9.3).

The central object for correlations in small-molecular mixtures are Scatchard's excess variables (Ref. 298) that were soon transferred to the polymer mixtures (Ref. 299). Assuming that for a given temperature T and pressure p the pure components are in the same liquid state as the mixture, or can be extrapolated to this state for reference, and denoting the mole fractions by x_B and the reference variables for the pure components by the index $0B$, we define a *mixing variable* by

$$\Delta V^{\mathrm{M}} = V(x, p, T) - \Sigma x_B V_{0B}(p, T), \tag{9.6a}$$

$$\Delta H^{\mathrm{M}} = H(x, p, T) - \Sigma x_B H_{0B}(p, T), \tag{9.6b}$$

and so on (V volume, H enthalpy). For the free enthalpy $\equiv$ Gibbs free energy G the mixing variable $\Delta G^{\mathrm{M}} = \Delta G_{\mathrm{m}}$ of Eq. 3.16 and 9.3 with the substitution $\phi_B \rightarrow x_B$.

An *excess variable* is defined by the difference: mixing variable minus the ideal combinatorical term for the mixing entropy,

$$\Delta S_m = -R\Sigma x_B \ln x_B, \tag{9.7}$$

where it is relevant. Therefore,

$$\Delta V^{\mathrm{E}} = \Delta V^{\mathrm{M}}, \Delta H^{\mathrm{E}} = \Delta H^{\mathrm{M}}, \tag{9.8a}$$

but

$$\Delta S^{\mathrm{E}} = \Delta S^{\mathrm{M}} - \Delta S_{\mathrm{m}} = S(x, p, T) - \Sigma x_B S_{0B}(p, T) + R\Sigma x_B \ln x_B, \tag{9.8b}$$

$$\Delta F^{\mathrm{E}} = \Delta F^{\mathrm{M}} - T\Delta S_{\mathrm{m}}, \tag{9.8c}$$

$$\Delta G^{\mathrm{E}} = \Delta G^{\mathrm{M}} - T\Delta S_{\mathrm{m}}. \tag{9.8d}$$

A mixture with zero excess variables ($\Delta S^{\mathrm{E}} \equiv 0$ and $\Delta H^{\mathrm{M}} \equiv 0$ and $\Delta V^{\mathrm{M}} \equiv 0$) is called *ideal*. This is a reasonable term for small-molecule mixtures because Eq. 9.7 for ΔS_{m} has a clear physical meaning, and because mixtures of similar molecules tend to the ideal case. But for polymer mixtures the term ideal is not so clear because the combinatoric term Eq. 3.15 is problematic, see Secs. 3.3 and 3.4. Nonetheless the Flory Huggins parameter χ is often discussed like an excess free energy or a mixing enthalpy.

Let us mention the following properties: Per definitionem all the mixing and excess variables are zero for any pure component ($x_B = 1$). According to the tendency Eq. 9.5 phase decomposition can be expected for $\Delta G^{\mathrm{E}} > RT/2$, see Eq. 3.23. From

$$T\Delta S^{\mathrm{E}} = \Delta G^{\mathrm{E}} - \Delta H^{\mathrm{M}} \tag{9.9a}$$

we get, using $S = -(\partial G/\partial T)_p$, the Gibbs-Helmholtz equation,

$$\Delta H^{\mathrm{M}}(p, T) = \Delta G^{\mathrm{E}} + T(\partial \Delta G^{\mathrm{E}}/\partial T)_p, \tag{9.9b}$$

expressing the enthalpy H in the natural variables p, T for G.

There is a second method from thermodynamics of small-molecule mixtures: the *equation-of-state method*. If for given p, T a pure component is far from the liquid state, e.g. a supercritical gas, then it is difficult to define a hypothetical liquid reference state. Then (and also for the normal case, of course) one can construct an equation of state $G = G(x, p, T)$ for the whole, or a large, fluid region, including metastable and locally unstable regions, from which the phase equilibrium can be calculated. High standards are required for the precision. Starting from $G(x, p, T)$, the pVT data come from a differentiation by p, $\partial/\partial p$, the caloric data from $\partial/\partial T$, and the chemical potentials from $\partial/\partial x$. Furthermore a resonable shape of the $G(x, p, T)$ function is also necessary for the environment of the region of interest since technical screening or optimization procedures can lose their way and shall come back.

For a mixture of "pure liquids" the two methods are, of course, not true alternatives; one can calculate excess properties from $G(x, p, T)$ and vice versa.

The following strategy outlines dominant activities of thermodynamics, and there are prominent masters for reference (e.g. Refs. 300, 301).

(1) Decision between equation of state or excess variable approach, the latter includes the appropriate choice of reference states.

(2) Construction of functions for $G(x, T, p)$ or $\Delta G^{E}(x, T, p)$ or $\chi(x, T, p)$.

(a) Number and character of constants (fewer constants mean better extrapolation, more constants higher precision, and higher theoretical input better prediction).

(b) Physical intuition in the sense of van der Waals.

(i) Generalization according to the correspondence principle of thermodynamics: All fluids behave similar and the remaining differences can well be correlated by few parameters such as acentric factor or dipole moment. Critical parameters (T_c, V_c, p_c) or potential parameters (ε, σ) are often used for scaling. (ii) How to modify the hard sphere equation of state, being a first approximation for the liquid structure, to describe the fluid behavior at large pressures and temperatures? (iii) How to mix the constants in the equations and how to combine the potential parameters to calculate (or estimate) binaries from pure, ternaries from binaries, and so forth? (iv) What about shape factors to take the mutual orientation of molecules into consideration? (v) Correction for nonadditivity of intermolecular potentials and for intramolecular contributions. (vi) Decomposition of larger molecules into groups to consider a mixture of group contributions instead of a mixture of molecules, or into segments, parts of surfaces, and so on. Can these contributions be correlated across mixtures of different substance classes? (vii) Nonrandomness of local concentrations up to association.

This list could be enlarged.

The result of point (2) is an equation with (often) many constants and ingenious structures to withstand the necessary differentiations and to allow a certain extrapolation in x, T, p and perhaps to higher numbers of components. It can only be handled with computers.

(3) Testing the function.

(a) Flexibility: Which complexity of phase diagrams (e.g. two miscibility gaps) can actually be represented?

(b) Precision of reproduction for the system of interest.

(c) Ability for extrapolation and/or prediction (new state regions, higher number of components, known systems that are not used for the construction). If the testing stage is positive, usually after some iterations of (2) and (3), then one has to make

(4) Programs to gain the constants from molecular properties, thermodynamic data, and so on. Usually several (or better, only a few) of the constants or parameters are released for adjustment. As a rule, they then lose, more or less, their original physical meaning (if they had one at all), so that no exact conclusion about the inter (and, for polymers, intra) molecular situation is possible even from an excellent reproduction, see also the discussion of computer simulation in Sec. 4.3.

But, nonetheless, good equations of states are extremely useful in the hand of their masters. For polymers, the state of the art is reviewed in Refs. 9 and 302. A few successful equations should be given by their names: Flory, Orwoll, Vrij (Ref. 303, hard core, interaction per segment, free volume), Koningsveld, Onclin, Kleintjens (Ref. 304, interacting segment surfaces, gas-lattice model), and Simha e.a. (Ref. 305, cell-hole model, free volume).

9.2 Small-molecule mixtures

To prevent exaggerated hopes for a too simple interpretation of mixing or excess functions we shall consider the simplest case, a liquid mixture from Lennard Jones molecules with different potential parameters, ε and σ (Lennard Jones mixture). The interaction is defined to be in pairs additive by the potential of Fig. 59a,

$$\varphi(r) = 4\varepsilon[(\sigma/r)^{12} - (\sigma/r)^{6}], \tag{9.10}$$

where the three possible pairs are labelled by the indices 11, 12, and 22. The mixing and excess variables are illustrated in Figs. 59b–e as a function of parameter ratios for a mole fraction $x = 0.5$ (symmetrical mixture), a temperature $T = 0.726\ \varepsilon_{12}/k$, and for the symmetrical combination rules

$$\varepsilon_{12} = (\varepsilon_{11}\varepsilon_{22})^{1/2},\ \sigma_{12} = (\sigma_{11} + \sigma_{22})/2. \tag{9.11}$$

The surfaces were obtained from computer simulations (Ref. 306). It is

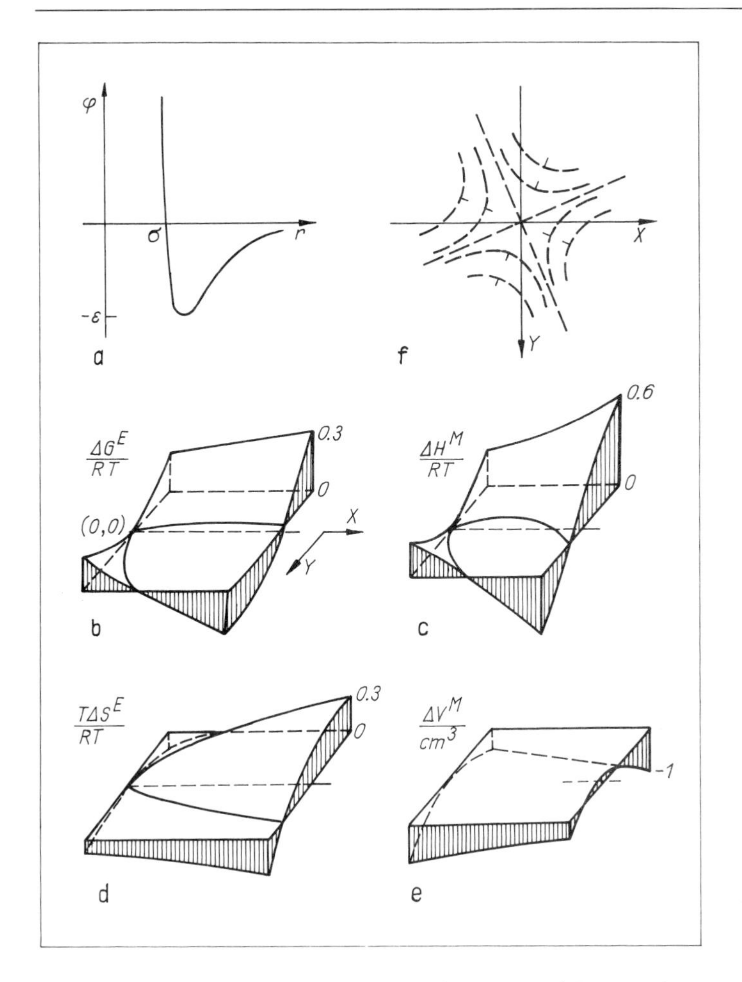

Fig. 59. Mixing and excess properties for Lennard-Jones mixtures.
a. Definition of potential parameters σ, ε. $\varphi(r)$ intermolecular potential as a function of distance between a pair of molecules. b. Reduced excess Gibbs free energy $\Delta G^{\mathrm{E}}/RT$ as a function of difference in diameter ($X = 1 - \sigma_{11}/\sigma_{12}$) and energy ($Y = 1 - \varepsilon_{11}/\varepsilon_{12}$). c. Ditto for mixing enthalpy $\Delta H^{\mathrm{M}}/RT$. d. Ditto for excess entropy $\Delta S^{\mathrm{E}}/R$. e. Ditto for mixing volume per mole ΔV^{M}. f. Contour lines for a saddle rotated against the frame.

surprising that they are so complex for small parameter variations in the order of 10 per cent,

$$\sigma_{11}/\sigma_{12} \equiv 1 + X = 1.00 \ldots 1.12,$$
$$\varepsilon_{11}/\varepsilon_{12} \equiv 1 + Y = 0.81 \ldots 1.235. \quad (9.12)$$

The first order contribution is zero for the symmetrical combination rules Eq. 9.11, so that we have saddles of second order in the lowest approximation for $\Delta A^E = \{\Delta G^E, \Delta H^M, \Delta S^E, \Delta V^M\}$,

$$\Delta A^E \approx aX^2 + bXY + cY^2. \quad (9.13)$$

They are rotated against the frame axes (Fig. 59f). For example it is a considerable difference whether the larger molecule (index 1) has the larger energy $Y > 0$ – then ΔH^M, $T\Delta S^E$ and $\Delta G^E < 0$ for $Y = 0.235$, better optimum – or whether it has the smaller energy $Y < 0$: then ΔH^M, $T\Delta S^E$ and $\Delta G^E > 0$ for $Y = -0.19$. This means that the b cross term in Eq. 9.13 is of considerable significance. Furthermore the quadratic saddles are modified by higher order terms even for the small X and Y values of Figs. 59b–e. The mixing volume ΔV^M is positive in a small XY region, against the expectation from packing effects that implies $\Delta V^M < 0$.

The theory of liquids and liquid mixtures was developed in the Sixties (see e.g. Ref. 307). Roughly, the properties of liquids are determined by two things. The steep repulsion potentials with their details in the orientation dependence make the structure, and the attraction potentials with their wells inside the first shell care for the general connection in the condensed-matter state. The dependence on X and Y corresponds, in a way, to the structure. The general condensation, in another way, can be expressed by an additive deviation from the symmetrical combination rules Eq. 9.11. Let this break of symmetry be denoted by $\delta\varepsilon_{12}$ and $\delta\sigma_{12}$, e.g. $\varepsilon_{12} = (\varepsilon_{11}\varepsilon_{22})^{1/2} + \delta\varepsilon_{12}$, then the surfaces of Figs. 59b–e are raised or lowered, approximately as a whole, for $x = 0.5$ without rotation or deformation. The computer simulation results in the following values per mole

$$\delta(\Delta G^E/RT) \approx -4.2\delta\varepsilon_{12}/\varepsilon_{12} - 1.5\delta\sigma_{12}/\sigma_{12} \quad (9.14a)$$

$$\delta(\Delta H^M/RT) \approx -6.0\delta\varepsilon_{12}/\varepsilon_{12} + 0.0\delta\sigma_{12}/\sigma_{12} \quad (9.14b)$$

$$\delta(\Delta S^E/\mathrm{RT}) \approx -1.8\delta\varepsilon_{12}/\varepsilon_{12} + 1.5\delta\sigma_{12}/\sigma_{12} \quad (9.14c)$$

$$\delta(\Delta V^M/\mathrm{cm}^3) \approx -5.7\delta\varepsilon_{12}/\varepsilon_{12} + 45\delta\sigma_{12}/\sigma_{12}, \quad (9.14d)$$

with $\sigma \approx 0.3\,\mathrm{nm}$ as for argon.

This means that the general structure of mixing or excess functions for $x = 0.5$ is

$$\Delta A^{E} \approx aX^2 + bXY + cY^2 + \cdots + \chi_1 + \chi_2 + \cdots \tag{9.14e}$$

where the first terms stand for "structure" variations and the second terms for "condensation" variations. Only the latter are approximately additive and can additively be interpreted to be "specific" interactions or long-reaching attractive potentials. The relation between the mixing properties at $x = 0.5$ and the quadratic structure terms can only be separated in case of dominance of one particularity, e.g. $|X| \gg |Y|$. In general, therefore, different origins cannot be superimposed additively to a mixing or excess function.

Using the equation-of-state method it was found (Ref. 306) that the pictures of Figs. 59b–e can well be reproduced by the so-called one-fluid model. One starts from a good equation of state for the pure Lennard Jones components scaled, according to the correspondence principle, by potential parameters σ_x and ε_x depending on the mole fraction x,

$$\Delta F^{E}(V, T, x) = N_A \varepsilon_x f(\tilde{V}, \tilde{T}) \tag{9.15a}$$

with

$$\tilde{V} = V/\sigma_x^3, \; T = T/(\varepsilon_x/k) \tag{9.15b}$$

where the so-called van der Waals *mixing rules* are applied,

$$\varepsilon_x = \Sigma x_i x_j \varepsilon_{ij} \sigma_{ij}^3/\sigma_x^3, \; \sigma_x^3 = \Sigma x_i x_j \sigma_{ij}^3. \tag{9.16}$$

These rules would be obtained if the van der Waals constants a and b of Eq. 3.12 were mixed like σ_x^3 in Eq. 9.16. Then the mixing and excess functions are calculated from Eq. 9.15a and compared with the pictures after calculating X, Y, $\delta\varepsilon_{12}$ and $\delta\sigma_{12}$ from Eq. 9.16. Usually, small post corrections for $\delta\varepsilon_{12}$ are necessary. [This paragraph was not intended to recommend a special model but to illustrate what is meant by the terms scaling (reduction) and mixing rules.]

All perturbations to Lennard Jones mixtures, e.g. originating from the orientation dependence of intermolecular potentials by fixed averages of X, Y, $\delta\varepsilon_{12}$, and $\delta\sigma_{12}$, lower the thermodynamic potential $\Delta G^{E}(p, T, x)$ because the linear terms are canceled in the excess and mixing terms and the second-order terms are generally negative. According to the tendency of Eq. 9.5 the miscibility is improved by such terms.

Let us make some outlook on polymer mixtures referring to the Flory Huggins χ parameter introduced in Secs. 3.3 and 3.4. For long chains, i.e.

large N, the combinatoric term Eq. 3.15 is small so that small χ's of order N^{-1} decide about miscibility, see Eq. 3.24. Small variations of χ are therefore of great interest for polymers. Even if a Lennard Jones mixture is not an excellent first order approximation for a polymer blend, the following comparison may be useful.

For a symmetrical polymer mixture the critical χ is $\chi_c \approx 2/N$, see Eq. 3.24 again. This means $\chi_c \approx 0.02$ for $N \approx 100$. Considering only one $\chi_1 = \chi$ for $\Delta G^E/RT$, then, according to Eq. 9.14e, the miscibility could be induced or prevented also by small variations X, Y, $\delta\varepsilon_{12}/\varepsilon_{12}$, or $\delta\sigma_{12}/\sigma_{12}$ of order a few per cent (see Fig. 59b).

We can draw three conclusions from this comparison: (i) it is not necessary to think immediately about specific interactions by a discussion about compatibility/incompatibility in polymers. (ii) There are surely (see Fig. 59d) also noncombinatoric contributions to the excess entropy of order $\pm$ some tenths of R for polymer blends. (iii) It is the smallness of the ψ level term Eq. 3.15 that enlarges the importance of ϱ level terms for the thermodynamics in polymer mixtures.

9.3 Flory Huggins again

In polymer science the Flory Huggins χ parameter plays a great role that goes far beyond the role of the Porter constant A in $\Delta G^E = Ax_1x_2$ for small-molecule mixtures. Roughly speaking, the reason is that in many cases χ can be considered as the only ϱ level mixing parameter, in the sense of $\delta\varepsilon_{12}$ in Eq. 9.14a, which can be transferred ("by the long chains") from the ϱ into a ψ level treatment of polymer mixtures. The other aspects of Eqs. 9.13 and 9.14a (X, Y, $\delta\sigma_{12}$, . . .) can sometimes be ignored or otherwise be corrected, e.g. by an adjustment of χ.

Historically, such an interpreptation goes back to a very rough treatment of the so-called *Flory Huggins model* as defined in Fig. 60. The segments of flexible chains of kind 1 (○) and 2 (●) are placed on the sites of a rigid lattice. Sometimes a certain number of voids is added as a third "component" to facilitate the mobility. Attractive forces are assumed to act between the nearest neighbors that are not connected by the chains, ε_{11}, ε_{12}, and ε_{22}, see Fig. 60 again. The repulsive forces and the chain stiffness are modelled by the lattice distance a of order the structure length Eq. 1.37.

This model is interesting and its thermodynamic properties can be cal-

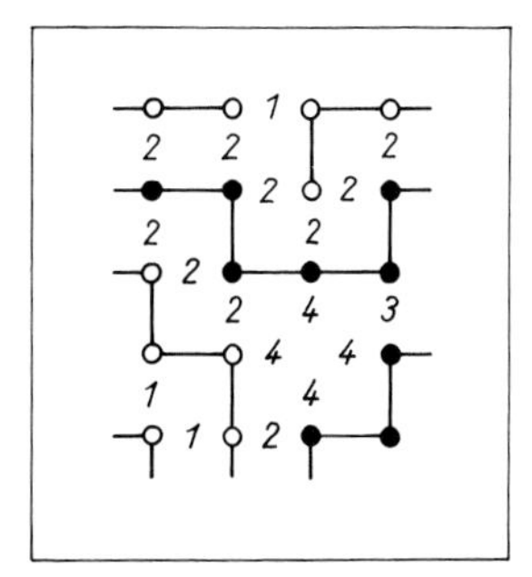

Fig. 60. Flory Huggins model for a two dimensional lattice. The figures indicate the kind of interaction energy: 1 ε_{11}(○○), 2 ε_{12}(○●), 3 ε_{22}(●●), 4 zero: (○ void) or (● void).

culated by computer simulation (see e.g. Ref. 308). The result is, for instance, a mixing Gibbs free energy that depends on the set of parameters, $\Delta G^M(\phi_1, T, \varepsilon_{11}, \Delta\varepsilon, \varepsilon_{11}/\varepsilon_{12}, N_1, N_2, \phi_V)$ and the type of lattice, where ϕ_1 is the (volume) fraction of chains 1, ϕ_V ditto for the voids, N_1, N_2 the length of chains 1, 2, and

$$\Delta\varepsilon \equiv (\varepsilon_{11} + \varepsilon_{22} - 2\varepsilon_{12})/2. \tag{9.17}$$

This function can be modelled by the Flory Huggins *equation* Eq. 3.17,

$$\frac{\Delta G^M}{RT} = \frac{\phi_1}{N_1} \ln \phi_1 + \frac{\phi_2}{N_2} \ln \phi_2 + \chi\phi_1\phi_2, \tag{9.18}$$

only in a very rough approximation, if χ is not allowed to depend on ϕ. Eq. 9.18 has the classical, mean-field critical exponents, whereas the Flory Huggins *model* gives the nonclassical critical exponents in Eq. 4.44.

The term *Flory Huggins approximation* is used for the rough treatment of the Flory Huggins model that just results in Eq. 9.18 with

$$\chi = \chi_0 \equiv -(z - 2)\Delta\varepsilon/RT \tag{9.19}$$

per mole segments + voids, with z the coordination number of the lattice. Comparing formally with Eqs. 9.14a,b this would correspond, for $x_1 x_2 = \phi_1\phi_2 = 1/4$, to

$$4.2\delta\varepsilon_{12}/\varepsilon_{12} \approx (z - 2)\Delta\varepsilon/4\,kT \tag{9.20a}$$

or

$$6.0\delta\varepsilon_{12}/\varepsilon_{12} \approx (z - 2)\Delta\varepsilon/4\,kT \tag{9.20b}$$

depending on the start either from ΔG^E or ΔH^M.

This corresponds to the standard interpretation of χ_0: smearing out of short-range potentials (ρ level) for applying to ψ level mean field approximations. Interesting details are suppressed by this approximation. Not only many possibilities of the Flory Huggins model (such as lowering of energy and entropy by a decrease of the number of 12 contacts, or clustering of voids and other correlations, see Refs. 309, 310), but also many possibilities from a variation of segment sizes and shapes, particularities of conformations, off-lattice effects, specific interactions etc. are not available in the χ_0 approximation. Obviously very delicate ϕ dependencies of a $\chi(\phi)$ function are necessary for modelling the details of thermodynamic properties of polymer blends and solutions.

A positive example for the χ_0 philosophy is the combined correlation of thermodynamic and scattering data in the ψ level that uses only average correlations from the ρ level, see Sec. 4.1.7. The RPA method that uses a quasi-thermodynamic treatment of the fictitious external fields anyhow deals exactly with a mean field interpretation of the difference in the reciprocal structure factors of Eq. 4.30, similarly as in Eqs. 9.14a–d and 9.19. This corresponds to the rough treatment of the ρ level by the a^2 term of Eq. 4.31, viz.

$$\frac{1}{S_{\text{coll}}(q)} \approx \frac{1}{\phi_1 N_1} + \frac{1}{\phi_2 N_2} - 2\chi_0 + \frac{1}{18} \cdot \frac{a^2}{\phi_1 \phi_2} \cdot q^2. \tag{9.21}$$

The segment length a (= structure length according to Eq. 1.37) and some contributions to χ_0 are the only indications of the ρ level, and no X, Y, . . . from Eq. 9.14e is reflected.

The set of Eqs. 9.18, 9.19 and 9.21 is therefore a conceptually consistent ψ level description of thermodynamic and structural mixture properties with a rough ρ level correction.

Improvements of the ρ level contributions are suggested, e.g. via the direct correlation functions $c_{AB}(q)$ of the pairs (Ref. 77). The equation obtained,

$$\chi_{\text{eff}}(q) \approx \frac{\rho}{2}\,(c_{11}(q) + c_{22}(q) - 2c_{12}(q)), \tag{9.22}$$

is in striking similarity to Eq. 9.17. The dependence on the scattering vector $\boldsymbol{q}$ reflects some nanometer structure input for the thermodynamic variables.

The way to Eq. 9.22 uses pair distribution functions $g_{AB}(q)$ for the scattering function. Using the Ornstein-Zernicke equation Eq. 1.29 (see also Eq. 1.28), $g_{AB}(q)$ is substituted by the short-ranging direct correlation function.

Corrections are made for intramolecular correlations along the chains, and less important parts of $c_{AB}(q)$ are modelled directly by the intermolecular potentials φ_{AB}. The connection to the free energy is made by RPA equations like Eq. 4.29.

Another example is described in Ref. 311, where off-lattice fluctuations are discussed in terms of the local structure asymmetries (like X, Y of Eqs. 9.12a,b).

9.4 Polymer-polymer mixtures and polymer solutions

This section is to illustrate some typical polymer aspects in the phenomenological treatment of mixtures and solutions.

Besides the complicated statistics further difficulties arise from the fact that an exact knowledge about the details of inter and intra (conformational) molecular potentials is not available.

The general approach is illustrated by an example selected from many similar ones by its relationship to the approach of Secs. 9.2 and 9.3 (Ref. 312). The aim is to introduce the energy and size ratios of segments (like X and Y according to Eq. 9.12) and a mean "specific" interaction (like as in Eqs. 9.14a) in a way compatible to the chain nature of the mixture components. On balance of the conceptions of Sec. 9.1 the authors decided to use mixing rules from small-molecule mixing theory (e.g. Ref. 313) applied now to segments. The chain nature is contained in the Flory Huggins combinatoric entropy (used for the definition of excess variables, instead of ΔS_{m}), in a (simplified) distribution function for segments in a chain, and in a Flory equation of state. All that is expanded into a Taylor series up to second order in the differences, with coefficients A, B, C coming now from the Flory equation. A saddle rotation ot the main axes $(X, Y) \rightarrow (X', Y')$ for the volume fraction $\phi = 0.5$ gives

$$\Delta G^{\mathrm{E}} \approx AX'^2 + BY'^2 + C\Delta\varepsilon \tag{9.23}$$

where, roughly, the A term describes size differences of segments, the B term energy differences (the coefficient B is calculated mainly from the difference in thermal expansion of the pure components – a free volume effect), and the C term describes the segmental interaction as $\delta\varepsilon_{12}$ in Eq. 9.14a or b. The structure of Eq. 9.23 corresponds to Eq. 9.14e.

As the mixing procedures for the parameters induce very complicated ϕ dependencies a great flexibility can be obtained. This is necessary to model phase diagrams similar to Fig. 61f, below. But the flexibility is controlled by two things: The correspndence principle used in the mixing rules, and the simple Flory equation of state, namely

$$\tilde{p} = \frac{\tilde{T}\tilde{V}^{-2/3}}{\tilde{V}^{1/3} - 1} - \frac{1}{\tilde{V}^2}, \tag{9.24}$$

where the reduction ($V \to \tilde{V}$ etc.) is made by the segment potential parameters.

The temperature dependence of the three terms in Eq. 9.23 is shown in Fig. 61a ($\phi = 0.5$). Such a diagram was first discussed in Ref. 313a. The parameter C is negative for specific attractive interactions. This corresponds to $\delta\varepsilon_{12} > 0$ in Eqs. 9.14a,b. The decrease of the amount $|C|$ follows from the Boltzmann scaling U/kT in the Gibbs distribution, enlarged by its exponential character. The temperature dependence is further steepened by frustration effects due to the orientation dependence of the molecular potentials.

Both A and B are obtained to be positive, against the expectation from the saddle surface Fig. 59b,f for Lennard Jones mixtures. The following interpretation is rather a speculation than a description of facts. The A term estimates the size differences, the structure. Only a weak effect in the X direction is observed for Lennard Jones mixture. But the structure is heavily modified by polymer chains. The steep repulsion terms can easily produce large positive contributions for incommensurable polymer components. At higher temperature the influence of the structure on thermodynamic variables generally decreases. This can be seen from the behavior of a hard-sphere fluid, $pV/kT \to 1$.

The B term estimates energy differences. At high temperatures, where the A term is small, the situation is dominated by the B term. Then, for $X \approx 0$, even the Lennard Jones pictures Figs. 59b and d show $\Delta G^{\mathrm{E}} > 0$ and $\Delta S^{\mathrm{E}} < 0$ which would mean $B > 0$ and $dB/dT > 0$, see Eq. 9.9b. [The argument for polymers can rest on the free volume. The thermal expansion is relatively large in polymers, i.e. the free volume difference between the components also becomes larger for higher temperature. This results in less energy contributions from optimization at larger distances because the intermolecular potential tends to zero for large distances: $B > 0$ and $dB/dT > 0$.]

In any case, the different behavior of the three terms ensures a high flexibility and opens a large field for application. Let us remark something positive and something negative. The succession in Figs. 61b,c,d (LCST, LCST + UCST, hourglass) can be explained, for instance, by variation of

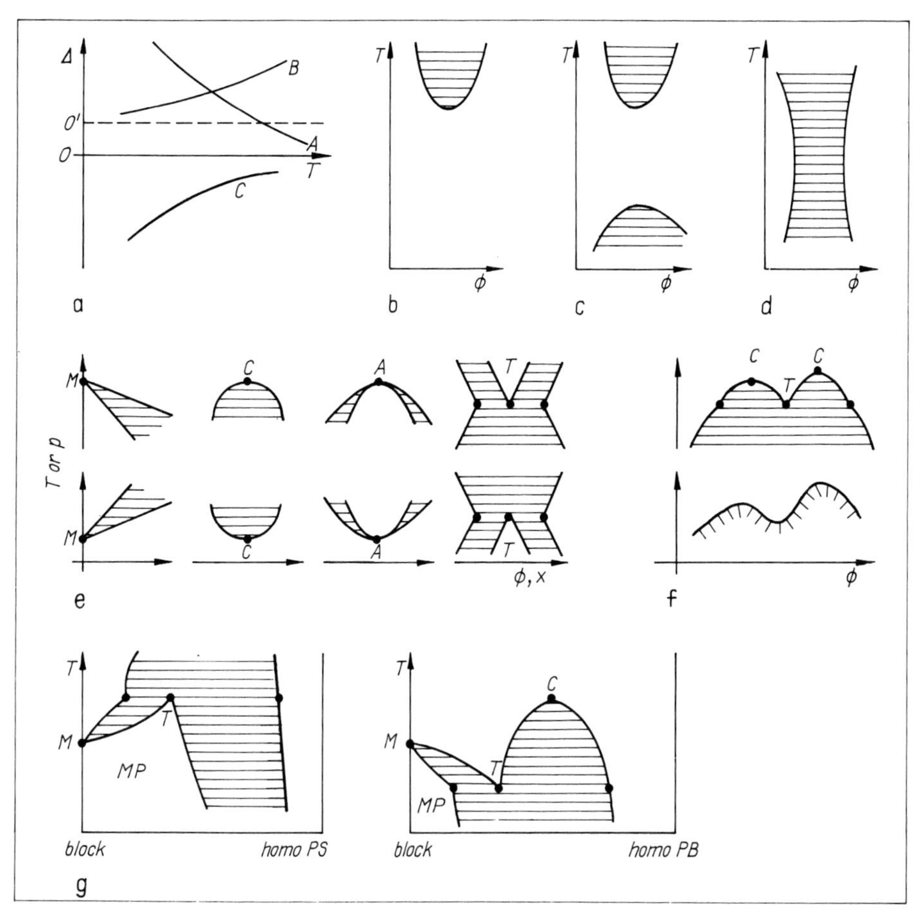

Fig. 61. Polymer mixtures (schematically).
a. Temperature dependence of the three contributions to Gibbs free energy at $\phi = 0.5$ according to Eq. 9.23. 0 zero for ΔG^{E}, 0′ zero for ΔG^{M} (the difference being the combinatoric part). b, c, d. Frequent phase diagrams for polymer–polymer mixtures. e. The four types of extrema in $T\phi$ or $p\phi$ phase diagrams of binary systems without chemical bonding of the components. M marginal, C critical, A "azeotropic", T three phase. All "points" are sections of lines in a $pT\phi$ diagram (marginal line ($\phi = 0$ or $\phi = 1$), critical line etc.). For polymers, binary means two monodisperse polymers; polydispersity leads to typical modifications of real diagrams (not shown). f. Binary phase diagram of the PB(2700) + PS(2100) system. Lower part as measured (e.g. by turbidity) in a real system. Upper part as analyzed by a combination of a three-phase line and two critical lines for the ideal binary. g. Binary phase diagrams for a styren-butadiene block copolymer with the cohomopolymers PS (l.h.s.) and PB (r.h.s.). MP macroscopically homogeneous phase with microphase decomposition in the 10 nanometer range (see Sec. 10.7).

only one parameter, the increase of X' (A term, size difference between polymer segments). On the other hand, from a parameter adjustment for real systems one can get parameters that caricature the original meaning, since the theoretical base for Eq. 9.23 is generally too small.

So far we have mainly considered the case $\phi = 0.5$. Let us now make some remarks to the composition dependence. There are vivid arguments that result directly in ϕ dependencies. Staverman (Ref. 314) suggested to set the number of the nearest neighbors of a molecule or segment proportional to its surface area. Application to polymer solutions and mixtures yields a concentration dependent interaction parameter

$$\Delta\varepsilon = \Delta\varepsilon_0 + \Delta\varepsilon_1/(1 - \gamma\phi_2) \tag{9.25}$$

where $\gamma = 1 - O_2/O_1$, with O_i the surface area of a molecule or repeat unit (Ref. 315). Eq. 9.25 is a good formula for screening and estimation. It can also be used at a higher level of sophistication.

Excluding the formation of chemical compounds there are four basic particularities in binary phase diagrams (see e.g. Ref. 45). They are explained in Fig. 60e. Combinations of them are well known in fluids at higher pressures, in liquid-solid equilibria, and in mixtures of liquid crystals. A polymer example for two critical points (or curves if pressure is varied) combined with a three-phase point is sketched in Fig. 60f. Especially copolymer components generate interesting phase diagrams. Two examples of a block copolymer mixed with the homo polymer component are shown in Fig. 60g (Ref. 316). Combinations of marginal, critical, and three-phase points explain the pictures. An interesting property of these diagrams is the occurrence of microphase (MP) separation that will be described in Sec. 10.7. MP forms a large ϕT region in these phase diagrams.

Mixtures of a statistical copolymer with a co-homopolymer (e.g. AABABB . . . + AAAA . . .) have, as a rule, better compatibility than the corresponding mixture (same ϕ_A, ϕ_B) of homopolymers, AAAA . . . + BBBB . . . This can be explained not only by the forced mixing inside the copolymer chain but also by the method to estimate mean-field energy differences. We have, on the Flory Huggins level of Eqs. 9.17 to 9.19, the following $\Delta\varepsilon$ contributions (Ref. 317),

$$\Delta\chi = g_{1A}\phi_A + g_{1B}\phi_B - g_{AB}\phi_A\phi_B, \tag{9.26}$$

with g_{1A} for 1 + homo A, g_{1B} for 1 + homo B, and g_{AB} inside the copolymer (the mixture is AABABB . . . + 1111 . . ., 1 = A or B, or a third component C). The minus sign is a consequence of the analogy with Eq. 9.17,

$-\Delta\varepsilon \sim (2\varepsilon_{12} - \varepsilon_{11} - \varepsilon_{22})$: the g_{1A} and g_{1B} terms correspond to the ε_{12} contribution that is exclusively from different chains.

Using such ingredients, like Eqs. 9.25 and 9.26, or results from complex mixing rules for the construction of more sophisticaed formulas, a lot of interesting phase diagrams can be reproduced. A sufficiently large ϕ variation can generate details in $\partial^2 G/\partial\phi^2$ ("several plaits") that can further restrain the effect of the small combinatoric entropy.

In polymer solutions the nanoheterogeneity (Sec. 1.4, Figs. 4a,b) gives rise to new complications in the ϕ dependence of mixing functions. The polymer chains run through regions with high (heap) and low (blob) polymer concentration. This results in a partial screening of the subtle long-ranging interaction along the chain. This effect can badly be represented by mean field methods (χ parameter). Only a concentrated polymer system behaves similar to polymer-polymer systems.

To give an example (Ref. 318) the nanonheterogeneity contribution to the excess function is

$$\Delta G^{\mathrm{E}} \approx g^*(T, Z) \cdot P, \tag{9.27}$$

where g^* denotes the free energy difference between the two regions of nanoheterogeneity, and

$$P \approx \exp\,(-\lambda_0 Z^{1/2}\phi_P) \tag{9.28}$$

is the probability that a given volume element in the solution does not fall in the concentrated regions (λ_0 being a parameter for molecular properties, Z the degree of polymerization, and ϕ_p the volume fraction of the polymer). The difference g^* can also contain intramolecular contributions when the distribution of chain conformations is different in the two regions.

An ansatz like Eqs. 9.27 and 9.28 reflects the art to make thermodynamics with only a few parameters. Of course one could also take a more rigorous *virial expansion* for small ϕ_p,

$$\Delta G^{\mathrm{E}} = \sum_n A'_n \phi_P^n, \quad n = 1, 2, 3 \ldots \tag{9.29}$$

with A'_n the viral coefficients (modified for this $\Delta G^{\mathrm{E}}(\phi_p)$ function). Being coefficients of a Taylor series expansion the A'_n can depend on the chain length, temperature and others, but not on ϕ_p. Several viral coefficients would be needed to represent the nanoheterogeneity situation, whereas the intuitive Eqs. 9.27 and 9.28 do with only two parameters, λ_0 and g^*. But even the virial expansion is not without physics: electrolytes, due to their Coulomb

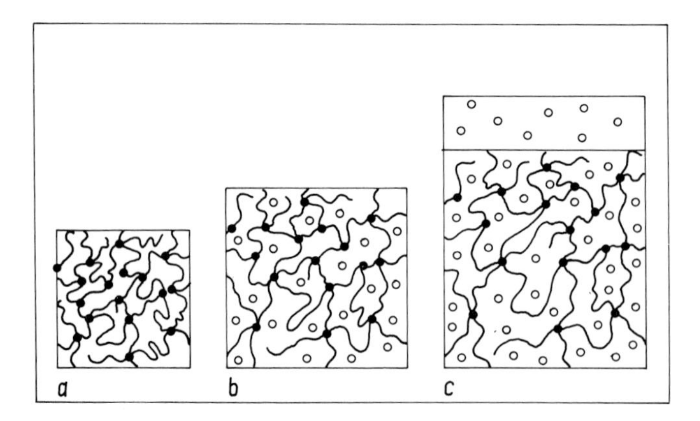

Fig. 62. Swelling of polymer networks.
a. Dry network. b. Swollen network, o solute particles. c. Swelling equilibrium solute + swollen network.

forces, would need Debye Hückel $\phi^{1/2}$ terms and can, therefore, not be represented by Eq. 9.29 (nor 9.28).

9.5 Swelling equilibrium of polymer network

Adding a solvent (volume fraction ϕ) to a dry network the solvent molecules can be incorporated up to a certain degree. The term swelling means that the network gets a larger volume, see Figs. 62a–c. The thermodynamic treatment includes the change of the free energy by virtue of the solvent and the elastic potential due to the swelling and, if necessary, due to an external deformation (Refs. 319 and 320, the presentation here is similar to Ref. 321).

We make three assumptions:

(i) Separability of elastic (see Eq. 3.6) and mixing free energy.

(ii) The mixing free energy (the Helmholtz free energy $\Delta F^{M}(V, T, \phi)$ is used to have the volume V explicitly) can be identified with the free energy of mixing of the uncross-linked polymer solution of infinity molecular weight. [This means we have no combinatoric part for the polymer component.]

(iii) For given V, the mixing free energy of the isotropically swollen network (the "gel") is invariant against anisotropic constraints. (This can only be valid as long as the orientation correlation between segments is small, e.g. we must be far enough from the onset of the up turn, see Sec. 3.2).

Using the dry polymer as a reference state we have

$$\Delta F^{\mathrm{M}} = \sum_n A'_n \phi^n + (RTv/2)[\lambda_1^2 + \lambda_2^2 + \lambda_3^2 - 3] \tag{9.30}$$

with v the dry concentration of elastic chains (see Eq. 3.6 again), λ_i the principal deformation ratios and A'_n the virial coefficients (per volume) of, now, the solute in the concentrated polymer system. The use of the model-free virial ansatz is advantageous for a test of assumption (ii). Three examples will shortly be discussed.

Swelling pressure
For the isotropic case (see Sec. 3.2) the swelling, according to assumptions (i) and (iii), is expressed by

$$\lambda_i = \phi^{-1/3}, \tag{9.31}$$

the extension is generated by the additional solute volume. After ϕ differentiation ("$p = -\partial F/\partial V$" with ϕ for V) the swelling pressure is obtained as

$$\tilde{\omega} = \sum_n A_n \phi^n - RTv\phi^{1/3} \tag{9.32}$$

with $A'_n = A_n n/\phi$, A_n being the conventional virial coefficients related to the pVT data.

Swelling equilibrium
Adding solvent enough we come to the true two-phase equilibrium solvent + swollen network, see Fig. 62c. This can be calcualted from the requirement of equal chemical potentials of the solvent in the two phases. Returning to the Flory Huggins form (with combinatoric part for the solvent component) we obtain from Eqs. 3.18 (for the solvent) and 9.30 (the virals replaced) the Flory Rehner equation,

$$RT[\ln(1 - v) + \chi v^2 + v] = \partial \Delta F^{\mathrm{M}}_{\mathrm{elast}}/\partial n_1, \tag{9.33}$$

corresponding to the v form Eq. 3.17b with v for the polymer and n_1 for the solvent. In the format of Eq. 9.33 the deformation ratios for the isotropic case read as

$$\lambda = \lambda_1 = \lambda_2 = \lambda_3 = [(n_1 V_1 + V_0)/V_0]^{1/3} = v^{-1/3}, \tag{9.34}$$

instead of Eq. 9.31, with the volumes V_0 for the dry polymer and V_1 for the pure solvent. Eqs. 9.33 and 9.34 define the swelling equilibrium in a format suitable for correlation e.g. with respect to the mean molecular weight M_c of chains between the network crosslinks, the identity carrier of the network.

Uniaxial deformation
Consider a swollen network under uniaxial extension given by the external deformation ratio λ_1, and use the format of Eq. 9.30 again. Then, from

$$\lambda_1 \lambda_2^2 = \phi^{-1}, \tag{9.35}$$

we obtain

$$\Delta F = \sum_n A'_n \phi^n + (RTv/2)[\lambda_1^2 + 2/\phi\lambda_1 - 3]. \tag{9.36}$$

The equilibrium is obtained from optimization,

$$\phi(\partial \Delta F^{\mathrm{M}}/\partial \phi)_{\lambda_1} = \sum_n nA'_n \phi^n - RTv/\phi\lambda_1 = 0. \tag{9.37}$$

Network experts use the term uniaxial "compression ratio" (of deformed gel in equilibrium) for

$$\Lambda \equiv \lambda_1 \phi^{1/3}. \tag{9.38}$$

Since the swelling pressure is zero now, $\tilde{\omega} = 0$, it follows

$$\Pi_{\mathrm{mix}} \equiv \sum_n A_n \phi^n = RTv\phi^{1/3}/\Lambda \tag{9.39}$$

with A_n the conventional virial coefficients explained in eq. 9.32.

Remark. The swelling equilibrium should not be confounded with the sol-gel transition. The latter term is used for the percolation stage of crosslinking, varying the number of crosslinks. Chemical crosslinking means generation of increasing clusters of the linked chains. At a certain stage the clusters percolate, which means a cluster becomes infinitely large. All segments that belong to this (or such) clusters are called the gel fraction, the rest is the sol fraction. The situation near this transition can be described with critical indices and with fractals. For the swelling equilibrium, however, the number of crosslinks is constant. This is an ordinary first-order transition for the solvent with no critical point for varying ϕ, even if the network phase is usually called a gel and the solute sometimes a sol as well.

10. Compatibility of polymers

The thermodynamic description of miscibility gaps and the corresponding potential is only one aspect. The actual realization of a phase decomposition also depends on structural and kinetic properties being the subject of this chapter. The red thread of the representation is twisted by spatial (structural) and temporal aspects. The aim is not a systematic development; rather some basic ideas are described by means of several illustrative examples. Polymer systems are also good examples for experimental work because of the convenient time scales for the flow transition or the vicinity of glass temperature.

10.1 Terminology

Some often used terms will be defined with reference to the miscibility gaps in Fig. 63.

The environment of the critical point C is dominated by long-reaching fluctuations (critical fluctuation, critical opalescence) that are generated by the vicinity of stability limits, the spinodals. These fluctuations also contribute to the thermodynamic properties, see Sec. 4.2.5. For instance, the values of critical exponents are universally determined by them; the remaining question is whether the molecular interaction is short or long reaching – not trivial for long chains – and what is the dimension.

In the metastable region K between binodals b and spinodals s the new phase can only be formed by overcoming an activation barrier connected with a small subsystem: *nucleation*. Metastability means that smaller fluctuations are stationarily present (local stability). Only if a larger fluctuation peak (zacke) corresponds to a large enough size and has a sufficient similarity to the new phase structure (critical nucleus), then the new phase can grow. In the locally unstable region S between the two spinodal branches of the plait no such an activation barrier is to be overcome. Nonetheless the new phase is not formed immediately. Rather we find a characteristic kinetics and structure that is called *spinodal phase decomposition* (or unmixing). The typical experiment is a quench from the stable into the unstable region S, e.g.

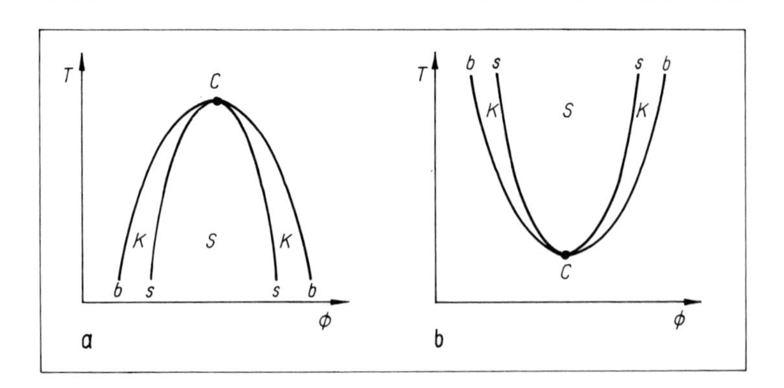

Fig. 63. Miscibility gaps.
a. For a UCST type. b. For a LCST type phase diagram. C critical fluctuation, K nucleation, S spinodal phase decomposition, b binodals, s spinodals.

by heating or cooling with an external rate faster than the internal kinetics. In the ideal case the unmixing starts from a homogeneous situation in S far from equilibrium. The expected chaos, however, develops structures in the earliest stage, we find a typical selforganization in space and time that passes gradually into the new equilibrium of two large-scale phase regions. The classical references are Refs. 322 and 323.

Of course, the usual terms of selforganization, such as evolution (succession of dissipative structures), nonlinearity, feed back (amplification), synchronization, symmetry breaking, historicity, and the principle of internal determination – a relative nondependence on the boundary conditions – can also be applied to this unmixing. But there is an essential difference to the dissipative structures usually considered, e.g. Bernard cells. There, beyond a critical distance from equilibrium, the cell size is mainly determined by phenomenological scales (and the layer thickness), the cell structure could also be realized by a true continuum with the same properties. It is not easy to reach small scales for such structures, because the nonequilibrium is generated from outside by, for instance, an external temperature gradient between the bottom and top of the layer. The continuum behavior is governed by partial differential equations, $\partial/\partial t \ldots = \Sigma\,(\partial/\partial x)^n \ldots$, that produce critical numbers (e.g. a critical Rayleigh number) which are proportional to L^α / t^β with exponents $\alpha > 0$ and $\beta > 0$, L being a length and t a time. This means that a given critical number (for a structure window, e.g.) cannot be reached for too small L.

In the spinodal region S, however, supposing a fast enough preceding quench, the state far from equilibrium is not characterized by an external gradient or an external length scale L. For the formation of a structure from the homogeneous state only molecular scales are available.

We shall try to understand which particularities come from the chain structure (coil radius R_0) when unmixing in polymers is compared with unmixing in small-molecule systems. The chains enlarge the effective range of interaction and the mean-field tendency. An interesting point is that, usually, the spinodal phase decomposition in polymers does not start in the segment scale but in the R_0 scale, i.e. in a spatial scale about ten times larger than in the small-molecular mixture although, for instance, the characteristic length of dynamic glass transition is of the same order in both systems.

10.2 Ornstein Zernicke (OZ) approach to critical phenomena

1. Classical treatment for small-molecule systems (Refs. 45, 324)

Consider the critical point C of the vapor-liquid phase equilibrium in a pure normal fluid. The density fluctuation becoming singular at C (critical opalescence) is described by an order parameter $\psi(r)$ suitable for the long-range correlations. In the Ornstein Zernicke (OZ) approach, their influence on the thermodynamic potential J is described by the first and third term of Eq. 4.10,

$$\Delta J = \int dV[\alpha t\psi^2 + K'(\partial\psi/\partial r)^2]. \tag{10.1}$$

This is a ψ-level mean-field ansatz, the linear t term ($t \equiv T - T_c$, $|t| \ll T_c$) follows from an analyticity at the critical temperature T_c with α the temperature coefficient. The ρ level influence is expressed by the coefficient K' in the gradient term, originating from the idea that the fluctuation can produce such steep gradients that the free energy must depend on them (see Sec. 10.8).

Spatial Fourier transformation in the scattering volume V yields

$$\Delta J = V\sum_q (\alpha t + K'q^2)|\psi_q|^2 \tag{10.2}$$

with $\boldsymbol{q}$ the scattering vector. The mean fluctuation of a ψ_q component follows from the dispersion of the Boltzmann-Einstein Gauss-function. This is half of the exponent denominator in the probability density for the occurrence of a given ΔJ,

$$w \sim \exp(-\Delta J/kT). \tag{10.3}$$

From Eq. 10.2 we find the mean squared fluctuation of ψ_q,

$$\overline{|\psi_q|^2} = \frac{kT}{2V(\alpha t + K'q^2)}. \tag{10.4}$$

This corresponds to the correlation function

$$G(\boldsymbol{r}) = \overline{\psi(\boldsymbol{r})\psi(0)} = \sum_q \overline{|\psi_q|^2}\, \mathrm{e}^{\mathrm{i}\boldsymbol{qr}} \sim \frac{1}{K'r}\, \mathrm{e}^{-r/\xi} \tag{10.5}$$

with the correlation length

$$\xi = (K'/\alpha t)^{1/2}, \tag{10.6a}$$

or, more general,

$$\xi \sim (T - T_c)^{-\nu}, \quad \nu = \tfrac{1}{2}. \tag{10.6b}$$

Eq. 10.6a can easily be understood from a dimensional analysis of the denominator of Eq. 10.4. The correlation length ξ is factorized into the molecular, ρ level gradient-term coefficient times the reciprocal temperature term. Eq. 10.6b shows the classical divergence of ξ at the critical (or spinodal) point $t \to 0$ $(T \to T_c)$ with the critical exponent $\nu = \frac{1}{2}$. This approach qualitatively explains the critical opalescence.

Introducing the short-ranging direct correlation function $c(r)$ one can substitute the gradient term coefficient by a more direct ρ-level structure term. Remember that the decomposition of $G(r)$ into a self and distinct part and the application of the OZ equation gives a relationship between the structure factor and the direct correlation function, Eqs. 1.21, 1.24 and 1.28. Using Fourier transforms we obtain for the structure factor

$$S(q) = n/(1 - nc(q)) \tag{10.7}$$

with n (formerly $\bar{n}$) the number density of particles. The relation to the order parameter is realized by the so-called *OZ approximation*, i.e. a Taylor expansion of the inverse structure factor up to second order in q, viz.

$$I^0(q)/I(q) = n/S(q) = 1 - nc(q) \approx R^2(\kappa^2 + q^2 + \ldots). \tag{10.8}$$

(cf. also Eq. 10.4 with $S^{-1}(q) \sim |\psi_q|^2$), $I(q)$ being the scattering intensity.

Small values of q^2 correspond to large distances of the long-range fluctuations. The Debye Hückel coefficients follow from the OZ equation,

$$R^2 = \text{const} \int r^2 c(r) dr, \qquad (10.9a)$$

$$\kappa^2 = (1 - nc(0))/R^2. \qquad (10.9b)$$

The integral is a spatial representation of the q^2 coefficient in the $c(q)$ series. A comparison of Eqs. 10.8 and 10.5 gives the desired relation, viz. between the gradient-term coefficient K' and the correlation length ξ on the one hand and the range R of direct correlation function and the isothermal compressibility κ_T (see Eq. 1.30) on the other. Therefore the OZ approximation is characterized by two lengths, a ρ level one,

$$K' \sim R^2, \qquad (10.10)$$

and a ψ level one,

$$\xi^2 \sim R^2 \kappa_T^{-1}, \qquad (10.11)$$

$(\kappa_T = (\partial n/\partial p)_T/n)$. The dampling length R is of order one nanometer, the ρ level is here characterized by a spatial integral over the direct correlation function. The ψ-level correlation length ξ is large by thermodynamic reasons. Eq. 10.11 has the same structure as Eq. 10.6a, of course, the length is enlarged by the divergence of the compressibility, $\kappa_T \sim t^{-\gamma}$, i.e. the critical exponent for ξ is ν with $\nu \approx \gamma/2$.

The situation of Eq. 10.8 is usually adjusted in a I^{-1} vs. q^2 diagram, Fig. 64a, the so-called *OZ diagram* (with a formal relationship to the Zimm diagram, see Eq. 1.15b in Sec. I). One finds a set of parallel straight lines parametrized by the temperature $t = T - T_c$. The slope is R^2 and gives information about the range of the direct correlation function $c(r)$ in the ρ level. The ordinate value for $q = 0$ gives, after thermodynamic input from the compressibility, the correlation length and, correlated with t, the critical exponent ν. The parallelism of the straight lines indicates a constant ρ-level length R, i.e. the direct correlation function remains short-ranged up to the critical point.

2. Adaptation for polymers

The adaptation to the critical point of unmixing in polymer systems is developed in Refs. 325–327, we shall partially follow the representation of Ref. 328. The order parameter is now defined by a concentration (volume

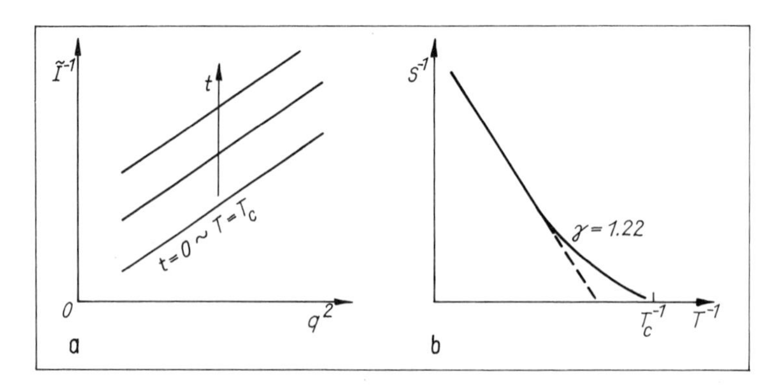

Fig. 64. Scattering near the critical point.
a. Ornstein–Zernicke (OZ) diagram for scattering in the homogeneous phase: I^{-1} vs. q^2, i.e. reciprocal intensity vs. the square of the scattering vector. Parameter $t = |T - T_c|$. b. Deviation from mean field behavior ($\gamma = 1$) in the polymer binary PS_d + PB, $\phi = 0.53$, from neutron scattering (Ref. 330). S^{-1} reciprocal structure factor.

fraction) fluctuation, $\psi = \phi(\mathbf{r}, t) - \bar{\phi}$. The gradient term equation, Eq. 10.1. or 10.2, is substituted by the RPA (random phase approximation) for $S^{-1}(q)$,

$$\frac{1}{S(q)} = \frac{1}{\phi_A N_A} + \frac{1}{\phi_B N_B} - 2\chi + \frac{1}{18}\frac{a^2}{\phi_A \phi_B} q^2 \tag{10.12}$$

(see Eq. 9.21 or 4.31). The ψ level is therefore specified by chain structure, since the $(\phi N)^{-1}$ factors stem from a simplified Debye function for Gaussian coils, see Eq. 1.15. The ρ level is only represented by the coarse $a^2 q^2$ term with a the structure length, $a \approx 0.7\,\text{nm}$ for vinyl polymers. The thermodynamic control ($\alpha t \psi^2$ in Eq. 10.1) is now undertaken by the Flory Huggins χ parameter. [Any reference to the details of the scattering experiment is neglected in Eq. 9.21 or 10.12, e.g. the scattering length for neutron scattering etc., see Ref. 238 again.]

Transformation of Eq. 10.12 into the OZ form,

$$S(q) \approx S(q = 0)/[1 + \xi^2(T, \phi)q^2], \tag{10.13a}$$

gives the correlation length

$$\xi^2(T, \phi) \approx (a^2/36)[\phi(1 - \phi)(\chi_s - \chi)]^{-1}, \tag{10.13b}$$

where the $(\alpha t)^{-1}$ dependence of the original approach to ξ^2 (Eq. 10.6a) is now transferred to the χ dependence, see Fig. 5a: The original $t = 0$ ($T = T_c$) condition is now called $\chi = \chi_s$, with $\chi_s = \chi_c$ at the critical point ($\phi = \phi_c$) or χ_s the χ value at the spinodal for $\phi \neq \phi_c$. The special form of Eq. 10.13b is obtained by the difference of Eq. 10.12 as presented minus its value at the spinodal, $S^{-1} = 0$ for $\chi = \chi_s$ (local stability limit). The substitution of the gradient term by the a^2 term makes sense since both the K' term of Eq. 10.2 and the a^2 term of Eq. 10.12 are proportional to q^2.

Eq. 10.13b corresponds exactly to the classical formula Eq. 10.11, or Eq. 10.6a, with the critical exponent $\nu = \frac{1}{2}$. The ρ level is now expressed by the structure length a ($\approx 0.7\,\text{nm}$) alone. The chain coil scale ($R_0 \approx aN^{1/2}$, $N = Z$) is not explicitly contained but enters by the N scaling of the thermodynamic singularity: For a symmetrical polymer mixture ($N_A = N_B = N$) the critical χ value is $\chi_c \approx 2/N$, see Eq. 3.24. This N is a consequence of the Flory Huggins combinatoric entropy, i.e. the occurrence of this N implies that the combinatoric term has an important influence on the critical parameters. [Strong ϕ-dependence of χ can alter the situation, see Sec. 9.4.] Since $\chi \approx \chi_s$ close to the singularity the correlation length scales with N. In formulas: We have a

$$\tilde{\Delta} = (\chi/\chi_s) - 1 \tag{10.14a}$$

so that $\chi - \chi_s = \chi_s \cdot \tilde{\Delta}$, with $\tilde{\Delta} \to 0$ for $\chi \to \chi_s$. Then, from Eq. 10.13b,

$$\xi^2(T, \phi) \approx (a^2 N/36)[\phi(1 - \phi)\tilde{\Delta}]^{-1} \cong (R_0/6)^2[\phi(1 - \phi)\tilde{\Delta}]^{-1}. \tag{10.14b}$$

The general form of the ξ^2 formula, it being a product of a length times a thermodynamic amplification factor, is preserved, but the singularity is now scaled by $R_0^2 \sim N$. In principle, far from critical points $\tilde{\Delta}$ could reach larger values so that the $\xi \ll R_0$ case is possible. This is, however, not very probable because the basic structure of the phase diagrams for polymers, Fig. 61c, prevents too large χ variations with temperature.

In summary, near the stability limit, where $\tilde{\Delta}$ is of order 1, the short-ranged ρ level correlation is transformed thermodynamically, via the combinatoric Flory Huggins entropy, into the ψ-level R_0 scale.

3. Discussion

The critical exponents of particle systems with short-range intermolecular potentials are independent of whether a fluid or a liquid binary is considered (universality). The values for $d = 3$ dimensions are

$$\begin{array}{ccccc} \alpha & \beta & \gamma & \delta & \nu \\ \hline 0.11 & 0.34 & 1.22 & 4.6 & 0.63 \end{array} \qquad (10.15)$$

The fluctuations of the order parameter – completely neglected in the OZ approach – modify the thermodynamic properties in the singularity region (usually of order a few Kelvins), the Ginsburg criterion is violated here.

Another universality class must be applied to the mean field case (long-range intermolecular forces)

$$\begin{array}{ccccc} \alpha & \beta & \gamma & \delta & \nu \\ \hline 0 & \frac{1}{2} & 1 & 3 & \frac{1}{2} \end{array} \qquad (10.16)$$

The Ginsburg criterion is valid everywhere so that the order-parameter fluctuations have no influence.

Consider now a real binary polymer system with a short-range interaction between the segments and a long-range chain structure (coil diameter R_0). If the thermodynamic R_0 scaling of ξ^2 Eq. 10.14b corresponds to an effective enlargement of the range of intermolecular forces in the critical region, then a tendency to the mean field behavior is expected. At a certain temperature $t' = T' - T_c \neq 0$ ($t' > 0$ for an UCST) where ξ is of order R_0 (or even larger, see Ref. 329), a crossover is expected from the mean field behavior of Eq. 10.16 for $t > t'$, $\xi < R_0$, i.e. far from the spinodal, to the non-classical behavior of Eq. 10.15 for $0 < t < t'$, $\xi \gg R_0$, i.e. near the spinodal (see Fig. 64b, Ref. 330). $T' - T_c$ was observed to be rather large, of order 10 K (Ref. 331), because the temperature dependence of χ is rather low. It is difficult to predict exactly the crossover temperature T' or the corresponding ξ because the thermodynamic scaling of the RPA formula does not mark a length, whereas the coil diameter is, of course, a length, but of another character than a potential range.

For a polymer solution with small-molecule solute the N^{-1} scaling gets lost (see Eq. 3.25). The nanoheterogeneity defines a new length smaller than R_0. If this length corresponds to the range of an effective potential then, relative to ξ, the crossover to the non-classical singularity is expected to occur at a relatively large distance from the critical point (see e.g. Ref. 332 for experiments).

A sharp critical point ($T = T_c$ at $\phi = \phi_c$) is, of course, the consequence of long-ranging correlation. Large fluctuations beyond the Ginsburg criterion do not change this statement. A sharp spinodal ($\phi \neq \phi_c$), however, is a mean-field construction. Beyond the Ginsburg criterion, sharp singular lines are not possible, if they would imply a finite exponent $\alpha > 0$. Such lines would correspond, when transformed to pure substances, to lines with singular specific heat at constant volume, $C_v \to \infty$. This is not possible from phenomenological reasons (Ref. 333). The spinodal is therefore smeared out when the fluctuations are too large. This is also indicated by experiments, the smearing-out is considerable, e.g. of order 10 K (Ref. 334).

The parallelism of slopes in the OZ diagram (Fig. 64a) is observed for many systems including polymer solutions and mixtures. As mentioned above, according to Eqs. 10.8 and 10.9a, this is caused by a nearly constant range of the direct correlation function(s) $c(r)$ in the critical region (see also Fig. 2b). Since $c(r)$ belongs to the ρ level (a few nanometers) the ρ and ψ level is "separated" in the critical region (cf. the $a^2 \cdot N$ factorization of the prefactor in Eq. 10.14b). Deviations from this parallism indicate serious changes in the ρ level, or its coupling to the ψ level, when the critical point is approached (see Sec. 10.9 below).

For kinetic aspects of critical states the reader is referred to the classical work of Kawaskai (e.g. Ref. 335).

10.3 Spinodal phase decomposition

1. Cahn Hilliard approach

We start from a spatially homogeneous situation, far from equilibrium, reached after a quench into the locally unstable region S of Fig. 63a, b between the spinodals of a plait. The main idea of the Cahn Hilliard approach is to describe the formation of an unmixing structure in the unstable region by means of a length scale introduced by a gradient term in the free energy (Ref. 336, cf. also Eq. 4.10).

Let us define the order parameter by the deviation of the local actual concentration of component A from the average c_0, $\psi(r, t) = c_A(r, t) - c_0$. Forming a structure means mass transport by microscopic currents, J_A and J_B, with $J_A = -J_B \equiv \tilde{J}$ for constant partial densities, ρ_A and ρ_B. Taking a thermodynamic ansatz: current = kinetic Onsager coefficient M times the difference of chemical potential gradients we have

$$\tilde{J} = -M\nabla\tilde{\mu} \tag{10.17}$$

where $\tilde{\mu} \equiv \mu_A - \mu_B$. We assume that the structure formation is so violent that the free energy $f(\psi)$ depends on it. If we confine ourselves to the first and third term of Eq. 4.10 we find

$$\tilde{\mu} = \partial f/\partial\psi - 2K\nabla^2\psi \tag{10.18}$$

with $K > 0$ the gradient term constant (or coefficient), see also Sec. 10.8 below. Using ψ conservation (mass conservation),

$$\partial\psi/\partial t + \nabla\tilde{J} = 0 \tag{10.19}$$

we obtain the *Cahn Hilliard equation* for the order parameter,

$$\partial\psi/\partial t = M(\partial^2 f/\partial\psi^2)\nabla^2\psi - 2MK\nabla^4\psi. \tag{10.20}$$

This is the mean field equation of the lowest nontrivial order. The first term of the r.h.s. describes a diffusion coefficient

$$\tilde{D} = M \cdot \partial^2 f/\partial\psi^2 \tag{10.21}$$

being a product: mobility times a thermodynamic factor. Inside the spinodal region S the local instability is expressed by the mean field (van der Waals) formula

$$\partial^2 f/\partial\psi^2 < 0 \text{ in S.} \tag{10.22}$$

Since $M > 0$ this means $D < 0$, diffusion *in* the gradient direction, i.e. amplification (instead of smoothing) of concentration differences.

The second term of the Cahn Hilliard equation Eq. 10.20 moderates the structure formation because too steep gradients are thermodynamically disadvantageous in Eq. 4.10.

Taking the separation ansatz $\psi_q = F(t) \cdot \cos(\boldsymbol{qr})$ we obtain from Eq. 10.20 for the spatial $\boldsymbol{q}$ component

$$\psi_q = e^{R(q)t} \cdot \cos(\boldsymbol{qr}) \tag{10.23}$$

with the *amplification factor*

$$R(q) = M(-(\partial^2 f/\partial\psi^2)q^2 - 2Kq^4) \tag{10.24}$$

that can be measured by dynamic scattering experiments with the scattering vector $\boldsymbol{q}$. Using Eq. 10.22 we see from Eq. 10.24 an amplification (*instability production*) of disturbances ψ_q at small q always present in a chaos. It is the gradient term ($K > 0$) that takes care of a decrease of amplification at large q. Above a critical q value

$$q_c^2 = -(\partial^2 f/\partial\psi^2)/2K \qquad (10.25)$$

(smaller for larger K), damping is obtained ($R(q) < 0$) instead of amplification. Large q values correspond to small distances: Approaching the ρ level (see Sec. 10.8 below) we expect the generation of finite fluctuations only.

The Cahn Hilliard equation describes the formation of spatial structures with exponentially increasing amplitudes in the starting length scale of

$$\lambda \approx 2\pi/q_c \qquad (10.26a)$$

or, more expressive, with typical gradients in the scale $\lambda' = \lambda/2\pi$,

$$q_c^{-1} = 2K/(-f'') \qquad (10.26b)$$

given by the ratio K over $f'' \equiv \partial^2 f/\partial\psi^2$. The general structure of this length-scale formula is, similar to the OZ formulas, defined by a product: gradient-term constant times a thermodynamic factor.

The Cahn Hilliard approach therefore describes a spatial structure formation with the following properties:
(1) Exponential growth of unstable amplitudes
(2) Constant length scale
(3) Linear decrease of R/q^2 vs. q^2 with a critical q value (see Fig. 65a).

Some examples for scattering experiments are sketched in Figs 65b and $65c_1$, c_2. The former is for a small-molecule mixture (Al + Zn) investigated by X ray scattering (Ref. 337), the latter are for polymer mixtures. Fig. $65c_1$ is for a mixture of deuterated and hydrogenated 1.4 poly butadiene with high molecular weights investigated by light scattering at low quench depths (i.e. near the spinodal, Ref. 338), and Fig. $65c_2$ is for the polymer mixture PS_D + PVME (Ref. 339).

In the small-molecule system the structure is observed in the spatial region of several nanometers. The corresponding distances are of order the length scales for the last die down of pair correlations ($g(r)$) in stable regions. This means that there is no instability production at the ρ level. Instability is produced in the transition region to large lengths and at the ψ level, i.e. in a quasi-continuum with thermodynamic properties generated at the ρ level. A detailed analysis shows that deviations from exponential growth of the amplitudes ψ_q are soon observed. The typical length $\lambda \approx q_c^{-1}$ (approximately defined by the intensity maximum) increases with time, and deviations from

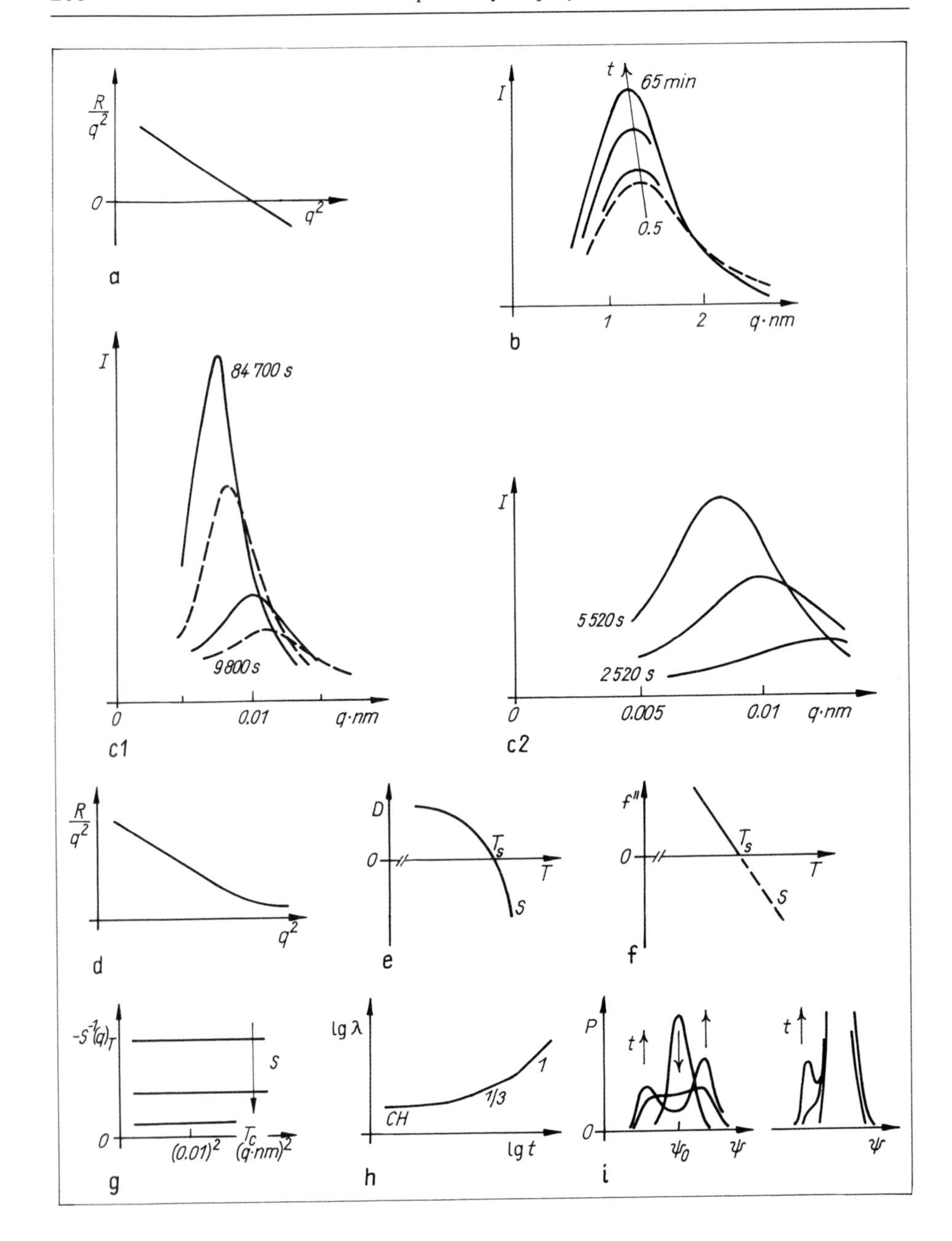
R/q²
0
q²
a
t
65 min
I
0.5
1
2
q·nm
b
I
84 700 s
9800 s
0
0.01
q·nm
c1
I
5520 s
2520 s
0
0.005
0.01
q·nm
c2
R/q²
q²
d
D
T_s
0
T
S
e
f''
T_s
0
T
S
f
$-S^{-1}(q)_T$
S
T_c
0
$(0.01)^2$
$(q·nm)^2$
g
lg λ
1
1/3
CH
lg t
h
P
t
0
ψ_0
ψ
t
ψ
i

the R/q^2-vs.-q^2 linearity are also observed, although negative values of R at large q are still met (i.e. the ρ level remains stable).

In the polymer systems, especially at low quench depths (i.e. $|\tilde{\Delta}|$ in Eq. 10.14a is of order one), the Cahn Hilliard conditions (1) and (2) are better fulfilled. This corresponds again (see Sec. 10.2) to the mean field tendency induced by the polymer chain structure. The general spatial scale of unmixing is larger by a factor of about 10 or even 100, i.e. in a ψ-level scale with the property as if the radius of interpenetrating coils (R_0) would be the particle diameter. On the other hand negative amplication factors $R < 0$, expected for the larger $q \geqslant R_0^{-1}$, i.e. for the smaller-than-R_0 scale, are not yet observed in polymer systems. Obviously, the polymer segments can produce some instability in the length region $a \ll \lambda \lesssim R_0$ similar to the particles in small-molecule systems.

2. Further ingredients

Before the length-scale discussion is continued some newer theoretical inputs should be mentioned (Refs. 325–327) that include higher-order fluctuations in unstable regions (Refs. 340–342), and polymer properties. The representation partially follows Ref. 339.

Fig. 65. Spinodal phase separation.
a. Cahn HIlliard amplification factor R as a function of squared scattering vector, q^2. b. X ray scattering intensity I for spinodal phase decomposition in the small-molecule Al + Zn binary. (ρ level, $T = 65°C$), parameter time. c. Spinodal unmixing in polymer systems. c1. Deuterated + hydrogenated 1,4 PB for low quench depth. c2. PS_D + PVE, $\phi_{PSD} = 0.19$, also for e, f, g. d. Typical behavior of the amplification factor R at the first stages in polymer systems ($R > 0$). e. Interdiffusion coefficient D in the PVME + PS_D system at both sides of the spinodal. S is the unstable region between the spinodals. T temperature, T_s spinodal temperature. f. $f'' = \partial^2 f/\partial\phi^2$, second ϕ derivative of free energy as a function of temperature. Full line from experiments, broken line extrapolation into the S region. g. Virtual structure factor $S(q)_T$ in the S region, parameter temperature $T \rightarrow T_c$. h. Typical time behavior of the length scale λ for spinodal unmixing in polymer systems. The exponent 1/3 corresponds to the Ostwald ripening stage, CH Cahn Hilliard (starting) stage. i. 1-point distribution function $P(\{\psi\}, t)$ for spinodal phase decomposition (l.h.s) and nucleation (r.h.s.). ψ order parameter = concentration of one polymer component, c.

Besides the structure function, $S(q, t) \equiv S_2(q, t)$, so-called higher moments are introduced for the description of nonlinearities,

$$S_n(q, t) = \int dr\, \overline{[\phi(\mathbf{r}, t) - \phi_0]^{n-1}[\phi(0, t) - \phi_0]}\, e^{iqr}. \tag{10.27}$$

The composition dependence is again parameterized by volume fractions ϕ. Eq. 10.27 enlarges the Cahn Hilliard formula, written for the structure function, to

$$\frac{\partial S}{\partial t}(q, t) = -2Mq^2\left[\left(\frac{\partial^2 f}{\partial \phi_0^2} + 2Kq^2\right) S(q, t) + \frac{1}{2}\frac{\partial^3 f}{\partial \phi_0^3} S_3(q, t) + \frac{1}{6}\frac{\partial^4 f}{\partial \phi_0^4} S_4(q, t) + \ldots\right] + 2MkTq^2. \tag{10.28}$$

The S_n terms, $n = 3, 4, \ldots$, come from a Taylor series expansion of the free energy functional $F[\psi(r, t)]$ into $\psi \sim \phi - \phi_0$ powers; F can depend on higher fluctuation in the S region. The last term is the Cook term (Ref. 343). This is a mean fluctuation in the sense of FDT. Formally it results from the $f(t)$ term in a Langevin equation (like Eqs. 4.18, 4.21) when Eq. 10.28 is considered to be the q representation of the corresponding Fokker Planck equation. Referring to an effective Hamiltonian, with Eq. 4.21, then the Cook term reflects more a ρ level than a ψ level fluctuation.

The $n > 2$ terms cannot be investigated by elementary means. Numerical calculations show that they can qualitatively model an increase of the typical length λ.

Eq. 10.28 with $S_3 = S_4 = \ldots = 0$ is called *Cahn Hilliard Cook equation* (CHC),

$$\partial S/\partial t(q, t) = 2R(q)S(q, t) + 2MkTq^2. \tag{10.29}$$

This equation is inhomogeneous by virtue of the Cook term and is usually a good approximation for (not too) large times t. The corresponding task is the reduction of two properties (gradient term, Cook fluctuation) to experimentally adjustable variables. A virtual structure factor $S(q)_T$ for the unstable region S is introduced (Ref. 344) and then RPA is used for polymers like as for critical phenomena in Sec. 10.2.

Assuming

$$\lim_{t\to\infty} \partial S/\partial t(q, t) = 0 \tag{10.30a}$$

for the extrapolation to large t, the solution is

$$S(q, t) = S_\infty + (S_0 - S_\infty)\, e^{2R(q)t} \tag{10.31}$$

where the thermal (Cook) noise is absorbed into S_∞. Both S_∞ and S_0 are functions of the scattering vector q.

$$S_0 \ = S_0(q) \ = \ \text{intensitity at } t = 0 \qquad (10.30b)$$

is the starting structure after a quench to the temperature T being so fast that S_0 = const, i.e. S_0 is a "false" S_0 for the new temperature T in the spinodal region S. The S_∞ parameter is the *virtual structure factor* arising from the thermal (Cook) noise at the unmixing temperature T, virtual because it can be negative in the spinodal region,

$$S_\infty \ = \ S(q, t \to \infty) \equiv S(q)_T. \qquad (10.32)$$

From CHC Eq. 10.29 we obtain, using the constant-limit condition Eq. 10.30a, a relationship between the virtual structure factor and the amplification factor,

$$R(q) \ = \ -MkTq^2/S(q)_T. \qquad (10.33)$$

There remains some obscurity with the limes $t \to \infty$ of Eq. 10.30a, because, according to Eq. 10.31, $\partial S/\partial t \to \infty$ for $R(q) > 0$.

For polymers the gradient term approach is again substituted by RPA, similar to the adaptation of the OZ approach to the critical point. Then the amplification factor is

$$R(q) \ = \ -Mq^2[\partial^2 f/\partial\phi^2 + kTa^2q^2/36\phi(1 - \phi)]. \qquad (10.34)$$

If the term $\partial^2 f/\partial\phi^2$ is also substituted by $\chi_s - \chi$ then the typical length (without ϕ factors) for spinodal phase decomposition is found in the typical OZ product form

$$\xi \approx (R_0/6)\tilde{\Delta}^{-1/2}, \qquad (10.35)$$

i.e. scaled by $R_0 \gg a$ for long polymer chains, as expected from the experimental results.

The pair of Eqs. 10.31 and 10.34 is well suited for the correlation and adjustment of the experimental data of Fig. $65c_2$ for the PS_D + PVME system with a low quench depth of few kelvins, see Figs. 65e–f (Ref. 339, the diagrams are for $\phi(PS_D) = 0.19 \approx \phi_c$). The adjustment reproduces the original data with the premise that $\partial^2 f/\partial\phi^2$ can simply be extrapolated from the stable region ($f'' = \partial^2 f/\partial\phi^2 > 0$, $T > T_c$) to the unstable region S ($f'' < 0$, see Fig. 65f). Then a smooth behavior of the Onsager coefficient M as a function of temperature is obtained, without change of sign, and the diffusion coefficient D is also smooth at the spinodal but with change of sign,

$D < 0$ inside the S region. The virtual structure factor of the S region, $S(q)_T$, is negative and white (independent of q) in the relevant q region (of order 1/100 nm) and becomes singular only at the mean-field spinodal itself. This means that the adjustment used is related to a finite, though negative parameter for the chaos in the spinodal region S with local thermodynamic instability. Otherwise the driving force would be infinite contrary to the experiments.

3. General discussion

The treatment suggests that the instability is transferred to large scales ($\gtrsim R_0$) by thermodynamic arguments. Connecting the relation $R_0 \gg a$ (Eq. 10.35) with a tendency to mean-field behavior we understand that CHC is a better starting equation for polymers than for small-molecule mixtures.

The following increase of the typical length for unmixing, $\lambda(t)$, is expected to be rather universal, see Fig. 65h. This increase is interpreted as an effect of the nonlinearities. The structure function and the scattering vector can often be scaled by $\lambda(t)$: $\lambda^3 S(\lambda q)$. Such a scaling is typical for structure formation on unstable situations. The take off from the ρ level, and later on from the R_0 level, corresponds to the general features of non-equilibrium thermodynamics that describe the structure evolution by macroscopic thermal properties. The so called Ostwald ripening of the final stage, $\lambda \sim t^{1/3}$, does not have any reference to the length scale of molecules.

As mentioned above, the fact that a negative amplification factor has never been observed for polymers as yet has to do with small mode lengths $a < \lambda \lesssim R_0$. The coil radius R_0 does not define a hard sphere but is formed by segments of interpenetrating chains. As discussed in Secs. 3.3–3.4 and Chap. 9, the segments are thermodynamically active and define, via the χ parameter for instance, the mixing gap properties. They cannot, therefore, a priori be excluded from the instability production in the first stages at small scales.

This view to the role of small scales in polymer unmixing is partially confirmed by the following finding (Ref. 345). Despite that the structures observed by electron microscopy after a deep quench in the PS + PVME system are, as usual, in the 50 . . . 100 nm scale ($\gtrsim R_0$) their development could be better correlated if one assumes that they are grown from structures in the 2 . . . 10 nm scale like as for small-molecule systems.

We can therefore ask the general question why observation of polymer unmixing, as a rule (Ref. 346), gives no or only scarce indications for a small-scale (ρ level) unmixing at small times. This problem will be considered in Secs. 10.8 and 10.9.

Finally we shall consider an illustrative example for the application of distribution functions and free energy functionals $F\{\psi(\boldsymbol{r})\}$ suggested by field theory approaches (Secs. 4.1.2–4.1.3). Divide the system into cells with volume L^3 and labelled by $\alpha = 1, 2, \ldots$. Be $P_L(\{\psi_\alpha\})$ the probability to find the order parameter value ψ_1 in cell number 1, ψ_2 in 2, . . . , ψ_α in α, i.e. ψ_α means $\psi(\boldsymbol{r}_\alpha)$ and the set of all ψ's is denoted by $\{\psi_\alpha\}$. Assume that the free energy of the cell system depends on this P_L via the effective Hamiltonian of Eq. 4.10 (or a shortened variant of it),

$$P_L(\{\psi_\alpha\}) \quad = \quad Z^{-1} \exp\left[-H_{\mathrm{eff},L}(\{\psi_\alpha\})/kT\right]. \tag{10.36}$$

The theoretical optimization is a hard exercise even for the equilibrium, and additional input is required for the kinetics in unstable regions.

The simplest version is the *1-point distribution function* obtained by integration over all the other cells,

$$P_{1,L}(\{\psi_\alpha\}) \quad = \quad \int \prod_{\beta \neq \alpha} P_L(\{\psi_\beta)\} d\psi_\beta . \tag{10.37}$$

This is the probability (frequency) to find an arbitrarily given ψ value among all the ψ in the different cells. Having a time series of photographs (a movie), $P_{1,L}$ can be obtained experimentally by digital evaluation. Fig. 65i shows an example from Refs. 334 and 347. Spinodal decomposition and nucleation in the PVME + PS system is compared in the length scale of micrometers and larger. A sharp starting peak splits into two peaks corresponding, at the end, to the equilibrium binodal concentrations ($\psi = c$). In the locally unstable S region a broad decay and double peak formation is observed, whereas in the metastable K region (Fig. 63a) the nucleus of the new phase grows at a sharp constant ψ value corresponding to the new binodal.

10.4 Homogeneous nucleation

Nucleation can be observed in the metastable region K between spinodal and binodal, see Figs. 63a and 63b. It is compared to the spinodal phase separation of the S region in Figs. 66a and 66b (Refs. 322, 338, 348). An attempt to map the stages of the two processes into the phase diagram is sketched in Figs. 66c and 66d.

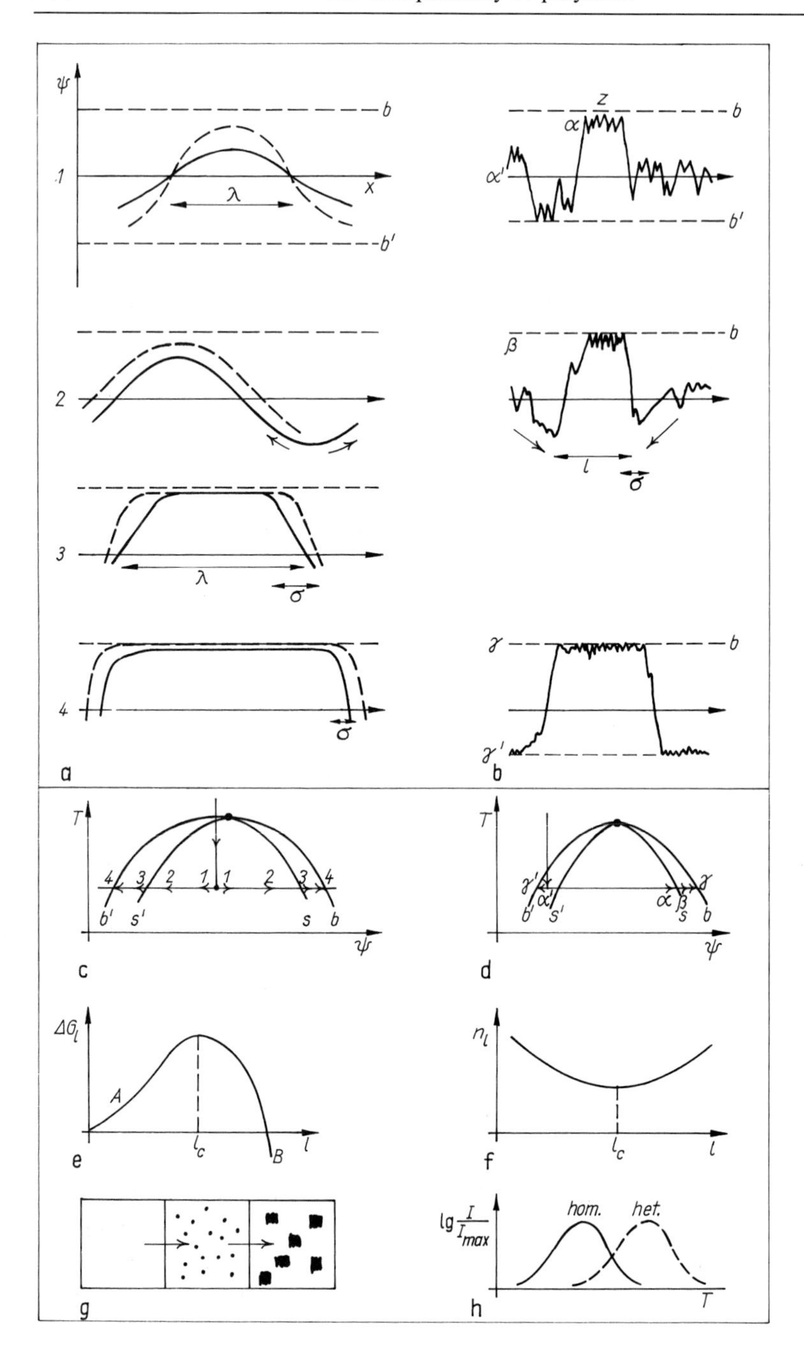
ψ
b
1
λ
x
b'
2
3
4
σ
a
z
α
α'
β
l
γ
γ'
b
T
4 3 2 1 1 2 3 4
b' s' s b
ψ
c
γ'
α
α β γ
d
ΔG_l
A
l_c
B
l
e
n_l
f
g
$\lg \frac{I}{I_{max}}$
hom.
het.
T
h

When the concentration (order parameter ψ) of the spinodal unmixing gets the vicinity of the binodal, then a second typical length appears that is diminished to the interfacial thickness during the growth stage 3 in Fig. 63a. In the ripening stage (4,γ) of *phase growth* no essential differences between nucleation and unmixing can be defined.

The main differences between the S and K region processes are observed in the first stages. In the metastable region a ψ-large ρ-level fluctuation peak (zacke) can not only be formed ($\alpha' \to \alpha$) but can also be stabilized (β) to the ψ of the new phase if the new structure has a chemical potential similar to that of the old phase. This is possible since the interdiffusion coefficient is positive and since the metastability allows the formation of interface layers with positive energy contribution.

In the spinodal region S the first stages (1, 2) are restricted to small ψ deviations far from the spinodals. The problem is that it is difficult to make oneself a clear picture of *unstable fluctuations*, especially with respect to a unique ψ amplitude. Small scale fluctuations, being present in any case, are therefore dropped in Fig. 66a. [In the sense of Sec. 10.3 one can speculate, perhaps, that the "waves" represent only the unstable ψ level fluctuation,

---

Fig. 66. Nucleation.
a. Spinodal phase decomposition. x spatial coordinate, ψ order parameter. 1, 2, 3, 4 succession of stages. In each stage, - - - is later than ——. λ "wave length", σ "surface thickness". The arrows mark the diffusion direction, b′, b binodal ψ's. b. (Homogeneous) nucleation in a similar representation. The succession of stages is now (α, α'), β, (γ, γ'). l size of nucleus. c. Map of stages 1, 2, 3, 4 in the phase diagram for spinodal unmixing. b, b′ binodals, s, s′ spinodals. d. ditto for the stages (α, α'), β, (γ, γ') of nucleation. e. Minimal work for nucleation as a function of nucleus size. l_c critical size of nucleus. A: surface term, B: bulk term of Eq. 10.38 dominates. f. Einstein–Boltzmann distribution for nuclei of different size l according to the local equilibrium formula Eq. 10.40. g. Vivid picture for the situation "stationary nucleation". Only the middle part is described by the theory. R.h.s. growth stage. h. Nucleation rate (normalized to the maximum value) as a function of temperature. hom. homogeneous nucleation (this case is described in the main text), het. heterogeneous nucleation.
Remark: The stages for the pictures a, b, c, d can shortly be characterized as follows.
1 Cahn Hilliard, constant λ,
2 increasing non-linearities, λ increases,
3 narowing of the interfacial, saturation,
4 Ostwald ripening with stabilized interface.
(α, α') large fluctuation (zacke z) near the critical size, β supercritical ($l > l_c$) nucleus (saturated, stabilized interface), (γ, γ') growth stage.

whereas the ρ level fluctuations indicate the ρ level stability in the sense of a Langevin noise so that the ψ waves arise after a ρ level preaveraging.]

Molecular thermodynamics is an art in science that attempts to apply thermodynamic concepts to situations of too small size. It is, in a sense, an extrapolation from large-scale experience to complicated molecular situations. Famous examples can be found in Ref. 349.

In the molecular thermodynamic approximation a nucleus is considered to be a small subsystem: *Droplet model*. The $\alpha' \to \alpha$ transition is ignored. The relationship to the environment is reflected by a term proportional to the surface of the subsystem. This term is added to the bulk term ($\sim \delta G$ of Chap. 9; Fig. 66e), so that

$$\Delta G_l = -Bl^3 + Al^2 \qquad (10.38)$$

with a bulk ($B > 0$) and a surface (area A) term constant, ΔG_l is the "free" energy difference between the new-phase nucleus of size l and the old-phase homogeneous system without nucleus. More precisely, ΔG_l is the minimal work $W_{\min}$ to create the nucleus. The minus sign at B in Eq. 10.38 means metastability ($\delta G \neq 0$ in Fig. 58b), the A term describes the additional energy expense for the structure in the border area, the "surface tension".

The Einstein formula for the probability of a fluctuation with $W_{\min} > 0$ reads

$$w \sim \exp(-W_{\min}/kT) \qquad (10.39)$$

(see, e.g., Ref. 45). It can, at the molecular-thermodynamic level, be interpreted as an activation barrier with $W_{\min} = \Delta G_l$, i.e. the distribution of nuclei of size l is given by

$$n_l = \text{const} \cdot \exp(-\Delta G_l/kT). \qquad (10.40)$$

The Gibbs free energy is to be used for isothermal-isobaric conditions. The distribution Eq. 10.40 has a *minimum* in the vicinity of the critical nucleus size l_c (see Fig. 66f) where ΔG_l is maximal,

$$l_c = 2B/3A. \qquad (10.41)$$

If Eq. 10.40 would describe an equilibrium situation, then all the other sizes are more probable (especially we would have $n_l \to \infty$ for $l \to \infty$). That is, Eq. 10.40 describes an unstable (metastable here) situation.

The problem to define a stationary situation with three ingredients: activation barrier (A term), driving force (B term), and unstable distribution, was solved by Becker and Döring (Ref. 350, see also, e.g., Ref. 342). This situation is called *stationary nucleation* and is sketched in Fig. 66g. Only the middle part of this figure is subject to the approach. The treatment of the large-n_l catastrophe remains local: metastable states decay via thermal activation of localized unstable fluctuations (l.h.s. of Fig. 66g), and the survivors growing up to phase regions (r.h.s. of Fig. 66g) are removed by hand from the calculation. In other words we shall consider the stationary droplet mass flow through the middle part of Fig. 66g accompanied by the boundary conditions

$$n_l \to 0 \text{ for } l \to \infty, \quad n_l^s = n_l \text{ of Eq. 10.40 for } l = l_c. \tag{10.42}$$

Consider first the construction of an evolution equation. Let $n_m(t)$ be the average number of droplets at time t containing m ($\sim l^3$) units or segments. Let us assume that any evolution $n_m(t)$ is only due to a phase-changing mechanism ($\alpha \leftrightarrow \alpha'$) in which a droplet loses or gains a single unit, $m \to m \pm 1$. Then

$$(\partial/\partial t)(n_m)(t) \quad = \quad J_{m-1}(t) - J_m(t) \tag{10.43a}$$

with J_m the growth rate per volume, $m \to m+1$. The master equation is

$$J_m(t) \quad = \quad R_m n_m(t) - R'_{m+1} n_{m+1}(t) \tag{10.43b}$$

with R_m, R'_{m+1} the kinetic coefficients depending on m. Consider, for large m, the variable m to be continuous. Then Eq. 10.43a reads $\partial n_m/\partial t = -\partial J_m/\partial m(t)$, and Eq. 10.43b leads to a Fokker Planck equation

$$(\partial n_m/\partial t)(t) \quad = \quad -(\partial/\partial m)\left[R_m\left(1 - \exp\left(-\frac{\partial \Delta G/\partial m}{kT}\right)\right) n_m(t) \right.$$
$$\left. - R_m \partial n_m/\partial m(t)\right] \tag{10.44}$$

where Eq. 10.40 is used in a continuous version for $m \to m+1$. The equilibrium condition would be $J_m = 0$, but this is not interesting for the metastable state.

The second step is to find whether there is a steady state solution of Eq. 10.44 suitable to the situation of Fig. 66g (Eq. 10.42). The answer is yes. Defining a *nucleation rate* for stable nuclei $m > m_c$ independent of the size m,

$$I \equiv J_m^s, \tag{10.45a}$$

(s for stationary) and assuming that $B = \text{const}$ (no depletion), the desired solution can be found after a Taylor series expansion around m_c. Large n_m

for $m \to \infty$ are suppressed by the diffusion type of the Fokker Planck equation and the minus sign in Eqs. 10.39 or 10.40 (for details see Ref. 342); the treatment remains, therefore, localized. The solution reads

$$I = I_0(R_{mc}, \varepsilon_c'') \exp(-\Delta G_{mc}/kT) \tag{10.45b}$$

where I_0 is a harmless function of R_{mc} and $\varepsilon_c'' \equiv \partial^2 \Delta G_m/\partial m^2$ for $m = m_c$, i.e. I_0 does not depend exponentially on its arguments. The result is a thermally activated steady state nucleation where the activation energy is determined by the critical minimal work alone, i.e. we have a preferential size (m_c, l_c) for nucleation. The prefactor I_0 is, of course, not accessible by such methods because all the details of structure and kinetics are suppressed by the R ansatz and the Taylor expansion.

A useful method for the adjustment of the temperature dependence is a *Volmer ansatz* (i.e. the exponential factor in Eq. 10.45b) where the additional kinetics is also approximated by an apparent activation term,

$$I = A \exp(-(\Delta G_{mc} + \Delta G_D)/kT). \tag{10.46}$$

The "diffusion term" ΔG_D is to characterize the mass transport, absorption, and rearrangement in and outside the droplet. Its temperature dependence can therefore be approximated by the slope of the relevant dispersion zone(s) in the Arrhenius diagram. Eq. 10.46 describes a maximum of I as a function of T. The high-temperature flank is determined by ΔG_{mc} with the "driving force" δG contained, the low-temperature flank is defined by the decreasing transport ability due to the increasing relaxation times.

In case of heterogeneous nucleation the activation energy is lower because the size of necessary rearrangements is smaller. The maximum of the nucleation shifts, according to Eq. 10.46, to higher temperatures when ΔG_{mc} is lowered.

From the standpoint of statistical physics the nucleation is a complex dynamic saddle-point problem with collective (droplet), cooperative (rearrangements) and fluctuative aspects in metastable materials. The molecular-thermodynamic approach indicates, however, that two conceptions are reasonable for a first-order approximation: diffusion and interface tension. These properties become dominating for the ripening stage leading to the equilibrium.

10.5 Diffusion

Diffusion is necessarily connected to the fact that the "chemical" environment of the components is changed. The hypothetical microflow, mentioned in Sec. 5.6, i.e. a fluctuative mutual motion of regions with no essential change in molecular or chain neighborhood (Fig. 67a) is more a shear activity and does not contribute to the diffusion, even if the amplitude would be large. Only the relative motion of a chain (segment) against its environment is important. If all chains could be differently colored, despite of their identity, then diffusion always means a change of the mixing color of a region around a given segment or chain part.

Some diffusion concepts for binary (polymer) systems are sketched in Fig. 67b–d. The D activity of Sec. 6.10, i.e. the Brownian motion of the chain center in a matrix of same (or similar) chains is called self-diffusion (Fig. 67b). This is nondissipative if the chain is "identical" with the other chains. Self-diffusion can be measured when some chains are temporarily marked (e.g. by nuclear spins in a magnetic field gradient; see e.g. Ref. 351) in such a way that the segment-segment interaction is not disturbed. If the marking of single chains is permanent, then we have the case of tracer diffusion, see Fig. 67c. This is only dissipative if we can define a changing concentration $c(\mathbf{r})$ of tracers. Otherwise tracer diffusion is a $c \to 0$ limiting case of mutual or *inter diffusion*: Fig. 67d. This diffusion is characterized by concentration gradients in mixtures of (permanently) different components. It measures, e.g., the rate at which concentration gradients are dissipated.

For polymer blends one can consider tube models with mixed tube walls. If the two components have different chain lengths then a lot of tube modifications become relevant (see e.g. Refs. 352–353). Constraint release (Fig. 31a) by means of short host chains, e.g., is an important process for larger differences in chain lengths. If one component consists of small molecules, then we have different diffusion regimes according to the different concentrations of Figs. 4a–c. The study of small-molecule concentration gradients in case of high polymer concentrations is called penetrant diffusion, see Fig. 61e. The inverse situation, polymer diffusion in dilute solution, is illustrated in Fig. 61f.

The fluctuation dissipation theorem FDT tells us that any diffusion coefficient

$$D(\phi, T, p) \tag{10.47}$$

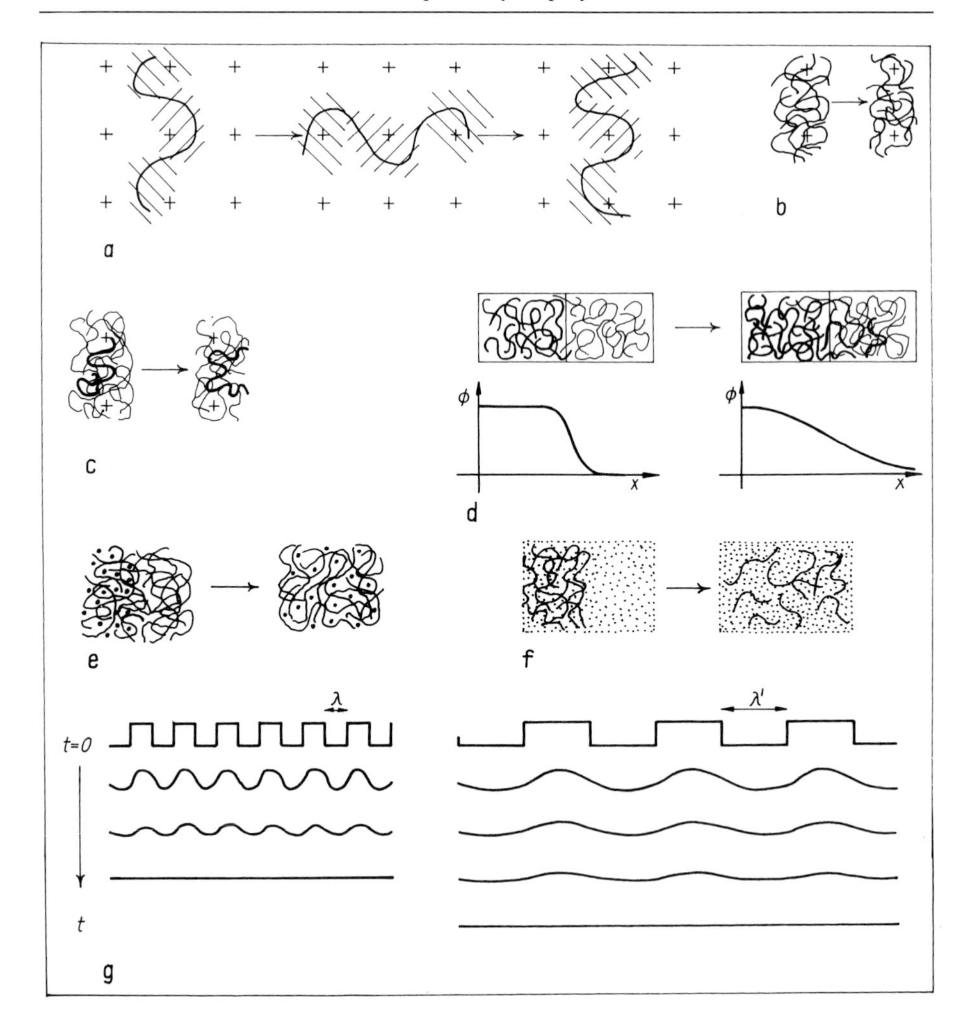

Fig. 67. Diffusion.
a. Microflow. This type of chain fluctuation does not contribute to the diffusion coefficient of the chain. b. Self diffusion. c. Tracer diffusion. d. Interdiffusion. Decay of a concentration gradient (ϕ volume fraction). e. Penetrant diffusion (the points are small molecules). f. Polymer diffusion in a dilute solution. g. Decay of a pinnacle concentration profile to measure a possible q dependence of D. q scattering vector.

is "generated" by concentration fluctuations in the equilibrium. From a dimension analysis of the diffusion equations we learn that

$$D = L^2/\tau \qquad (10.48)$$

where, for low enough temperatures, the length scale L is given by the size of the relevant natural subsystem and the time scale τ by the D activity of the relevant dispersion zone. It is the ϕ and T sensitivity of τ (Figs. 52b, c) that makes the large changes in D. A typical order of magnitude near the glass temperature is $D \approx (\mathrm{nm})^2/\mathrm{s} = 10^{14}\,\mathrm{cm}^2/\mathrm{s}$.

The multifarious diffusion phenomena mentioned above are further ramified by additional parameters and aspects. We shall consider two examples.

(i) To be precise we shall consider penetrant diffusion. The diffusion time related to a phenomenological diffusion situation will be denoted by τ_D, the relaxation time of the matrix by τ_R. Then we can define a Deborah number

$$\mathrm{De} = \tau_R/\tau_D \qquad (10.49)$$

(details see e.g. Ref. 354). The Kubo case of normal, "viscous" diffusion is observed for $\mathrm{De} < 0.1$. The "elastic" diffusion in the glass state is observed for $\mathrm{De} > 10$. In the case of "viscoelastic" diffusion, $\mathrm{De} \approx 1$, the diffusing molecule influences T_g with a feed-back to the diffusion. This implies that τ_R strongly depends on ϕ and we find nonlinear effects that modify the linear diffusion equation Eq. 1.6 (*anomalous diffusion*). Examples are front formation up to solvent swelling fracture or solvent crazing (failure) of the material.

(ii) For long chains the D activity of segments or chain parts may also be of interest as for nucleation or crystallization. According to the general scaling principle (GSP, Sec. 4.2.1) the long chain modes are slow and the shorter, segment modes are fast. This means that shorter chain parts can alter their chemical environment (new arrangement) in times where the chain center has been little displaced. In case of self diffusion, for instance, we observe a length-dependent diffusion coefficient ($D(q, \omega)$ after Fourier transformations) that is usually defined via the scattering function,

$$S(q, \omega) = S(q)/(\omega^2 + q^2 D(q, \omega)), \qquad (10.50)$$

see also Eq. 4.23. Such D's have, of course, influence on the diffusion profile in Fig. 67d.

An example how such length dependent diffusion coefficients can be measured is shown in Fig. 67g (Ref. 355). Consider the decay of pinnacle concentration profiles with different length scales λ in the relevant q^{-1} region.

Then the length scale of the wash-out is fixed by the given λ. According to the GSP the decay is expected to be faster (larger $D(\lambda)$) for shorter λ (also after λ^2/τ reduction to D). This can also be observed by light scattering as long as the method is specified to pursue the decay of the pinnacles, e.g. by interference. For nonspecified methods, of course, the usual scattering for "$\lambda \to \infty$" (Eq. 10.52, below) is observed after the amplitudes are small enough.

Let us conclude this section with a reference to a newer issue of interdiffusion in polymer mixtures. In principle the diffusion coefficient is a product of two factors: kinetics times driving force. The kinetic factors Λ_{ij} in the linear flow ansatz for a binary ($A + B$)

$$
\begin{aligned}
j_A &= -(\Lambda_{AA}\nabla\mu_A + \Lambda_{AB}\nabla\mu_B)/kT \qquad (10.51)\\
j_B &= -(\Lambda_{BA}\nabla\mu_A + \Lambda_{BB}\nabla\mu_B)/kT,
\end{aligned}
$$

are called *Onsager coefficients*. They obey the Onsager cross relation, $\Lambda_{AB} = \Lambda_{BA}$. In the transport zone (see Sec. 6.10.1) the Λ's are coefficients that can depend on ϕ but not on q.

The simplest way from the Λ's of Eq. 10.51 to the Cahn Hilliard coefficient M of Eq. 10.17 is to equalize all four Λ's, $\Lambda_{ij} \equiv M$. In the dispersion zone (see Eq. 10.50) and for the case of viscoelastic diffusion the Λ's can also depend on q.

For larger scales ($qR_0 \ll 1$) the interdiffusion of chain molecules can be pursued by light scattering, see e.g. Ref. 356. Defining the order parameter to be the Fourier component ϕ_q of the difference between the actual local volume fraction and the equilibrium ϕ, we obtain the structure function in the transport zone from the formulas of Sec. 2.8,

$$S(q, t) \equiv \overline{\phi_q(t)\phi_{-q}(0)} = S(q)\exp(-Dq^2t), \qquad (10.52a)$$

where $\phi_q \equiv \phi_{qA} = -\phi_{qB}$ for incompressible melts. [Motivated by the collective character of a diffusion experiment in a ϕ gradient, not restricted to the transport zone, a *collective diffusion coefficient* is sometimes defined by

$$D_{\text{coll}}(q, t) = -q^{-2}\partial \ln S(q, t)/\partial t \qquad (10.52b)$$

with the first cumulant

$$D_r(q) = \lim_{t\to 0} D_{\text{coll}}(q, t). \qquad (10.52c)$$

The self-diffusion coefficient is then

$$D_s(t) = \lim D_{\text{coll}}(q, t) \qquad (10.52d)$$

where the limes sign means extrapolation to large q.]

The thermodynamic factor of D is calculated by RPA, Eq. 4.31, see also Eq. 10.12. Elimination of noninteractive terms by additive reference to the spinodal, $1/S(0) = 0$ for $\chi = \chi_s$, again gives the OZ form for small q^2,

$$S^{-1}(q) = 2(\chi_s - \chi) + A^2 q^2 \phi(1 - \phi) \tag{10.53}$$

with $A = A(\phi)$ a length of order the structure length ($a \approx 0.7\,\text{nm}$) and ϕ the mean volume fraction of one component. Scaling of D with the correlation length ξ according to Eq. 10.13b yields a reasonable compromise of all motives by the formula

$$D = 2D^0(\chi_s - \chi)\phi(1 - \phi), \tag{10.54}$$

see also Fig. 65e. The diffusion coefficient $D^0(\phi)$ is usually taken for an unknown average of the tracer diffusivities D_A^0, A in B, and D_B^0, B in A, for $\phi \to 0$ and 1, respectively. It is not probable, however, to get a general formula like $D^0 = D^0(D_A^0, D_B^0, N_A, N_B, \phi)$ with only two D^0 coefficients because, in general, the three Onsager coefficients cannot be reduced to only two D_0's (see, e.g. Refs. 357, 358).

10.6 Interface

An interface is formed between two polymer phases in the vicinity of their stable binodal equilibria, the different compositions are denoted by volume fractions ϕ, see Fig. 68a. The phenomenon will be described in a simplified

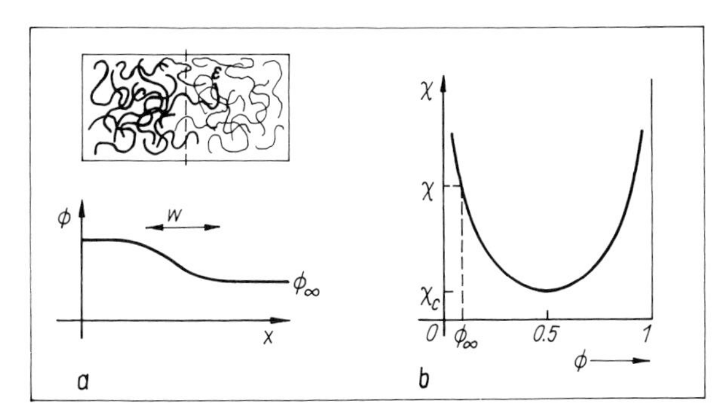

Fig. 68. Polymer–polymer interface.
a. Concentration gradient. ϕ volume fraction of the bold line component, x the spatial coordinate in direction normal to the interface. b. Binodal in a binary phase diagram of a symmetric mixture (for reference, cf. also Fig. 5a).

version of Ref. 359. Two aspects are important: the energy associated with unlike or missing bonds at the interface, and a loss of conformational entropy associated with the need to turn the long chain around as it approaches the interface.

Let us estimate, at the χ parameter level, how long (Z) a "foreign" chain end (ε in Fig. 68a) could be. The energy deficits of order $Z\chi kT$ should be of order the thermal fluctuation, kT. Therefore,

$$Z\chi \approx 1. \tag{10.55}$$

If the foreign chain end is Gauss coiled then the penetration depth λ is of order

$$\lambda \approx a\sqrt{Z} \approx a\chi^{-1/2} \approx w. \tag{10.56}$$

This is a measure of the interface thickness w. With $a \approx 0.7\,\text{nm}$ for the structure length we find $w = 1 \ldots 20\,\text{nm}$ for Flory Huggins parameters χ in the interval $10^{-3} \ldots 1$.

The interface tension σ can be estimated as the product of an energy ($\approx \chi kT$) times the number of units in the interface of thickness w, i.e.

$$\sigma \approx \chi^{1/2}\rho akT \approx kT\rho a^2/w \tag{10.57}$$

with ρ the unit number density.

The $\phi(x)$ profile across the interface can be obtained from a thermodynamic variation exercise. The rather sharp gradient should be reflected by a gradient term in the free energy,

$$\Delta F = kT \int dx[f(\phi(x)) + \lambda(\phi)(\nabla\phi)^2] \tag{10.58}$$

where f has the dimension of a reciprocal length, and λ is the gradient term constant of order the reciprocal interface thickness w. For steeper gradients the gradient term would be larger, for smaller gradients the free energy changes per length would be larger than it should be according to the estimations of Eqs. 10.56 and 10.57. From variation of the $\phi(x)$ function in Eq. 10.58 ($\Delta F = \min$), with a Flory Huggins formula for f, and Gaussian chains with a structure factor a, the optimum profile for a symmetric mixture ($N_A = N_B = N$, see Fig. 68b) is obtained to be a tangens hyperbolicus,

$$\phi(x) \approx \tfrac{1}{2}\{1 + (1 - 2\phi_\infty)\tanh(x/w)\}. \tag{10.59}$$

with w of order Eq. 10.56 but modified by a numerical and by a thermodynamic factor describing the distance from the critical point (or the mean field spinodal),

$$w = (a\sqrt{2}/3\sqrt{\chi_c})\{\chi/\chi_c - 1\}^{-1/2}. \tag{10.60}$$

The interface thickness diverges by approaching the critical state. It is larger than expected for small-molecule systems ($w \approx a$). The scaling factor for polymer mixtures is again (see Eqs. 10.14a and b) the coil diameter R_0 because $a\chi_c^{-1/2} \approx R_0$ from $\chi_c \approx 2/N$. Therefore,

$$w \approx (R_0/3)\tilde{\Delta}^{-1/2}. \tag{10.61}$$

As above, the coil structure enlarges the thermodynamic length because of the coupling between χ_c and N in Eq. 3.24.

Experiments in a nPS + dPS system with long chains (UCST $\approx$ 200°C) confirm Eq. 10.60 (Ref. 360). One can assume that the time scale τ for the establishment of the gradient is given by the flow transition, because the spatial scale is controlled by the coil diameter R_0. But the experiments show that $\tau \gg \tau_{\text{rept}}$ where the reptation time was calculated for the case of self diffusion. This points to a complicated microscopic motion within the flow transition including longer modes than reptation.

10.7 Microphase decomposition of block copolymers

Microphase decomposition is a good example for the small subsystems of Sec. 3.8. Consider a flexible diblock copolymer with components that are incompatible in the melt, see Fig. 69a. The incompatibility must lead to the formation of microphases with at least one domain dimension of order the coil diameter R_0 of blocks A or B. Otherwise the block connection would be incorporated into the phases – a contradiction to the presumed incompatibility – or the chains must be extended to larger lengths $> R_0$ which is not optimal for high extension ratios.

For a given total chain length the composition is defined by the block length ratio. Increasing the volume fraction ϕ_A we find regular domain superstructures (supercrystals) in the succession A spheres, A cylinders, layers (inversion), B cylinders, and B spheres, see Fig. 69b (Ref. 361). The point is that this structure succession cannot be explained by pure geometric reasons (e.g. area to volume ratios). Instead the thermodynamic properties of domains must depend on size and form as is typical for the small subsystems.

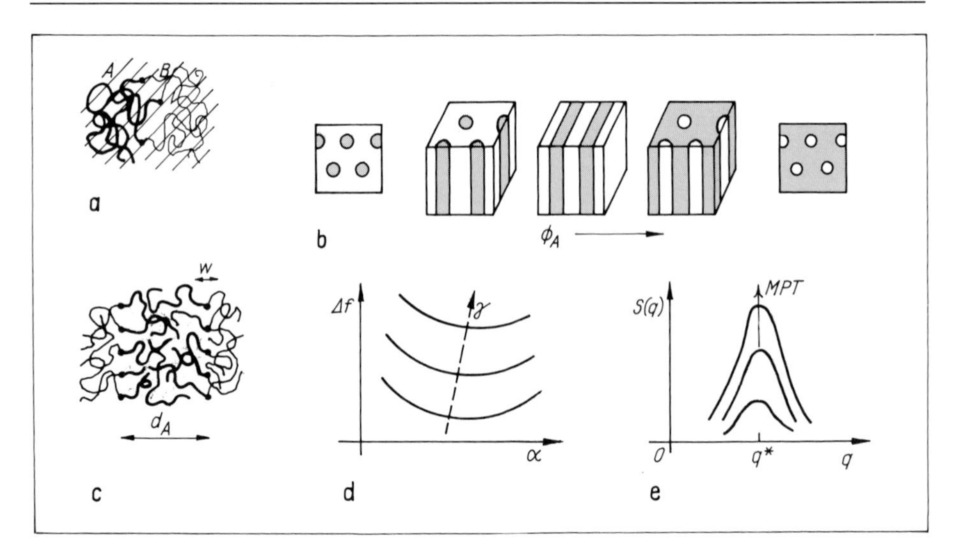

Fig. 69. Microphase decomposition of block copolymers.
a. Some block copolymers AB in the melt. b. The superstructure succession with increasing A content (gray). The changes are at about 5 and 20 per cent. c. Layer superstructure. d_A layer thickness (*A* domain size), *w* interface thickness. d. Free energy of a layer as a function of reduced layer thickness α (Eq. 10.63). Parameter γ is the reduced interface thickness (Eq. 10.64). e. Structure factor as a function of scattering vector when a micro phase transition MPT is approached; q^* corresponds to the expected layer thickness d_A.

Consider a layer structure Fig. 69c for $\phi \approx 0.5$ from this point of view (see, e.g. Ref. 362). Let be the chain length large enough to open the possibility for Gaussian coils, $R_{0A} \sim Z_A^{1/2}$, which also ensures having segments enough to form a thermodynamic subsystem across the layer. Each microphase is a *small* subsystem because the smallest length scale is smaller than the diameter of the total coil of the A + B chain. Suppose that the interface where the block bonds are collected is thin (large χ). Gaussian coils with the constraint that the block bond is located in one of the interfaces cannot form a phase with a (nearly) constant density. But this cannot be disadvantageous from thermodynamic reasons if the density deviation is larger than the natural fluctuation. The constraint, i.e. the existence of the other phase with the block bond in the interface, therefore leads to changes in the phase itself: A certain extension of the chains can provide for a constant density. This, however, requires an additional "elastic" energy.

Thus for thin interfaces we have a free energy formula for the A layer (and analogously for the B layer),

$$\Delta f \approx \frac{B'}{\sqrt{\gamma}} \alpha^{\mu} + \frac{A'}{\gamma\alpha} + c, \tag{10.62}$$

with two terms sensitive to the smallness of the layer subsystem (area term with A', elastic bulk term with B'). The variation parameter α is the layer thickness reduced by the Gaussian coil diameter,

$$\alpha = d_A / a Z_A^{1/2}, \tag{10.63}$$

and γ is the analogously reduced interface thickness

$$\gamma = w / a Z_A^{1/2}. \tag{10.64}$$

The elastic bulk term of Eq. 10.62 is the larger the larger the extension (d_A, α) and the higher the concentration of block links (small w, γ) are. The interface area term is the larger the smaller the layer thickness (more layers make more area) and the smaller the interface thickness (see Eq. 10.57) are. The exponents of the bulk term are not trivial ($\gamma^{-1/2}$, α^{μ} with $\mu \approx 2$), their calculation requires the expenditure of statistical mechanics (see e.g. Ref. 363). A diagram of Eq. 10.62 is shown in Fig. 69d. The equilibrium for a given interface thickness parameter γ is obtained from $\partial \Delta f / \partial \alpha = 0$, i.e.

$$d_A \sim Z_A^{0.6} > Z_A^{0.5}. \tag{10.65}$$

We find a chain extension due to the constraints of the small system. Without constraints the Gaussian 0.5 exponent would be obtained, of course.

Another question is: What are the conditions that during a change of temperature a homogeneous melt of block copolymers decomposes into microphases? This so-called micro phase transition was treated by Leibler (Ref. 74) with a Landau-type analysis. Two parameters are proved to be important: The relevant order parameter is $\psi(\boldsymbol{r}) = (1 - f)\rho_{\mathrm{A}}(\boldsymbol{r}) + f\rho_{\mathrm{B}}(\boldsymbol{r})$, where f is the fraction of A monomers in a chain, and the second parameter is χZ that is known from the phase diagram of polymer blends. In the latter $Z\chi_{\mathrm{c}} \approx 2$, for symmetrical mixtures, whereas $Z\chi_{\mathrm{c}} \approx 10$ for the critical point of the microphase transition.

As usual, the relevant fluctuations become large at the critical point, but one length remains fixed to the domain size d expected after the transition. If the total correlation length ξ is restricted to the domain size ($\xi \approx d$) then the transition remains relatively broad (ΔT of order 1 K). If d is only a modulation for an infinite correlation, $\xi \to d$, then the transition would be sharp in temperature ($\Delta T \to 0$). The structure factor is represented in Fig. 69e, the peak at the starred scattering vector $\boldsymbol{q}^*$ corresponds to the domain (layer) size (see Refs. 364 and 365). The theoretical situation was briefly described in the first example of Sec. 4.1.4. The actual, experimental realization of the domain structure is difficult for melts not far above T_g because relatively long times are required for the large rearrangements needed at a microphase separation. The normal way to ripe domain structures are via block copolymer solutions.

10.8 Gradient term and natural subsystem

Taking the free energy in the form

$$\Delta F = kT \int d^3\boldsymbol{r}[f\{\phi(\boldsymbol{r})\} + n\lambda^2(\nabla\phi(\boldsymbol{r}))^2], \tag{10.66}$$

then λ, the square root of the gradient term constant, has the dimension of a length (dim $f = 1/L^3$, n the mean number density of particles). Eq. 10.66 is a ψ level equation, the order parameter is a volume fraction ϕ with dimension 1,

$$\psi(\boldsymbol{r}) = \phi(\boldsymbol{r}) - \bar{\phi}, \tag{10.67}$$

where $\bar{\phi}$ is the average value. A general validity limit for Eq. 10.66 is the Ginsburg criterion Eq. 4.15, $(\Delta\psi)^2 \ll \psi^2$, where $\Delta\psi$ is the fluctuation of ψ. Larger ψ fluctuations cause ψ itself to be uncertain (see Fig. 70a) and can only be handled after renormalization (see Sec. 4.1.4).

This section is to discuss two aspects of Eq. 10.66: (i) λ cannot arbitrarily be chosen, and (ii) there is a relation to the ρ level.

Fig. 70b is for a moderate gradient. The point is that the functional $f\{\phi(r)\}$ alone can also describe gradients as far as they can be modelled by a series of equilibrium subsystems. Taking a gradient term here would mean to give the gradient a weight additional to $f(r)$. This is not correct. The situation changes for steep gradients when the difference $\Delta_x\phi$ (Fig. 70c) between neighbored *natural* subsystems (Fig. 20 of Sec. 3.6) is larger than the mean fluctuation $\Delta\phi$ in the natural subsystem. Therefore

$$\Delta_x\phi \lesssim \Delta\phi\text{: no gradient term,} \tag{10.68a}$$

$$\Delta_x\phi > \Delta\phi\text{: gradient term necessary.} \tag{10.68b}$$

For numerical estimations we should remember that $\Delta\phi$ is relatively large for natural subsystems of small size ξ.

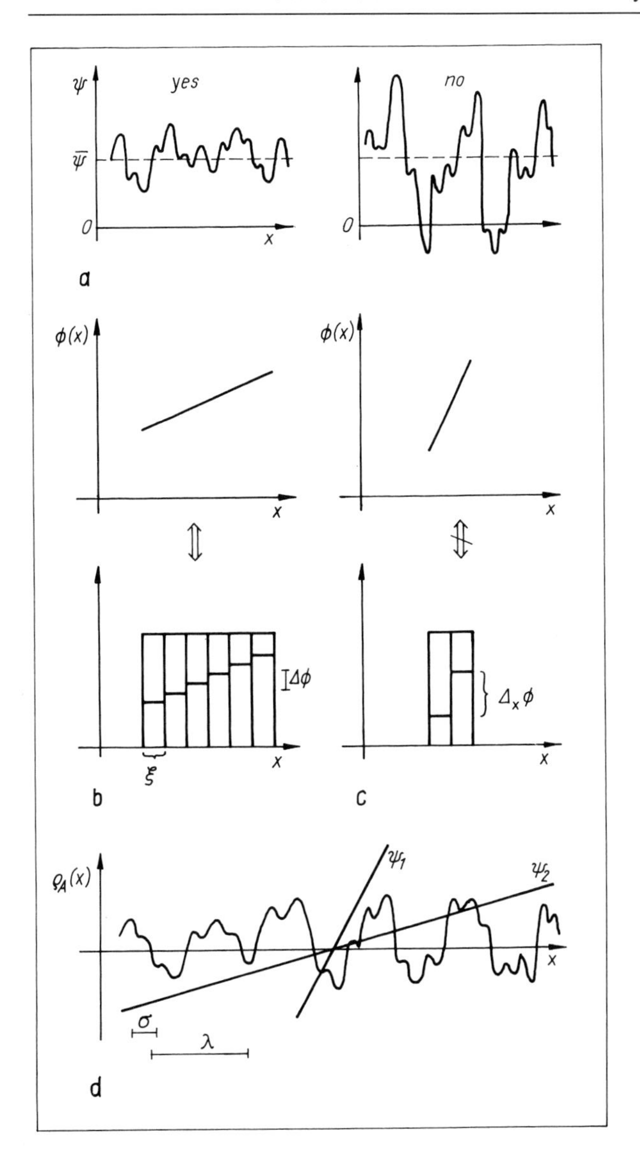

Fig. 70. Gradient terms.
a. Ginsburg criterion. b. Moderate gradients are equivalent to a series of natural subsystems. ξ diameter of a natural subsystem ($\xi \approx 2\xi_a$, ξ_a the characteristic length), $\Delta\phi$ ϕ fluctuation of a natural subsystem. c. Steep gradients are not equivalent of such a series. d. Comparison of ψ-level gradients ψ_1, ψ_2 to ρ-level fluctuation. For the ψ_2 situation the gradient term is forbidden. σ range of intermolecular potentials, λ gradient-term constant (length).

Two conclusions can be drawn from this picture. (i) The gradient term constant is of order ξ^2, i.e.

$$\lambda \approx \xi, \tag{10.69}$$

or somewhat larger, where ξ is the diameter of the relevant natural subsystem. In small-molecule glass-forming liquids between glass and splitting temperature ($T_g < T < T_s$) this subsystem has the characteristic length ξ_a according to Sec. 6.4. Therefore, $\xi \approx 2\xi_a$, which means several nanometers. If for polymer problems the flow transition is relevant (in the flow zone), then ξ is of order ξ_F or R_0, which means λ is larger, i.e. from several 10 nm up to 100 nm. This order was also obtained by the RPA analysis for the gradient-term problems in the foregoing sections. For the description of the faster main-transition kinetics (in the plateau zone), however, one has to take the corresponding smaller functional subsystem (Sec. 3.9) so that λ is again of order several nanometers, e.g. of order the entanglement spacing if the confined flow is relevant.

As, in general, the natural subsystem is larger than the range σ of intermolecular potentials or of the direct correlation function, we expect

$$\lambda > \sigma. \tag{10.70}$$

(ii) From Figs 70b and c and Eq. 10.66 we see that in case of moderate gradients Eq. 10.68a the use of gradient terms is forbidden because the thermal fluctuations can model such gradients with no additional expenditure. Since we cannot be sure that this prohibition is automatically done by Eq. 10.66 in the right way, this equation cannot be used without examination of situations with too moderate gradients. In brief, the gradient term must be switched off for too small gradients.

In contrast with the situation of too large fluctuations beyond the Ginsburg criterion, the small-gradient situation cannot consistently be discussed at the ψ level alone. At this level, Eq. 10.66 would be the zeroth and first order term of a generally valid Taylor series expansion for moderate $\psi(\boldsymbol{r})$ dependencies. The switch-off, however, makes a sense if the ρ level is taken into account, see Fig. 70d. This picture is an illustration of the conditions of Eqs. 10.68a, b if $\Delta\phi$ denotes the ρ-level fluctuation. The gradient term is clearly needed for the steep-gradient case ψ_1. In case of ψ_2 the gradient term is switched off because it is already contained in the variation that can be modelled by the functional $f\{\phi(\boldsymbol{r})\}$.

10.9 Granulated phase decomposition

Recall the difference mentioned several times above: On the one hand, the spinodal phase decomposition of polymer mixtures usually starts at the ψ level, i.e. in the spatial scale of the coil diameter $\lambda \approx R_0$ or larger, ξ_F. On the other hand, the main part of the thermodynamic excess function is "generated" in the ρ level by thermodynamic activities of chain segments. In this section we consider the question if the first stages of unmixing for polymers can also start at the ρ level, or, whether there are other reasons for the R_0 scale.

[Instead, one can also refer to the Flory-Huggins tendency of Fig. 18: the longer the chains are the worse (thermodynamically) the mixing in the total system gets. The long chains can dominate the unmixing phenomena, because in large scales of order R_0 the mixing entropy is heavily damped by a factor $1/N$. This opens, in principle, the way for small segment activities to act directly. Although they could be killed by a possibly large segment mixing entropy, the large scale damping can transfer their action to the large scale, i.e. to the total behavior.]

More consequently, a thermodynamic treatment of this problem is to base on the thermodynamics of small subsystems (Sec. 3.8). Does the small system only feel the chain part in it or does it feel the whole chain, partly outside? Does a smaller subsystem feel a smaller chain part? And: the outer chain parts form the constraints to the small subsystems. Do they act in another way than the inner parts? We shall discuss this problem by means of a Sec. 3.8 hypothesis (independent small subsystems, Ref. 366). The level of treatment is given by Eq. 9.5. [Of course, critical points are also possible for special excess functions $\chi(\phi)$. The following discussion, however, is not restricted to that case.]

To make the hypothesis more precise consider a small subsystem in the spatial scale $\lambda \lesssim R_0$, (see Fig. 71a, from Fig. 22a), being in the stable region of a homogeneous mixture in the vicinity of a critical state. The constraints of the small subsystems are assumed to be defined by the many chains going through their walls. Assume that the self-experiment hypothesis of Sec. 3.7, Figs. 21d, e, is valid. Then all (including nonextensive) variables of the small subsystem do, in principle, depend on the size λ. In a consequent version of the small-subsystem concept this statement includes the chemical potentials of the components, $\mu_B(\phi, \lambda)$. Such subsystems are called *independent small subsystems.*

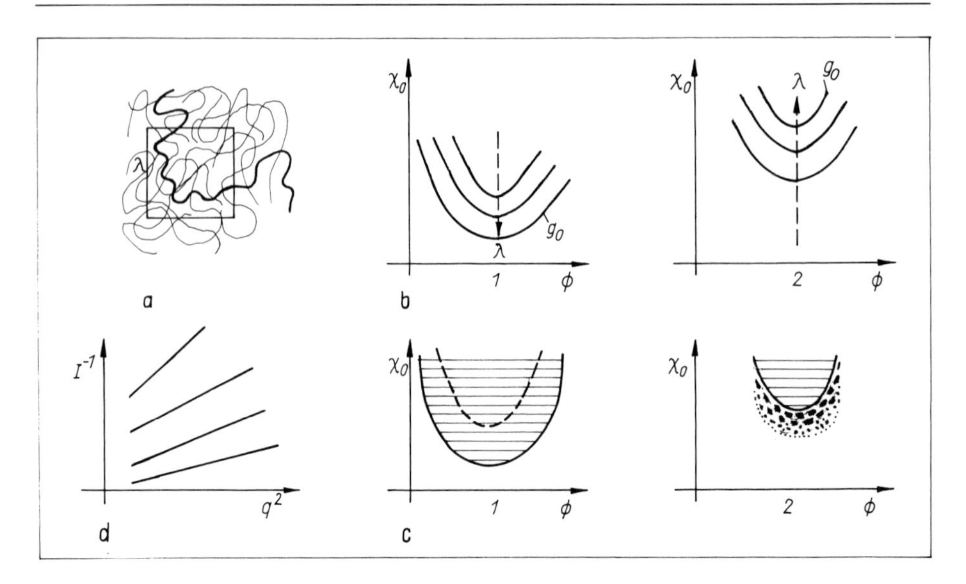

Fig. 71. Scale-dependent incompatibility for independent small subsystems.
a. Small subsystem of size $\lambda < R_0$ in a homogeneous mixture near critical point, see also Fig. 22a. (R_0 chain coil diameter). b. Formal λ dependence of miscibility gap in a $\chi_0\phi$ phase diagram. χ_0 bulk Flory Huggins parameter, ϕ volume fraction of one polymer component (not normalized). 1: case of Eq. 10.72a, 2: of Eq. 10.72b. c. 1 forced unmixing, 2 granulated unmixing. d. Modification of the Ornstein Zernicke (OZ) diagram for granulated unmixing with large excess enthalpy contribution $h^M(\lambda)$. I scattering intensity, q scattering vector.

The Gibbs free energy per thermodynamically active segment will be denoted by g, with g_0 its value for $\lambda \to \infty$ (e.g. $\lambda \gg R_0$). Then

$$g = g_0 + g(\lambda) \qquad (10.71)$$

where $g(\lambda)$ is the part depending on the size with $g(\lambda) \to 0$ for $\lambda \to \infty$. We can consider two cases:

$$1.\ \partial g/\partial\lambda > 0, \quad \text{or} \quad g(\lambda) < 0, \qquad (10.72a)$$

$$2.\ \partial g/\partial\lambda < 0, \quad \text{or} \quad g(\lambda) > 0. \qquad (10.72b)$$

Connecting an increasing $g(\lambda)$ with the tendency to unmixing (see Eq. 9.5) then we have the situation that the polymer compatibility in the vicinity of C depends on the size λ. The formal construction of mixing gaps depending on the spatial scale λ of the small subsystems is shown in Figs. 71b1 and b2:

1. $g(\lambda) < 0$, Fig. 71b1; larger λ correspond to larger g values, the unmixing starts at large scales.

2. $g(\lambda) > 0$, Fig. 71b2; g is larger for smaller λ, the unmixing starts at small scales.

The thermodynamic consequences are illustrated in Figs. 71c1 and c2.

Case c1, $g(\lambda) < 0$. The unmixing starts (and wins) in the largest scale, of order R_0. The larger scale is always more unstable than the smaller scale. This case is called *forced phase decomposition*. The time scale is dominated by typical relaxations in the R_0 scale, the smaller time scale of segment motion cannot be "seen" because their stability is undermined.

Case c2, $g(\lambda) > 0$. The unmixing starts in the small scale (order of segment size) and results in phase decomposition inside the χ_0 region of the state diagram corresponding to the large-scale g_0. In the other region, corresponding to the mixing gap of smaller scale $g(\lambda)$, the situation remains stable in larger scales. We obtain, therefore, a stable equilibrium modulation [at least in the time scale $\tau < \tau_R$. This restriction corresponds to a special interpretation of Fig. 21e, where the "external" quanta, going through the walls, influence the thermodynamics. For independent small subsystems the modulation is stable for all times.] The modulation length λ is small for small χ_0 and increases with increasing χ_0, see Fig 71c2. The modulation is superposed by the general critical opalescence. This case c2 is called *granulated phase decomposition*. The temperature region of granulated unmixing is of order 10 K because $d\chi_0/dT$ is rather small for typical polymer systems.

Writing $g(\lambda) = g_{\text{comb}}(\lambda) + g^{\text{E}}(\lambda)$ it depends on the scale-dependent additional excess contribution $g^{\text{E}}(\lambda)$ which case is actually observed, because the combinatoric contribution $g_{\text{comb}}(\lambda)$ is systematic. If the relevant self-experiments really occur only inside the independent small subsystems then the chain identity is also defined inside the subsystem. The smaller the subsystem is the smaller the identity carriers are. The identity is "screened", i.e. the chain lengths N_1 and N_2 in Eq. 3.17 become shorter for decreasing λ. This corresponds to the Flory-Huggins tendency, i.e. $\partial g/\partial\lambda > 0$, and favors forced phase decomposition.

This would explain why, as a rule, the unmixing in polymer systems is not observed in the segment scale. The noncombinatoric, "segmental" contribution to $g(\lambda)$, the excess part $g^{\text{E}}(\lambda)$, is not systematic in its sign. Large $g^{\text{E}}(\lambda)$ are necessary for granulated unmixing.

These phenomena may be described by a q-dependent χ parameter. (A $\chi(q)$ in the angstrom scale was introduced in Ref. 367, see also Eq. 9.22 that is from Ref. 77). But now the length scale λ corresponds to the ψ level, and not to the ρ level,

$$q \approx \lambda^{-1}. \tag{10.73}$$

In a crude approximation we can $\chi(q)$ model by

$$\chi = \chi_0 + \alpha R_0^2 q^2 \tag{10.74a}$$

(or ξ_F instead of R_0) with a constant α ($|\alpha|$ of order 1), where $\alpha > 0$ is for granulated and $\alpha < 0$ for forced unmixing. This corresponds to a free-energy density with, formally, two gradient terms,

$$f \approx \text{FH terms} + \alpha R_0^2 (\nabla\phi)^2 + b^2 (\nabla\phi)^2 \tag{10.74b}$$

The first (α) gradient term is the ψ level term. The b^2 term is the inevitable ρ level gradient term switched off for too small gradients (see Sec. 10.8).

For $\alpha > 0$, Eq. 10.74b can model the modulated ψ level structure, because the variation of the first two terms leads to a hyperbolic equation,

$$\alpha R_0^2 \Delta\phi' + \text{const}\ \phi' = 0, \quad \text{const} > 0. \tag{10.75}$$

with Δ the Laplacian and a ϕ' counted from the average composition, $\phi' = \phi - \phi_{av}$. The solution is proportional to sin (x/λ) or cos (x/λ) where x is a spatial coordinate. For $\alpha < 0$ the negative gradient term of Eq. 10.74b indicates a global (λ scale) instability of the free energy.

In the ρ level unmixing, with steep gradients, both gradient terms are important. This corresponds to one effective gradient term with a constant

$$K' \sim \alpha R_0^2 + b^2 \tag{10.76}$$

that determines, according to Eq. 10.10, the slopes in the Ornstein-Zernicke (OZ) diagram (and modifies the correlation length Eq. 10.11). For large values of $\partial\alpha/\partial T$ – large values of the λ dependent excess enthalpy $h^M(\lambda)$ – the gradient term constant K' depends on the temperature. This could explain the anomalous situation of an OZ diagram with slopes depending on temperatures, Fig. 71d. The correlation length meets the λ scale (R_0) in the region of granulated unmixing (a few ten K's beyond C).

Such an OZ diagram is observed by neutron scattering in the PS_D + PTMC system (Ref. 367). Eq. 10.9a indicates that the granulated phase decomposition ($h^M(\lambda) \neq 0$) would induce changes in the direct correlation function of the ρ level.

An experimental verification of the granulated unmixing would have a fundamental importance for general thermodynamics of small systems, even if the details of $g(\lambda)$ could not be explained in all molecular details. Furthermore, the conception of independent small subsystems would modify many theoretical treatments in the K and S regions of Fig. 63.

11. Semicrystalline polymers

The typical 10-nanometer structure elements in semicrystalline polymers are layer crystallites (Sec. 1.7, Fig. 8) with a lamellar thickness of about 5–20 nm. The relaxation phenomena connected with them are described in Sec. 8.5 and Fig. 54. With respect to the question of the thermodynamic activity of segments at the ρ and ψ level, discussed in Chaps. 9 and 10, a look to the chain structure of the lamellas (Fig. 8c) unambiguously leads to the conclusion that segments or short chain parts participate in controlling the thermodynamic processes.

Lamellas are not the equilibrium structures, see Fig. 72a and b. Equilibrium crystals are possible only under special conditions – e.g. the application of high pressure in linear PE to fill the whole sample with extended-chain crystals. This is the equilibrium structure because the free energy is much lower without foldings, see Sec. 11.5 below. However, even PE with moderate chain length crystallizes under normal conditions, from melt or solution, in the layer form.

Usually stacks of lamellas are organized, where the material between the lamellas is amorphous and entangled. The stack period is called long period (L in Fig. 72b), it is of order R_0, the mean chain end-to-end distance (or, in short hand, the coil radius or diameter). The lamellear cyrstallites are rather imperfect, the smaller their thickness l is the more imperfect they are, see Fig. 8h. This includes conformational disorder and cooperative segment motion, e.g. the α_c glass transition in Fig. 54c. [We shall not discuss that and how disorder in the chemical configuration (perturbation, or breakdown, of translation symmetry along the chain) hinders and finally prevents crystallization in polymers.]

This chapter is to discuss some nonequilibrium phenomena such as nucleation and crystal growth at a very primitive level of molecular thermodynamics. In the last section we shall consider the general reason for development of lamellas. A more thorough representation can be found in Refs. 8, 26, and 368.

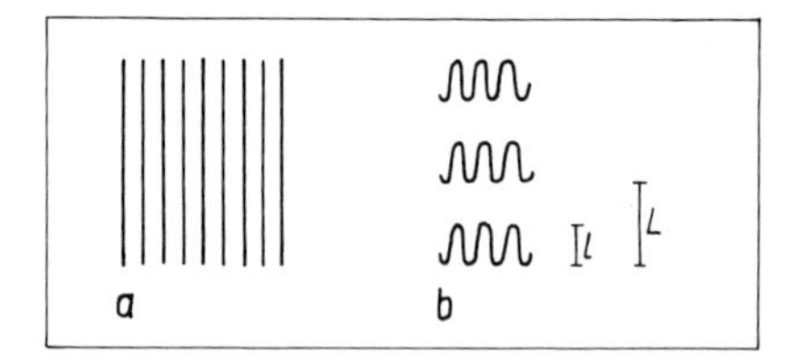

Fig. 72. Layer crystallites.
a. Extended-chain crystals are the equilibrium structure (helices included). b. Lamellas are not in thermodynamic equilibrium. l lamellar thickness $\approx$ fold length, L long period of a stack.

11.1 Terminology and salient facts

A chain folded lamella is schematically represented in Fig. 73a. The fold surface (area X^2) is characterized by the *fold surface energy* σ_e, an interfacial free energy, the four lateral surfaces (area each $X \cdot l$) by the lateral surface energy σ. The value of σ amounts only to a few ten per cents of σ_e, the latter is much larger due to the fold energy, even if the folds are not so regular as depicted in Fig. 73a. Examples: PE, $\sigma \approx 15$, $\sigma_e \approx 93$ (± 10); PEO, $\sigma \approx 10$, $\sigma_e \approx 20$ (or 40 according to other references, all values in erg/cm^2).

The crystal growth direction ($\rightarrow$) is normal to the lateral surfaces, the lateral

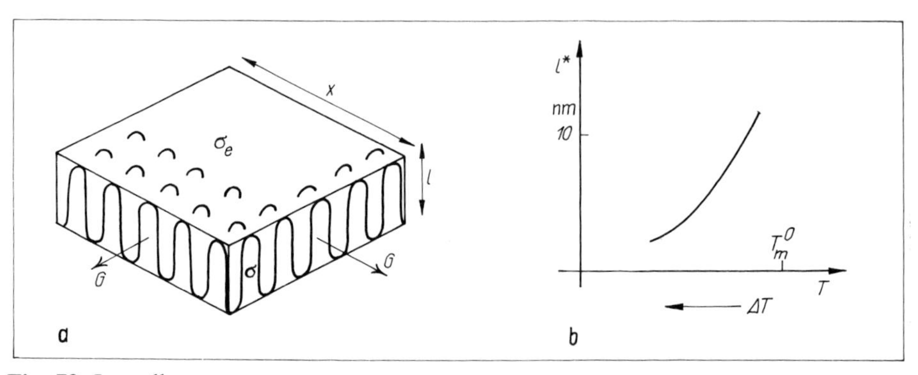

Fig. 73. Lamella.
a. Sketch of a chain-folded lamella. σ_e fold surface energy, σ lateral surface energy (both are interfacial tensions), G crystal growth, l fold length $\approx$ laminar thickness. b. Primary fold length l^* as a function of temperature. T_m^0 ideal melting temperature ($l \rightarrow \infty$), $\Delta T = T_m^0 - T$ undercooling.

growth rate is denoted by G (m/s). An increase of the fold length $l(t)$ is called thickness growth.

The free energy difference between a layer crystallite and the melt can be estimated from the thermodynamics of small systems. Using the molecular-thermodynamic concepts σ, σ_e we obtain

$$\Delta G = 4Xl\sigma + 2X^2\sigma_e - X^2 l\Delta g \tag{11.1}$$

where Δg (Joule/volume) describes the bulk contribution. Denoting the heat of fusion per unit volume by Δh, then, with $\Delta s = \Delta h/T_m^0$,

$$\Delta g = \Delta h - T\Delta s = \Delta h^0 - T\Delta h/T_m^0 = \Delta h^0 \cdot \Delta T/T_m^0 \tag{11.2}$$

where the upper index ^{0} refers to the ideal melting temperature T_m^0 of the equilibrium extended-chain crystal Fig. 72a. At $T = T_m^0$ we have $\Delta g = 0$, of course. The under or supercooling below T_m^0, for crystallization at T, is denoted by

$$\Delta T = T_m^0 - T > 0. \tag{11.3}$$

Using this terminology we can state the salient experimental facts (Ref. 369)

(i) Flexible chains have an astonishingly persistent tendency to fold; they will fold even when the chain length is only marginally longer than a primary fold length l^*.

(ii) The primary fold length and the lateral growth rate is uniquely determined by the supercooling.

(iii) The chains may tend to rearrange immediately after the primary chain folded deposition, with a tendency to bring the chain ends to the surface and to increase the overall fold length.

These facts are described in terms of a special kind of nucleus: (prefolded) pieces of extended chain parts, so-called primary folds of length l^*, that can be deposited as a whole at the lateral growth surface (Ref. 370). Molecular thermodynamic concepts of nucleation (see Sec. 10.4, and Sec. 11.2 below) yield

$$l^* = 2\sigma_e/\Delta g + \delta l = 2\sigma_e T_m^0/\Delta h\Delta T + \delta l. \tag{11.4}$$

The typical order is 10 nm, l^* decreases with increasing undercooling ΔT, Fig. 73b, down to a few nanometers for ΔT of order 50 K. The correction term $\delta l > 0$ is added to ensure an actual growth by means of a driving force that must be gained from ΔG of Eq. 11.1 by virtue of a slightly larger length. The lateral deposition indicates the first term of Eq. 11.1 to be the source of δl; one can therefore expect $\delta l \sim 1/\sigma$ for adjustments. Formally, Eq. 11.4 would lead to $l^* \rightarrow \infty$ for approaching T_m^0 ($\Delta T = 0$). This means that particularities are expected to occur near T_m^0 ($\Delta T \lesssim 10\,\mathrm{K}$).

11.2 Primary and secondary nucleation

Simple formulas for the activation energies of homogeneous (see Sec. 10.4, and Figs. 74a and b) and secondary nucleation are described in this section by means of molecular thermodynamics of small systems.

Consider a crystal nucleus to be a square column of length l, parallel to the stems, with a sectional area a^2. According to Eq. 11.1 the free activation energy is given by

$$\Delta G = -al^2\Delta g + 4al\sigma + 2a^2\sigma_e. \tag{11.5}$$

The difference $\sigma \neq \sigma_e$ indicates an anisotropy of the nucleus, surely amplified by some chain folding. The size of the critical nucleus (l^*, a^*) is obtained from $\partial\Delta G/\partial l = 0$ and $\partial\Delta G/\partial a = 0$, viz.

$$l^* = 4\sigma_e/\Delta g, \qquad a^* = 4\sigma/\Delta g, \tag{11.6}$$

i.e. $l^*/a^* = \sigma_e/\sigma$ (≈ 6 for PE). This means that the anisotropy also determines the (extended) form of the nucleus For small undercooling Δg can be substituted by Eq. 11.2. Then

$$l^* \sim a^* \sim (\Delta T)^{-1}. \tag{11.7}$$

For large undercooling the temperature dependence of the thermophysical properties must be taken into account. From adjustments a decrease of Δh with lower temperature was found, e.g. $\Delta h \approx \Delta h^0(T/T_m^0)$, then

$$\Delta g(T) \approx T\Delta T\Delta h^0/T_m^{02} \tag{11.8}$$

where the index ^{0} again indicates the equilibrium according to Fig. 72a.

The critical barrier height, $\Delta G^* = \Delta G(a^*, l^*)$, is obtained for small ΔT as

$$\Delta G^* = 32\sigma^2\sigma_e/(\Delta g)^2 = 32\sigma^2\sigma_e(T_m^0)^2/(\Delta h^0\Delta T)^2 \sim (\Delta T)^{-2}. \tag{11.9}$$

Heterogeneous nucleation at a foreign surface is only advantageous if there is a gain of area energy, i.e. for small enough $\Delta\sigma(<2\sigma)$. From Fig. 74c one finds

$$\Delta G = -abl\Delta g + 2ab\sigma_e + 2bl\sigma + al\Delta\sigma \tag{11.10a}$$

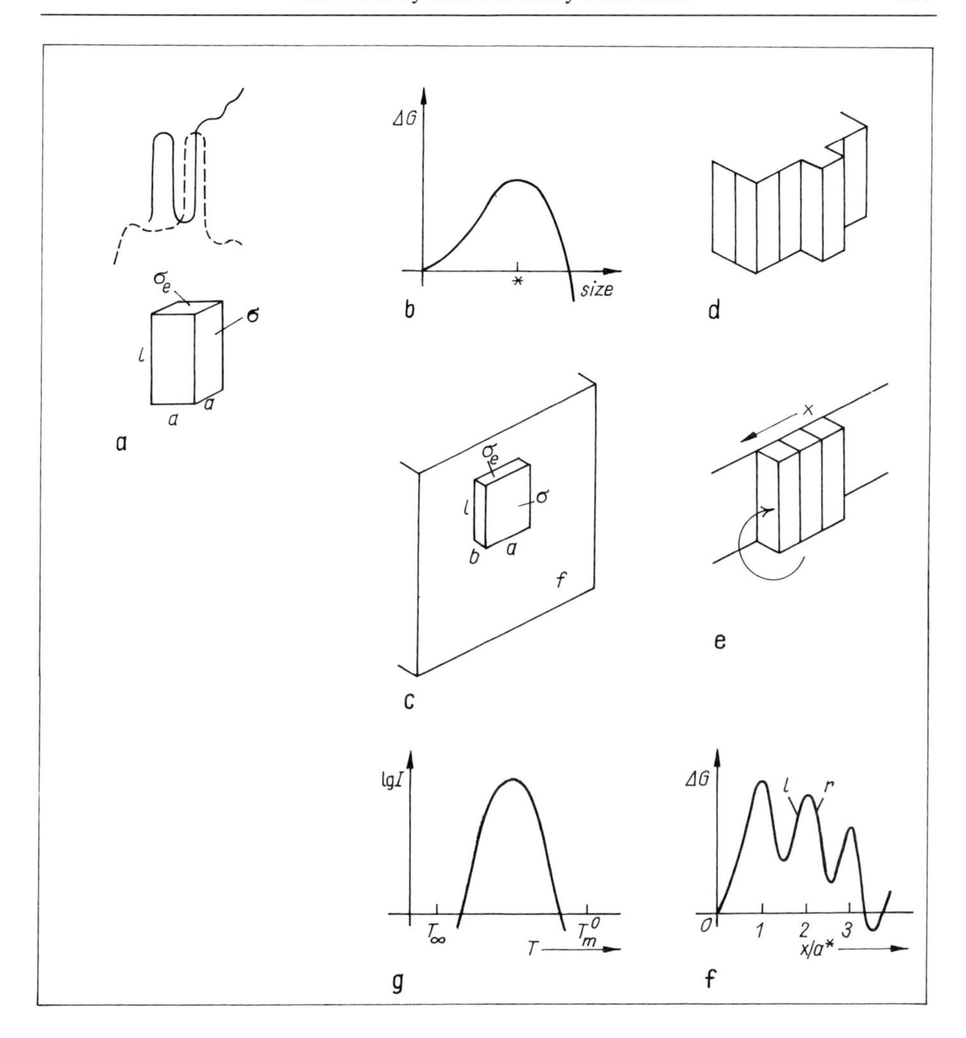

Fig. 74. Polymer nucleation.
a. "Homogeneous" nucleus. Upper part: a hypothetical structure in the melt (schematically). Lower part: molecular-thermodynamic simplification. σ, σ_e are the appropriate surface energies. b. Energy barrier for nucleation. ΔG additional free energy, * critical nucleus size. c. Model for heterogeneous nucleation, f foreign surface. d. Secondary nucleation (first step). e. Successive deposition in the niche. X direction for completion. f. Activation barrier system for successive deposition in niches (schematically). g. Maximum of nucleation rate $I(T)$. T_∞ Vogel temperature, T_m^0 equilibrium melting temperature.

where

$$\Delta\sigma = \sigma + \sigma(\text{foreign/crystal}) - \sigma(\text{foreign/melt}). \quad (11.11)$$

For large undercooling all dimensions of the nucleus become smaller, and b can reach the stem diameter b_0 ($\approx 0.5\,\text{nm}$). Then b is dropped from the optimization, and we have (without the correction Eq. 11.8)

$$\Delta G = -al(b_0\Delta g - \Delta\sigma) + 2b_0(a\sigma_e + l\sigma). \quad (11.10b)$$

Optimization for the critical nucleus yields

$$l^* = 2\sigma_e/(\Delta g - \Delta\sigma/b_0), \quad a^* = 2\sigma/(\Delta g - \Delta\sigma/b_0) \quad (11.12)$$

and

$$\Delta G^* = 4b_0\sigma\sigma_e T_m^0/(\Delta T\Delta h^0 - T_m^0\Delta\sigma/b_0) \simeq \Delta T^{-1}. \quad (11.13)$$

The barrier height of heterogeneous nucleation does not so sharply depend on the undercooling when compared with the homogeneous case of Eq. 11.9.

The so-called *secondary nucleation* (Refs. 370, 371) is the deposition of a nucleus formed in the vicinity of the lateral surface of the lamella. The surface is assumed, in the original version, to be rather smooth, i.e. not too rough. Secondary nucleation can then be considered as an heterogeneous nucleation with $\Delta\sigma = 0$. From Eq. 11.12 we obtain

$$l^* = 2\sigma_e T_m^0/\Delta T\Delta h, \quad a^* = 2\sigma T_m^0/\Delta T\Delta h. \quad (11.14a)$$

Eq. 11.14a substantiates the main term of Eq. 11.4 for the primary fold length. The energy barrier is formally obtained as

$$\Delta G^* = 4b_0\sigma\sigma_e T_m^0/\Delta T\Delta h \sim \Delta T^{-1}. \quad (11.14b)$$

The term "formally" was used because the activation barrier has not the form of Fig. 74b. The niches of Fig. 74d favor an advantageous deposition of further stems in their vicinity, see Fig. 74e. The consequence of this picture is a barrier system for secondary nucleation, Fig. 74f. As long as the surface roughness is not too high, the first (the largest) maximum is followed by further maxima. The left flank (l) of them is the energy expense for the next fold or "deposition", and the right flank (r) describes the remainder of the free energy from the "crystallization". This is a collective phenomenon, the stability ($\Delta G < 0$) is not reached before a certain number of barriers is deposited.

To describe the temperature dependence of the nucleation rate at the level of the enlarged Volmer equation Eq. 10.46,

$$I = I_0 \exp(-\Delta G^*/kT - \Delta G_D/kT), \quad (11.15)$$

we must add a transport term ΔG_D. The temperature dependence of ΔG_D is determined by the slope of the relevant dispersion zone in the Arrhenius diagram, e.g. by

$$\Delta G_D/kT \approx \text{const}/(T - T_\infty) \tag{11.16}$$

with T_∞ the relevant Vogel temperature. Taking l^* to be the relevant (mode) length, then for large undercooling we must consider the glass transition as the relevant motion, because their characteristic length ξ_a is of order l^* there. For lower undercooling (larger temperature) l^* increases (whereas ξ_a decreases) and we must consider modes of the confined flow, the hindering, and finally the flow transition, if the nucleation requires a "reeling out" (Ref. 372) of chains from the entanglement. This means that the situation is rather complicated, and we must additionally refer to modifications of molecular modes near the surface. An extreme example will be discussed, for crystal growth, in the next section.

Inserting the ansatz Eq. 11.16 into Eq. 11.15 we obtain an equation that is called *Turnbull Fisher equation* (Refs. 373, 374). The value of ΔG^* is to be specified for the different cases. For the secondary nucleation, e.g., with the modification of Eq. 11.8, we would obtain

$$I \approx I_0 \exp(-A/T\Delta T - B/(T - T_\infty)) \tag{11.17}$$

where A, B, and I_0 are parameters with low dependence on temperature. The parameter A originates from the thermodynamic driving forces, B from the appropriate WLF equation, and for I_0 there is at present only a little chance to understand its numerical value, since the details of molecular/segmental deposition in the nanometer scale are unknown and cannot, of course, quantitatively be calculated from molecular thermodynamics.

Eq. 11.17 describes a typical maximum in the temperature dependence, see Fig. 74g. The right flank is described by the thermodynamic driving force, increasing with ΔT, the left flank is determined by the transport ability of the material, decreasing for large ΔT. A similar curve is also obtained for homogeneous nucleation. The maximum of homogeneous nucleation is, as a rule, below the maximum of heterogeneous or secondary nucleation, mainly by virtue of the larger ΔG^* values and the sharper temperature dependence (e.g. ΔT^{-1} vs. ΔT^{-2}).

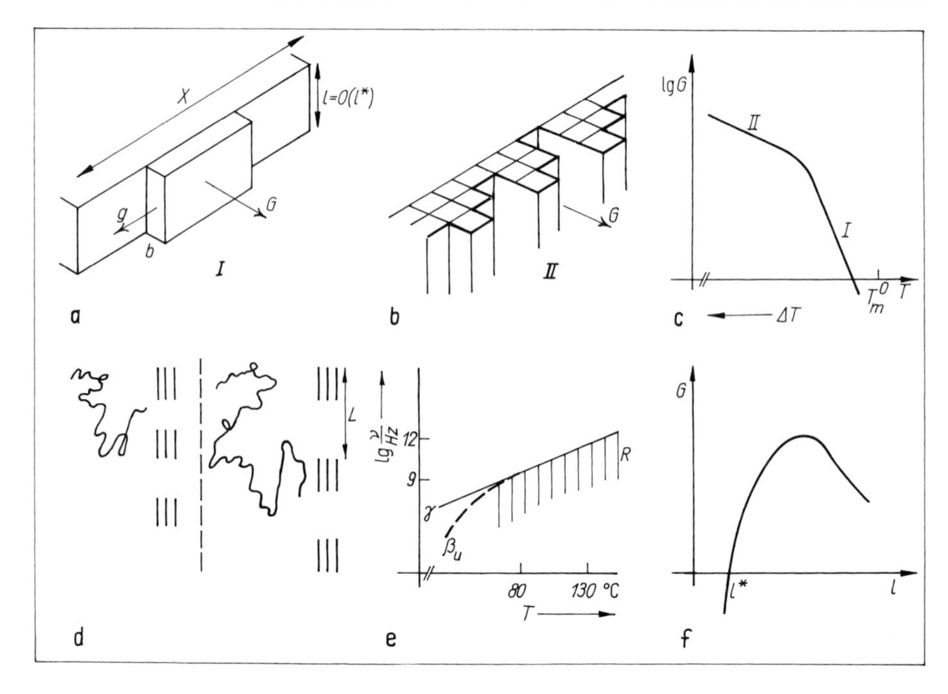

Fig. 75. Lateral growth of polymer layer crystals.
a. Regime I growth. X dimension of lamella, l thickness, b stem diameter, g completion velocity, G growth rate. b. Regime II growth. c. Temperature regions for the two regimes in polymers like PE. ΔT undercooling, T_m^0 ideal melting temperature. d. Chain-length dependence of long period L at the start of crystallization. e. Detail of the relaxation card for polyethylene. γ local mode, β_u upper glass transition (in the interface), R Rouse modes. f. Optimum growth rate at a certain lamellar thickness.

11.3 Nucleation-controlled crystal growth

Consider the possibility that the repeated secondary nucleation determines the lateral growth rate G. In other words we shall consider the secondary nucleation of Fig. 74e with respect to the possibility that it controls the crystal growth normal to a lateral surface, see Fig. 75a. Obviously the problem is influenced by three variables, i, g and X.

i, the surface nucleation rate (dimension 1/ms), is the number of chain straightenings per time and per length in the X direction.

g is the completion velocity (m/s) in the X direction, well imaginable if the surface is smooth enough; g is the velocity of a niche.

X is the available length (m), also well defined for not too rough lateral surfaces.

Consider a decreasing temperature (i.e. increasing undercooling ΔT). Then we can distinguish between at least two regimes (introduced in Ref. 375 for nonchain molecules; for polymers see e.g. Refs. 376–379).

Regime I. Small i allow the growth step, following nucleation, to sweep completely across the lateral surface of the crystal before the next layer of thickness b is formed. Then we have

$$G = biX. \tag{11.18}$$

Regime II. For larger i new growth steps are allowed to nucleate before the previous layer is completely filled, see Fig. 75b. This situation is not longer influenced by the length X. As long as G (m/s) remains linearly proportional to the stem diameter b, we find from dimensional analysis

$$G = b(ig)^{1/2}. \tag{11.19}$$

The crossover between I and II is determined by the ratio i/g and the length X available. The dimensionless characteristics is

$$X^2 i/g. \tag{11.20}$$

This crossover (the roughening transition) is found at $X^2 i/g$ values of order 1.

The temperature dependence of this lateral growth is described by a Turnbull Fisher equation Eq. 11.17,

$$G = G_0 \exp\left(-A_j/T\Delta T - B/(T - T_\infty)\right), \quad j = \text{I or II} \tag{11.21a}$$

where G_0 depends on X, the regime, and so on. Using the different i exponents in Eqs. 11.18 and 11.19 we observe $A_\text{I} = 2A_\text{II}$, where A_I is determined by Eq. 11.15 with ΔG^* according to Eq. 11.14b ($b = b_0$). Thus

$$A_\text{I} = 4b_0\sigma\sigma_\text{e}T_\text{m}^0/k\Delta h. \tag{11.21b}$$

For PE the crossover is at the high temperature flank of the maximum, and the behavior is sketched in Fig. 75c.

In the vicinity of the G maximum the delivery of "appropriate" segments (or stems) from the melt becomes important, see Ref. 372. The maximum value found in PE was $G \approx 2\,\text{m/s}$ at $T \approx 80°\text{C}$ ($\Delta T = 50\,\text{K}$), Ref. 380. The main problem is the question which melt modes can supply segments for a deposition at so high a rate.

The spatial situation for crystallization in polymer melts will be sketched for two limiting cases. (i) It was found (Ref. 381) that at the beginning of crystallization the long period L Fig. 72b approximately coincides with the coil radius R_0, $L \approx R_0$, see Fig. 75d. The longer the chain is, the larger is L,

$$L \approx aZ^{1/2}. \tag{11.22}$$

Neglecting all the details we may perhaps conclude that, at least for the first stage, individual chains play a role. Applied to the stationary case of nucleation-controlled crystal growth this would mean that the chain which has just supplied a stem has a certain advantage to deposit the next stem in the niche. This so-called *molecular nucleation* (Ref. 8) can also explain the clustering of stems in Fig. 8c, and the separation of chains with different length during the crystallization, as sometimes observed.

(ii) The very large G value of 2 m/s in PE at $T \approx 80°C$ can, seen from the standpoint of glass transition multiplicity, be procured only by local modes. The relaxation card of PE (Figs. 54a and 75e) shows secondary relaxation at $\nu \approx 1.4$ GHz for 80°C followed by slower and longer Rouse and flow-transition modes. A "crystallization length" (an average between stem diameter b and fold length l) can be calculated by

$$l_G \equiv G/\nu \approx (2\,\mathrm{m/s})/(1.4 \times 10^9/\mathrm{s}) \approx 1.5\,\mathrm{nm}. \tag{11.23}$$

The frequency of the Rouse modes would result in an even smaller length. This means $l \leqslant l^*$, i.e. the folding can also occur if only rather local modes are available (e.g. supported by much free volume, possibly concentrated by the niche geometry, by local raising of temperature, or by a stowage of chain ends). Local modes favor molecular nucleation, of course.

In the normal case G is of order nm/s or μm/s and there are more possibilities to deliver the stems from the melt. Perhaps we can find stems preformed in melt by fluctuations which are rearranged during crystallization (Ref. 382, the erstarrungsmodell), or a "spinodal defect decomposition" near the growth surface (Ref. 383). Many pictures for delivering chain parts from the melt and rearrangements after deposition, for the special role of surface roughness, and for special forms of or in lamellas are developed and are documented in the literature (e.g. Refs. 8, 26, 384, 385). It may also be of interest that the layer thickness (or primary fold length l^*) and the entanglement spacing are about of the same order of magnitude.

The main idea of the nucleation-controlled crystal growth is that secondary nucleation marks a length l^* that wins in competition to other lengths. But there are also other pictures, e.g. Ref. 386.

An optimum growth at such a length, independent from the Hoffman

Lauritsen mechanism, is discussed in Ref. 387: Let f be the additional thermodynamic driving force. Assume that f linearly controls both δl in Eq. 11.4,

$$\delta l \sim f/\Delta g, \tag{11.24}$$

and the growth rate on the high-temperature flank of the $G(T)$ maximum,

$$G \approx (a/\tau) \cdot (f/kT), \tag{11.25}$$

with τ^{-1} proportional to the deposition rate of segments and a being a constant. According to the general scaling principle any motion is the slower the larger its spatial scale is. Assume that the deposition rate depends on the sequence (stem) length l by an "activation" mechanism,

$$\tau^{-1}(l) \sim \exp(-\alpha l), \tag{11.26}$$

where α is the activation per segment, perhaps modified by the distance from the surface or niche. Eq. 11.26 means that the chain part straightening is more seldom for longer sequences. Eq. 11.26 can also be interpreted, as viewed from the activation picture, by statistical independence of equal-weight probabilities to deposit parts of l, since $\exp(-\alpha l) = \exp(-\alpha l_1) \cdot \exp(-\alpha l_2) \ldots$ with $\Sigma l_i = l$. Combining Eqs. 11.24–11.26 we obtain the growth rate G as a function of $l = l^* + \delta l$,

$$G \sim (a \cdot \delta l \cdot \Delta g/kT) \exp(-\alpha(l^* + \delta l)) \tag{11.27}$$

where l^* is now defined to be the zero in Fig. 75f. The maximum of G is now at $l = l^* + 1/\alpha$. The new l^* is, of course, approximately equal to the old l^* according to Eq. 11.4 without δl, because chain part straightening, chain folding and polymer nucleation are related phenomena, see the upper part of Fig. 74a.

11.4 Crystallization and melting

Some typical phenomena in polymer melts are described in this section. Consider the following temperature-time program. We start at a high temperature well above the equilibrium melting temperature T_{m}^0, then the crystallizable polymer material is cooled down to T_{g}, and heated again. The heat flux of a DSC (dynamic scanning calorimeter) is recorded and depicted in a thermogram ($\dot{Q}/\dot{T} \equiv C_p$ vs. T). Fig. 76a shows two typical limiting cases (there are, of course, transitions between them). In case of a large heterogeneous nucleation we find many nuclei in the cooling stage so that a lively

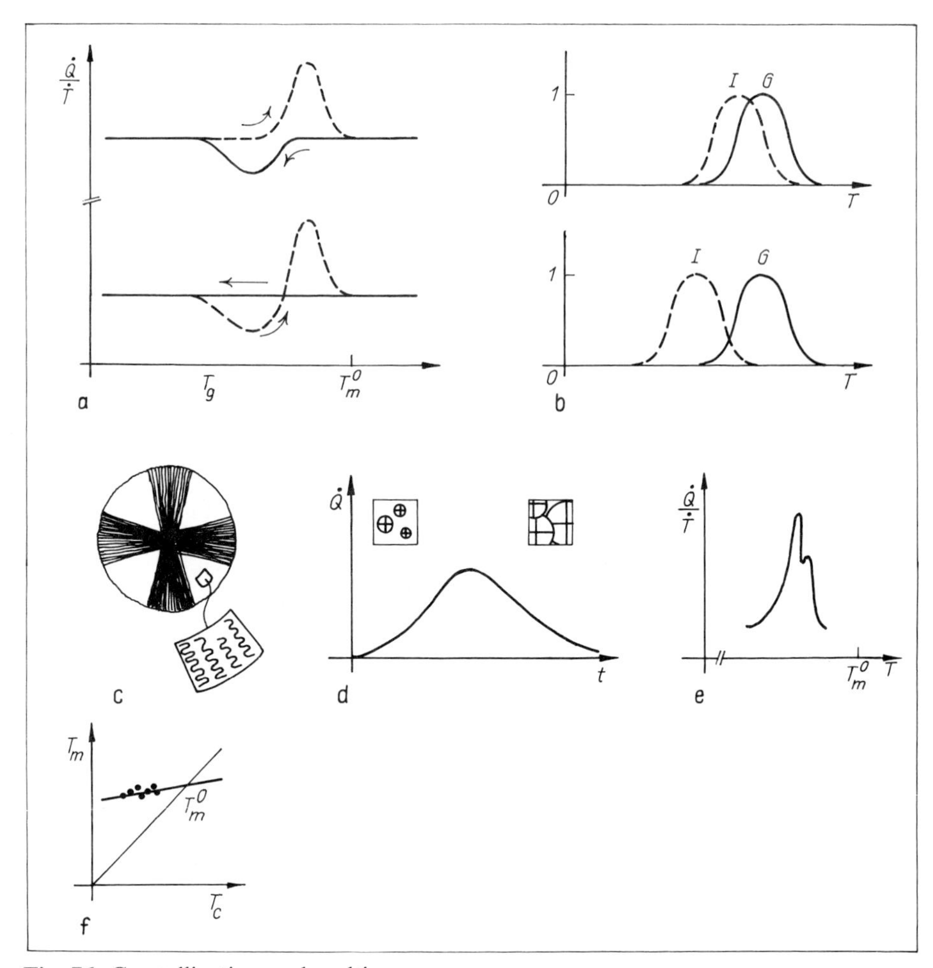

Fig. 76. Crystallization and melting.
a. Thermograms ($C_p \equiv \dot{Q}/\dot{T}$ vs. temperature T) for a cooling–heating cycle with rate $|\dot{T}|$. Upper part: typical for heterogeneous nucleation. Lower part: typical for homogeneous nucleation. The maximum corresponds to melting, the minimum to crystallization. b. The corresponding nucleation (I) and crystal growth (G) rates as a function of temperature T (both normalized to 1). c. Spherulite in polymers as seen with crossed Nicols. The detail shows the arrangement of lamellas in a spherulite. d. Basic conception for understanding the Avrami equation. $\dot{Q}$ heat flow, t time. e. From such a melting thermogram one can conclude to a bimodal distribution of lamella thickness. The high-temperature peak corresponds to larger l. f. Hoffman Weeks plot for extrapolation to the equilibrium melting temperature T_m^0.

crystallization can be observed during cooling (minimum in the thermogram). On the other hand, for the case of homogeneous nucleation, in particular if the maximum of the nucleation rate is at a relatively low temperature (Fig. 76b), the nuclei are formed at low temperature where the growth rate is too low to observe any considerable crystallization. Then crystallization is observed only in the heating stage where the growth rate is high enough. The following maximum is the melting peak at $T_m < T_m^0$.

The isothermal crystallization can often be adjusted by the *Avrami equation* (or Avrami Kolmogorov equation). This equation includes morphological features, especially the mutual hindering of growing crystallites. Most typical are spherulites in the 100 nm up to the 1 mm range, which are organized in such a way that the lateral growth of lamellas corresponds to the radial growth of spherulites, see Fig. 76e. [The growth rate G can then be observed by a light microscope.] In the later stages the crystallization can only fill the remaining gores between the spherulites, see Fig. 76d. This is approximately taken into account by a factor $(1 - \alpha)$ in the evolution equation for the degree of crystallization α, $0 \leqslant \alpha \leqslant 1$ (or $\leqslant \alpha_{max} < 1$)

$$d\alpha/dt = k(t)(1 - \alpha). \tag{11.28a}$$

Putting $k(t) = nK_n t^{n-1}$ for the Avrami growth rate we obtain after integration

$$1 - \alpha = \exp(-K_n t^n). \tag{11.28b}$$

The so-called Avrami exponent n is sometimes connected with the dimension of morphological units: $n = 1$ for linear structures, $n = 2$ for flat discs, and $n = 3$ for spherulites. Nucleation during crystallization also enlarges n ($\Delta n = 1$ for the case of "thermal nucleation").

Broken exponents are usually obtained from the experimental adjustments, and they can also depend on temperature. Other aspects (see e.g. Refs. 8, 372, 387a), besides Eq. 11.28a, should be included in a reasonable interpretation of the Avrami constants. (i) Insertion crystallization: displacement of ill crystallizable material, a differentiation process in the amorphous layers that can lead to additional lamellas of smaller thickness. (ii) Second stage crystallization: Lamellar thickening and other perfection or completion processes inside the spherulites. Such processes are assumed to be mainly responsible for deviations from the Avrami equation at large times. (iii) Surface melting. There is some dynamic instability connected with the lamella-thickness dependence of the α_c relaxation (see Sec. 8.4, Fig. 54c). If the thickness is diminished by some premelting, then $T_{\alpha c}$ is lowered which means that the mobility is enlarged which favors further melting. (iv) Recrystallization, and so on. [It should be remarked that structure variations in the time scale of minutes can now be pursued by X ray scattering at synchrotrons.]

The melting temperature T_m, e.g. the maximum of the curves in Fig. 76a, is usually considerably (e.g. 10 K) higher than the crystallization temperature T_c (the minimum of the curves). From the standpoint of fluctuations in a dynamic equilibrium, however, the melting is similar to crystallization, only seen "in the inverse direction". The difference between T_m and T_c comes from the necessity that small subsystems of different size are to be considered (e.g. larger thickness l for melting) or that larger-scale irreversibilities must be included. From the inversion argument we see, however, that it is the same barrier maximum function $\Delta G^*(T)$ that stabilizes both the nucleus (embryo) for crystallization and destabilizes the lamella for melting. Therefore, from Eq. 11.14 (with $\delta l = 0$) we now obtain for the melting temperature ($\Delta T = T_m^0 - T_m$)

$$T_m = T_m^0(1 - 2\sigma_e/l\Delta h) \tag{11.29}$$

where now, instead of l^*, l is the lamellar thickness enlarged by thickness growth in the history. In polymer science Eq. 11.29 is called *Thomson equation*. The melting temperature T_m increases with the lamellar thickness l, the latter can be estimated from T_m. The Thomson equation can also be used to determine l distributions from the broadness and structure of the melting peak. Thus, one can deduce a bimodal l distribution from a double melting peak, see Fig. 76e. In spite of the molecular thermodynamic approximations for small-system thermodynamics leading to Eq. 11.29 a solid parallelism was established between l distributions from this equation and those from other methods (electron microscopy, Raman spectroscopy, etc.).

The equilibrium melting temperature T_m^0 is usually obtained from extrapolation in a *Hoffman Weeks plot*, Fig. 76f. Allied values of T_m and T_c, that depend on a parameter such as the cooling/heating rate $|\dot{T}|$, are linearly extrapolated to the 45° straight line $T_m = T_c$. The intersection T_m corresponds to $l \to \infty$ in the Thomson equation.

Crystallization, and melting, in compatible polymer systems are similarly described. The kinetic part is now modified by the composition dependence of the glass or Vogel temperature, $T_g(\phi)$ or $T_\infty(\phi)$, and by the fact that the barrier height ΔG^* increases with the decreasing fraction of the crystallizable component. The latter effect diminishes the growth rate G and lowers its temperature maximum. The easier nucleation results in smaller spherulites. The value of T_{m}^{0} is also modified: The melting point depression informs us about the free mixing enthalpy of the melt, like as for small-molecule mixtures. Mixed crystals are not very probable for polymers, even if the possibilities for conformal disorder (see Sec. 1.7) are taken into consideration. Equalizing the chemical potentials of the crystallizable component in the melt and in the crystal one finds, after additive normalization for $\phi_{\mathrm{p}} = 1$ at $T_{\mathrm{m,pure}}^{0}$, by using the Flory Huggins equation Eq. 3.17a

$$(\Delta h V'/RV_p)(1/T_{\mathrm{m}}^{0} - 1/T_{\mathrm{m,pure}}^{0}) = -(\ln \phi_p)/Z_p + [1/Z_p - 1/Z']\phi' - \chi\phi'^2 \approx -\chi\phi'^2 \quad (11.30)$$

where the index p is for the crystallizable component, the prime for the other, V is the segment volume, and Z the degree of polymerization. The approximation is for large Z' and Z_{p}. Eq. 11.30 is called Nishi Wang equation (Ref. 388). This equation is related to the complete thermodynamic equilibrium and therefore contains no direct reference to a length scale.

11.5 Why lamellas?

This section is to collect arguments why layer crystals are the typical structure in polymers.

Firstly we have to show that the extended-chain crystal is the thermodynamic optimum for homopolymers. We start from the molecular-thermodynamic formula Eq. 11.1. Consider a given volume V for the crystal. Then the layer size X (Fig. 73a) can be eliminated. Since

$$X^2 l = V \quad (11.31)$$

we obtain

$$\Delta G = 4(Vl)^{1/2}\sigma + V(2\sigma_{\mathrm{e}}/l - \Delta g) \quad (11.32)$$

with the optimum, from $\partial\Delta G/\partial l = 0$,

$$l_{\mathrm{eq}} = V^{1/3}(\sigma_{\mathrm{e}}/\sigma)^{2/3}, \quad (11.33)$$

i.e. the equilibrium crystal is stretched into the l direction, $l_{\mathrm{eq}} > X$, and $l_{\mathrm{eq}} \to \infty$ for $V \to \infty$. This proves the statement that the lamella with thickness l^* is not a general equilibrium habit. On the contrary, in a layer structure of block copolymers with incompatible components, l is restricted and the lamellae can be an equilibrium structure for crystallizable blocks, see Ref. 389.

Now some reasons will be listed that favor layer crystals. [The basis is a general geometric property: straightening of a part of a coiled chain to a stem enlarges the probability for a larger curvature at the stem ends.]

1. There is a certain thickness $l \approx l^*$ that ensures a most effective lateral growth, e.g. by a nucleation controlled growth mechanism. The layer structure would then be a question of competition (Fig. 75f, Sec. 11.3).

2. According to the general scaling principle the polymer mobility decreases with increasing mode length. This means that large-scale rearrangements would consume long time intervals. So the lateral growth rate making lengths X larger and larger prevents (or restricts) the possibility to thicken the layer.

3. Dealing with layers we have to apply the thermodynamics of small systems. The interface, and other constraints due to the chains going through the layers, can modify the thermodynamic properties of all concerned (layer) phases. This can stabilize the layer structure (Ref. 390).

4. The thickness growth requires a special chain motion within the layers, see Ref. 384. But there is a special braking amplification by the thickness dependence of the α_c relaxation in the crystal layers (see Fig. 54c). $T_{\alpha c}$ increases with increasing layer thickness, higher $T_{\alpha c}$ means lower mobility (see, e.g. Refs. 391, 392). The thickening is stopped for $T_{\alpha c} > T_c$.

These four arguments are mainly addressed to the melt crystallization. In a weakened form they can also be applied for solutions, in particular when the local polymer concentration is enlarged in the vicinity of the growth surface.

Of course, the argumentation is not absolute. This is evident from the fact that extended chain crystals can be observed in the nature, e.g. for PE under high pressure. The high pressure lowers the influence of the attractive potential in the thermodynamic play. All energies and structures are then mainly determined from the hard van der Waals cores of segments. There may be new crystal structures with changing mobility possibilities in the different directions, the extension ratio σ_e/σ can be changed, and so on.

In summary, any attempt to explain the preferential layer structure in semicrystalline polymers provides an interesting example for combining arguments from relaxation and thermodynamics in polymers.

Appendices

Appendix 1

Fourier and Fourier Laplace transforms as used in this book (Ref. 49)

The mutual transforms are denoted by the same symbols, the distinction is only made by the symbols for the independent variables in the brackets.

Let $\phi(\boldsymbol{r}, \mathrm{t})$ be a function with a spatial and time behavior similar to a correlation function and assume that there are no convergence problems. The bare integral $\int$ without limits always means $-\infty \cdots +\infty$. Then we take

$$\phi(\boldsymbol{q}, t) = \int d\boldsymbol{r} \phi(\boldsymbol{r}, t) \mathrm{e}^{-i\boldsymbol{q}\boldsymbol{r}}, \tag{A.1}$$

$$\phi(\boldsymbol{q}, z) = \int_0^\infty dt \phi(\boldsymbol{q}, t) \mathrm{e}^{izt}, \tag{A.2}$$

where z is called complex frequency (advantage: easy arithmetic, disadvantage: need to take care of the cut along the real axis in the complex plane). The link to the real frequency ω is given by

$$\phi(\boldsymbol{q}, z) = \int \frac{d\omega}{2\pi i} \frac{\phi(\boldsymbol{q}, \omega)}{\omega - z} \tag{A.3}$$

and, inversely, by

$$\phi(\boldsymbol{q}, \omega) = 2 \operatorname{Re} \phi(\boldsymbol{q}, z = \omega + i\varepsilon), \tag{A.4}$$

with a small real $\varepsilon > 0$ labelling the cut.

Example 1 (Diffusion). Let ϕ be a solution of the following partial differential equation

$$\partial_t \phi - D\nabla^2 \phi = 0 \tag{A.5}$$

(∇ being the gradient, $\nabla^2 = \Delta$ the Laplacian) with the initial condition

$$\phi(\boldsymbol{q}, t = 0) = \phi_{\boldsymbol{q}} \tag{A.6}$$

where ϕ_q depends only on q. Then we have

$$\phi(\boldsymbol{q}, t) = \phi_{\boldsymbol{q}} \exp(-Dq^2 t), \tag{A.7}$$

$$\phi(\boldsymbol{q}, z) = i\phi_{\boldsymbol{q}}/(z + iDq^2), \tag{A.8}$$

and

$$\phi(\boldsymbol{q}, \omega) = 2\phi_{\boldsymbol{q}} \cdot Dq^2/(\omega^2 + Dq^2)^2 \tag{A.9}$$

where $q = |\boldsymbol{q}|$, the amount of $\boldsymbol{q}$. For a Dirac-delta initial condition at the origin $\boldsymbol{r} = 0$,

$$\phi(\boldsymbol{r}, t = 0) = \phi_0 \delta(\boldsymbol{r}), \tag{A.10}$$

we have constant $\phi_q \sim \phi_0$ and

$$\phi(\boldsymbol{r}, t) = \phi_0 (4\pi Dt)^{-3/2} \mathrm{e}^{-r^2/Dt} \tag{A.11}$$

for three dimensions ($d = 3$).

Example 2 (Convolution). Let us assume that we have some memory $D(t - t')$. Let a current be defined by

$$\boldsymbol{j}(\boldsymbol{r}, \mathrm{t}) = -\int_0^t dt' D(t - t') \nabla \phi(\boldsymbol{r}, t'). \tag{A.12}$$

Assuming conservation (a continuity equation)

$$\partial_t \phi + \nabla \boldsymbol{j} = 0 \tag{A.13}$$

and an exponential decay for the memory,

$$D(t - t') = (\tilde{D}/\tau) \exp[-(t - t')/\tau] \tag{A.14}$$

with a characteristic time τ and a coefficient $\tilde{D}$, we then have

$$\phi(\boldsymbol{q}, z) = i\phi_{\boldsymbol{q}} \frac{1}{z + \dfrac{iq^2\tilde{D}}{1 - iz\tau}}. \tag{A.15}$$

The second fraction bar (in the denominator) comes from the convolution theorem, viz. that the transform of a convolution $M * \nabla\phi$ gives a product of the transforms for M and $\nabla\phi$.

Example 3 (Homogeneous Langevin equation). This example is to show that the typical structure of a continued fraction can sometimes be related to a linear differential equation. Let $\phi(t)$ be a solution of the following homogeneous equation

$$v\dot{\phi}(t) + \Omega^2 \int_0^t M(t - t')\dot{\phi}(t')dt' + \Omega^2\phi(t) = 0 \tag{A.16}$$

with ν a "friction", Ω^2 an "eigenfrequency" ("mass operator"), M the memory, and $\dot{\phi}(t) \equiv d\phi/dt(t)$. Then, for the initial condition

$$\phi = 1 \quad \text{for} \quad t = 0 \tag{A.17}$$

we find the continued fraction structure

$$\phi(z) = \frac{1}{z + \frac{\Omega^2}{\nu + \Omega^2 M(z)}}. \tag{A.18}$$

Appendix 2

Molecular mass and chemical configuration of the monomeric unit for several polymers

PE	28	polyethylene	H H –C–C– H H
PP	42	poly propylene	H Me –C–C– , Me: $-CH_3$ H H
PS	104	polystyrene	H Ph –C–C– , Ph: 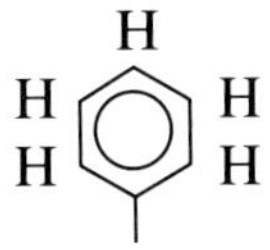 benzene ring with H, H, H, H, H H H
PVC	62.5	poly vinyl chloride	H Cl –C–C– H H
PMA	86	poly methyl acrylate	H AMe –C–C– , AMe: OCOMe (bond from C) H H
PMMA	100	poly methyl methacrylate	H AMe –C–C– H Me

PVAC	86	poly vinyl acetate	H Ac –C–C– , Ac: COOMe H H
cis PI	68	cis poly isoprene	Me H –C C– H C=C H H H
PVME	58	poly vinyl methyl ether	H OMe –C–C– H H
PIB	56	polyisobutylene	H Me –C–C– H Me
PTMC	298	poly tetra methyl carbonate (from bisphenol A)	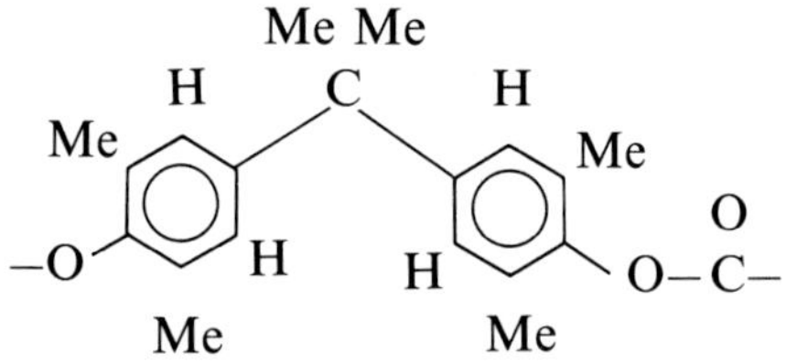 Me Me H C H Me Me O –O H H O–C– Me Me
PEO	44	poly ethylene oxide	H H –C–C–O– H H
PDMS	74	poly dimethyl siloxane	Me –Si–O– Me
PPO	58	poly propylene oxide	H Me –C–C–O– H H
PET	192	poly ethylene terephthalate	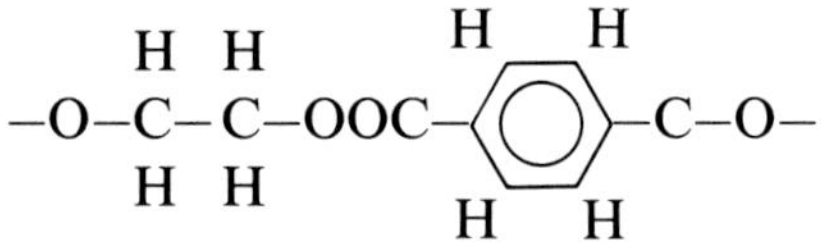 H H H H –O–C–C–OOC– –C–O– H H H H
PTFE	100	poly tetra fluoro ethylene	F F –C–C– F F

List of tables

Frequently used symbols, acronyms, and synonyms

Space and time

a	structure length ($\approx 0.7\,\text{nm}$ for flexible vinyl polymers)
b	general: segment length, stem diameter, a length between 0.5 and 1.5 nm
d_E	entanglement spacing ($\approx 7\,\text{nm}$), E or e index for entanglement
l, l_K	Kuhn segment length ($\approx 1.5\,\text{nm}$)
l, l_f	crystal fold length, stem length (2 . . . 20 nm)
L	extended chain length
L_t	chain contour (or tube) length
λ	general length scale (to be modified in the text, e.g. mode length)
R, R_0	mean end-to-end distance of polymer chain coil; shortly coil radius or coil diameter
σ	(van der Waals) diameter of molecules, chain diameter ($\approx 0.5\,\text{nm}$)
$g(r)$	radial distribution function
$G(r, t)$	van Hove correlation function
Z, M, N	chain length (degree of polymerization, molecular mass, number of segments)
M_c	critical m.m. for entanglement onset (physical M_c); also average m.m. between chemical crosslinks (chemical M_c)
ξ	general correlation length
ξ_a	characteristic length for a cooperativity region (with N_a particles contained in a volume of V_a) = size (radius) of natural subsystems
ξ_F	$= \xi_a$ for the flow transition
τ	relaxation or retardation time; or $t - t'$ ($d\tau = -dt'$).

Linear response and correlation functions

$G(\tau)$, $G^*(\omega) = G'(\omega) + iG''(\omega)$ shear modulus, general modulus
$J(\tau)$, $J^*(\omega) = J'(\omega) - iJ''(\omega)$ shear compliance, general compliance
J_e^0 steady state compliance
K^*, $B^* = 1/K^*$ bulk modulus, bulk compliance
ε^* dielectric permittivity (related to a compliance)
x extensive variable for compliance
f intensive variable for modulus
$x^2(\omega)$, $f^2(\omega)$ spectral density
$x^2(\tau)$, $f^2(\tau)$ (or $\varphi_x(\tau) = \varphi_{xx}(\tau)$, $\varphi_f(\tau) = \varphi_{ff}(\tau)$) correlation function

Scattering

q $= |\boldsymbol{q}|$, scattering vector
ω $(=2\pi\nu)$ frequency
z complex frequency
$S(q)$ structure factor
$S(q, t)$ structure function
$S(q, \omega)$ scattering function
"$S(\boldsymbol{r}, t)$" $\sim$ distinct part of $G(\boldsymbol{r}, t)$ [seldom used]

Thermodynamics and others

k Boltzmann constant
ψ order parameter
$\varphi(\boldsymbol{r})$ intermolecular potential
ρ general density
n number of units, also $n = \bar{n}$, number density
N_A, n_A Avogadro number
$n(\boldsymbol{r})$ number density ($\bar{n}$ its mean value)
ν_B number of B chains
T temperature
V, v volume
P, p pressure
S entropy

ΔF	Helmholtz free energy
ΔG	Gibbs free energy = free enthalpy
ΔH	enthalpy
G	crystal growth rate
H	Hamiltonian
σ	(inter, sur) face tension
E	index: excess
M	index: mixing
δT^2	mean square temperature fluctuation of a natural subsystem (volume V_a)
T_g	conventional glass temperature
T_∞	Vogel temperature
$\dot{T}_K$, $\dot{T}_H$	cooling, heating rate
μ, μ_B	chemical potential (of B component)
ϕ	volume fraction, $0 \leqslant \phi \leqslant 1$
x	mole fraction
η	shear viscosity
D	diffusion coefficient

Acronyms

PE, PS, . . .	polyethylene, polystyrene, . . . see Appendix 2
nPS, dPS, . . .	normal and deuterated PS; PS_D or PS_d is also used for the latter
FDT	fluctuation dissipation theorem
FT	flow transition
GLE	general Langevin equation
GSP	general scaling principle
GT(Z)	glass transition (zone)
MT	main transition
PZ	rubbery plateau zone
KWW	Kohlrausch Williams Watts
MPT	microphase transition
PLE	principle of local equilibrium
RPA	random phase approximation
SAXS (SANS)	small angle X-ray (neutron) scattering
WAXS	wide angle X-ray scattering

VFT(H)	Vogel Fulcher Tammann (Hesse)
WLF	Williams Landel Ferry

List of new terms introduced in this book

ρ level: Sec. 1.1
ψ level: Sec. 1,1
general scaling principle GSP: Sec. 2.8
mode length: Sec. 2.8
natural subsystem: Sec. 3.6
thermokinetic structure: Sec. 3.9
functional subsystem: Sec. 3.9
ideal dynamic glass transition: Sec. 6.4
WLF scaling: Secs. 4.2.3, 6.3
short and long glass transition: Sec. 6.5
survival of large structures: Sec: 6.8
thermokinetic diffusion: Sec. 8.8
granulated phase decomposition: Sec. 10.9

Frequently used synonyms

main transition MT	=	glass-to-rubber transition in polymers
flow transition FT	=	terminal zone (or transition)
glass transition GT	=	glass–liquid transition (or transformation)
plateau zone PZ	=	rubbery plateau zone
dispersion zone	=	relaxation or transition zone (or region)
Arrhenius diagram	=	relaxation card, lg ω vs. T^{-1} or
	=	$\pm$lg τ vs. T^{-1} diagram
glass temperature T_g	=	glass point, transformation temperature
time-temperature equivalence	=	tt superposition, or shift

Remark

The terms "cooperative" (and "cooperativity") and the difference to the term "collective" are described in the last paragraphs of the introduction to Chap. 6 and in Sec. 6.2.

References

1 P.J. Flory, Principles of Polymer Chemistry, Cornell University Press, Ithaca NY, 1953 (2nd Ed. 1962).

2 J.D. Ferry, Viscoelastic Properties of Polymers, 3rd Ed., Wiley, New York, 1980. For a historical review: J.D. Ferry, Macromolecules *24* (1991) 5237; H. Morawetz, Polymers – The Origins and Growth of a Science, Wiley, New York, 1985.

3 N.G. McCrum, B.E. Read, G.Williams, Anelastic and Dielectric Effects in Polymeric Solids, Wiley, London, 1967.

4 P.G. de Gennes, Scaling Concepts in Polymer Physics, Cornell University Press, 1979.

5 V.P. Privalko, Molekulyarnoe Stroenie i Svojstva Polymerov, Chimiya, Leningrad, 1986 (in Russian).

6 M. Doi, S.F. Edwards, The Theory of Polymer Dynamics, Clarendon, Oxford, 1988.

7 Some experimental techniques are described in: S. Dattagupta, Relaxation Phenomena in Condensed Matter Physics, Academic, New York, 1987; B.J. Berne, R. Pecora, Dynamic Light Scattering, Wiley, New York, 1976; B. Wunderlich, Thermal Analysis, Academic, Boston, 1990; G.D. Patterson, Photon Correlation Spectroscopy of Bulk Polymers, Adv. Polym. Sci. *48* (1983) 125; G.D. Wignall, Neutron Scattering, in Ref. 11, Vol. 10 (1987); F. Mezei (Ed.), Neutron Spin Echo, Springer, Berlin, 1980; D. Campbell, J.R. White, Polymer Characterization. Physical Techniques, Chapman & Hall, London, 1989.

8 B. Wunderlich, Macromolecular Physics (Three volumes), Academic, New York, 1973, 1976.

9 L.A. Utracki, Polymer Alloys and Blends: Thermodynamics and Rheology, Hanser, München, 1989.

10 L.R.G. Treloar, Introduction to Polymer Science, Wykeham, London, 1970.

11 Encyclopedia of Polymer Science and Engineering, 2nd Ed. (17 volumes), H.F. Mark, e.a. (Ed.), Wiley, New York, 1985–1989.

12 Comprehensive Polymer Science (7 volumes) G. Allen e.a. (Ed.), Pergamon, Oxford, 1989.

13 Enciklopediya Polimerov (V.A. Kabanov e.a. Ed.). Three volumes, Izd. Sovetskaya Enciklopediya, Moscow, 1972, 1974, 1977 (in Russian).

14 H. Dominghaus, Die Kunststoffe und ihre Eigenschaften, 3rd Ed., VDI-Verlag, Düsseldorf, 1988; W.D. van Krevelen, Properties of Polymers, 2nd Ed., Elsevier, Amsterdam, 1976.
15 R.L. Stratanovich, M.S. Polyakova, Elementy Molekulyarnoj Fiziki, Termodinamiki i Statisticheskoj Fiziki, Moscow University, Moscow, 1981 (in Russian).
16 W.J. Orr, Trans. Faraday Soc. *43* (1947) 12.
17 C.A. Croxton (Ed.), Progress in Liquid Physics, Wiley, New York, 1977. [An example for $g(r)$ in a molten polymer (PE) is: G.R. Mitchell, R. Lovell, A.H. Windle, Polymer *23* (1982) 1273].
18 P.J. Flory, Statistical Mechanics of Chain Molecules, Interscience, New York, 1969.
19 M.V. Vol'kenštejn, Konfiguracionnaya Statistika Polimernych Cepej, Izd. Akademiya, Moscow, 1959.
20 Yu.Ya. Gotlib, A.A. Darinskij, Yu.E. Svetlov, Fizicheskaya Kinetika Makromolekul, Chimiya, Leningrad, 1986 (in Russian).
21 W.W. Graessley, Polymer *21* (1980) 258.
22 W.W. Graessley, Adv. Polym. Sci. *16* (1974) 1.
23 W.W. Graessley, Adv. Polym. Sci. *47* (1982) 67.
24 S. Wu, J. Polym. Sci. *B27* (1989) 723.
25 Y.H. Lin, Macromolecules *20* (1987) 3080; see also Ref. 119.
26 L. Mandelkern, Crystallization of Polymers, McGraw-Hill, New York, 1964; J. Polym. Sci. *C50* (1975) 189.
27 A. Keller, Phil. Mag. *2* (1957) 1171.
28 E.W. Fischer, Z. Naturforsch. *12a* (1957) 754.
29 P.H. Till, J. Polym. Sci. *24* (1957) 301.
30 D.G.H. Ballard, J. Schelten in J.V. Dawkins (Ed.), Developments in Polymer characterization, Vol. 2, Appl. Sci. Publ., London, 1980; J. Schelten, D.G.H. Ballard, G.D. Wignall, G. Longman, W. Schmatz, Polymer *17* (1976) 751. For the semidilute regime see M. Daoud, J.P. Cotton, B. Farnoux, G. Jannink, G. Sarma, H. Benoit, R. Duplessix, C. Picot, P.G. de Gennes, Macromolecules *8* (1975) 804.
31 E.W. Fischer, Macromol. Chem. Macromol. Symp. Ser. *12* (1987) 123; E.W. Fischer, K. Hahn, J. Kugler, U. Struth, R. Born, M. Stamm, J. Polym. Sci. Polym. Phys. Ed. *22* (1984) 1491; see also B. Crist, J.D. Tanzer, W.W. Graessley, J. Polym. Sci. *B25* (1987) 545.
32 E.R. Walter, F.P. Reding, J. Polymer Sci. *20* (1956) 561; see also Ref. 8, Vol. 1, p. 154.
33 C.G. Vonk, H. Reynaers, Polym. Comm. *31* (1990) 190.
34 S.J. Opella, J.S. Waugh, J. Chem. Phys. *66* (1977) 4919.
35 H.G. Olf, A.J. Peterlin, J. Polym. Sci. *A-2, 8* (1970) 753.

36 B. Wunderlich, M. Möller, J. Grebowicz, H. Baur, Adv. Polym. Sci. *87* (1989) 1; D.W. Noid, B.G. Sumpter, B. Wunderlich, Macromolecules *23* (1990) 664; see also W. Pechhold, Makromol. Chem. Suppl. *6* (1984) 163.

37 E. Helfand, J. Chem. Phys. *54* (1971) 4651.

38 W. Kuhn, O. Künzle, A. Preissmann, Helv. Chim. Acta *30* (1947) 307, 464; R.D. Andrews, N. Hofman-Bang, A.V. Tobolsky, J. Polym. Sci. *3* (1948) 669.

38a J. Brandrup, E.H. Immergut (Eds.), Polymer Handbook, 3rd Ed., Wiley, New York, 1989.

38b Spravochnik po Fizicheskoj Chimii Polimerov (Three Volumes), Yu.S. Lipatov e.a. (Ed.), Naukova Dumka, Kiev, 1984, 1985 (in Russian).

38c S.M. Aharoni, Macromolecules *16* (1983) 1722; *18* (1985) 2624; *19* (1986) 426.

39 A.C. Pipkin, Lectures in Viscoelastic Theory, Springer, New York, 1972; F.R. Schwarzel, L.C.E. Struik, Advan. Mol. Relax. Proc. *1* (1967–1968) 201; F. Schwarzel, in Ref. 11, Vol. 17, p. 587 (1989).

40 J. Jäckle, H.L. Frisch, J. Polym. Sci. Polym. Phys. Ed. *23* (1985) 675; J. Jäckle, Z. Physik *B64* (1986) 41.

41 L.R.G. Treloar, Introduction to Polymer Science, Wykeham, London, 1970.

42 G.V. Vinogradov, A.Ya. Malkin, Reologiya Polimerov, Chimiya, Moscow, 1977; A.S. Kraus, H. Eyring, Deformation Kinetics, Wiley, New York, 1974.

43 H. Bateman, A. Erdelyi e.a., Tables of Integral Transforms, Vol. II, McGraw-Hill, New York, 1954, Chap. XV.

44 N.W. Tschoegl, The Phenomenological Theory of Linear Viscoelastic Behavior. An Introduction, Springer, Berlin, 1989.

45 L.D. Landau, E.M. Lifshits, Statistical Physics, Pergamon, Oxford, 1980.

46 G. Kluge, S. Bark-Zollmann, Wiss. Z. Univ. Jena, NR (1985) 81; G. Kluge, private comm. 1990.

46a S.W. Provencher, Comput. Phys. Comm. *27* (1982) 213; J. Honerkamp, J. Weese, Macromolecules *22* (1989) 4372; S.-L. Nyeo, B. Chu, Macromolecules *22* (1989) 3998.

47 B.V. Gnedenko, Kurs Teorii Veroyatnostej, 3rd Ed., Gosud. Izd., Moscow, 1961.

48 H. Nyquist, Phys. Rev. *32* (1928) 110.

49 D. Forster, Hydrodynamic Fluctuations, Broken Symmetry, and Correlation Functions, Benjamin, London, 1975; R. Kubo, M. Toda, N. Hashitsume, Statistical Physics II. Nonequilibrium Statistical Mechanics, Springer, Berlin, 1985.

50 D. Pines, P. Nozieres, The Theory of Quantum Liquids, Benjamin, New York, 1966.

51 e.g. F. Noack, in P. Diel, E. Fluck, R. Kosfeld (Eds.), NMR Basic Principles and Progress, Vol. 3, Springer, Berlin, 1971, p. 83.

52 J.D. Ferry, E.R. Fitzgerald, J. Colloid Sci. *8* (1954) 224; M.L. Williams, J.D. Ferry, J. Colloid Sci. *9* (1954) 479; J.D. Ferry, S. Strella, J. Colloid Sci. *13* (1958) 459; M. Miyake, Rep. Progr. Polym. Phys. Japan *4* (1961) 36.

53 M. Kakizaki, T. Kakudate, T. Hideshima, J. Polym. Sci. Polym. Phys. Ed. *23* (1985) 809; see also G.D. Patterson, Ref. 7; R.-J. Roe, J.J. Curro, Macromolecules *16* (1983) 428.
54 E. Donth, K. Schneider, Acta Polym. *36* (1985) 213, 273; see also Ref. 195.
55 J.K. McKinney, H.V. Belcher, J. Res. Nat. Bur. Stand. *A67* (1963) 43; C.H. Wang, B.Y. Li, R.W. Rendell, K.L. Ngai, Ref. 129, p. 870.
56 E.H. Lee, Quart. Appl. Math. *13* (1955) 183.
57 e.g. H.W. Spiess, J. Non-Cryst. Solids *131–133* (1991) 766.
58 W. Kuhn, Angew. Chem. *51* (1938) 640.
58a P.J. Flory, Proc. Roy. Soc. London *A351* (1976) 351; see also W. Burchard, Ber. Bunsenges. Phys. Chem. *89* (1985) 1151.
59 S. Kästner, Faserforsch. Textiltechn. Z. Polymerforsch. *27* (1976) 1; Acta Polymerica *31* (1980) 444.
60 B. Erman, P.J. Flory, Macromolecules *15* (1982) 800.
61 S.F. Edwards, Proc. Roy. Soc. London *91* (1976) 513; S.F. Edwards, T. Vilgis, Rep. Progr. Phys. *51* (1988) 243.
62 B. Erman, L. Monnerie, Macromolecules *18* (1985) 1985.
63 R. Oeser, B. Ewen, D. Richter, B. Farago, Phys. Rev. Lett. *60* (1988) 1041 [For the fluctuations of entanglements see e.g. D. Richter e.a., Phys. Rev. Lett. *47* (1981) 109].
64 H.G. Kilian, Polymer *22* (1981) 209; H.G. Kilian, T. Vilgis, Colloid Polymer Sci. *262* (1984) 15; T. Vilgis, H.G. Kilian, Polymer *24* (1983) 949.
65 P.J. Flory, J. Chem. Phys. *9* (1941) 660; *10* (1942) 51.
66 M.L. Huggins, Ann. N.Y. Acad. Sci. *43* (1942) 1; J. Chem. Phys. *9* (1941) 640.
67 A.J. Staverman, J.H. van Santen, Rec. Trav. Chim. *60* (1941) 76.
68 E. Donth, Wiss. Z. TH Leuna-Merseburg *24* (1982) 475.
69 F. Reif, Statistische Physik und Theorie der Wärme, 3. Aufl. (W. Muschik, Ed.), de Gruyter, Berlin, 1987, p. 686.
70 T.L. Hill, J. Chem. Phys. *36* (1962) 3182; T.L. Hill, Thermodynamics of Small Systems, Part I and II, Benjamin, New York, 1963, 1964.
71 W.W. Graessley, S.F. Edwards, Polymer *22* (1981) 1329.
72 E.W. Fischer, E. Donth, W. Steffen, Phys. Rev. Lett. *68* (1992) 2344.
73 F.W. Wiegel, Introduction to Path-Integral Methods in Physics and Polymer Science, World Scientific, Singapore, 1986.
74a K.F. Freed, Renormalization Group Theory of Macromolecules, Wiley, New York, 1986.
74b L. Leibler, Macromolecules *13* (1980) 1602.
75 G.H. Fredrickson, E. Helfand, J. Chem. Phys. *87* (1987) 697; *89* (1988) 5890; K. Binder, G.H. Fredrickson, J. Chem. Phys. *92* (1990) 6195; S. Stepanow, private comm. 1991.
76 K.G. Wilson, Rev. Mod. Phys. *55* (1983) 583; R. Balescu, Equilibrium and Nonequilibrium Statistical Mechanics, Wiley, New York, 1975.

77 K.S. Schweizer, J.G. Curro, Phys. Rev. Lett. *60* (1988) 809; J.G. Curro, K.S. Schweizer, J. Chem. Phys. *88* (1988) 7242; D. Chandler, H.C. Anderson, J. Chem. Phys. *57* (1972) 1930.
78 D. Bohm, D. Pines, Phys. Rev. *92* (1953) 609; D. Pines, The Many Body Problem, Benjamin, Reading MA, 1962.
78a S.F. Edwards, Proc. Phys. Soc. London *85* (1965) 613; *86* (1966) 265.
78b G. Jannink, P.G. de Gennes, J. Chem. Phys. *48* (1968) 2260.
79 G.D. Wignall, R.W. Hendricks, W.C. Koehler, J.S. Lin, M.P. Wai, E.L.T. Thomas, R.S. Stein, Polymer *22* (1981) 886; H. Benoit (1980) as quoted therein.
80 A.Z. Akcasu, M. Benmouna, H. Benoit, Polymer *27* (1986) 1935; see also A.Z. Akcasu, M. Tombakoglu, Macromolecules *23* (1990) 607.
81 B. Widom, J. Chem. Phys. *43* (1965) 3894.
82 N. Metropolis, A.W. Rosenbluth, M.N. Rosenbluth, A.H. Teller, E. Teller, J. Chem. Phys. *32* (1953) 1087.
83 S. Nosè, F. Yonezawa, J. Chem. Phys. *84* (1986) 1803; S. Nosè, J. Chem. Phys. *81* (1984) 511.
84 K. Binder, Colloid Polym. Sci. *266* (1988) 871; K. Kremer, K. Binder, Comp. Phys. Rep. *7* (1988) 259; see also I. Carmesin, K. Kremer, Macromolecules *21* (1988) 2819; H.P. Wittmann, K. Kremer, K. Binder, J. Chem. Phys. *95* (1991) 3476.
85 J. Heijboer, in R.A. Pethrick, R.W. Richards (Eds.), Static and Dynamic Properties of the Polymeric Solid State, Reidel, London 1982, p. 197, see also Ref. 128a (1976), p. 104.
86 H.A. Kramers, Physica *7* (1940) 284.
87 H. Eyring, J. Chem. Phys. *3* (1935) 107; H. Eyring, E.M. Eyring, Modern Chemical Kinetics, Van Nostrand Reinhold, New York, 1965.
88 J. Koppelmann, Progr. Colloid Polym. Sci. *66* (1979) 235.
89 F. Simon, Ergebnisse exakt. Naturwiss. *9* (1930) 222.
90 M. Goldstein, J. Chem. Phys. *51* (1969) 3728.
91 G. Adam, J.H. Gibbs, J. Chem. Phys. *43* (1965) 139.
92 W. Kauzmann, Chem. Rev. *43* (1948) 219.
93 D. Turnbull, Contemp. Phys. *10* (1969) 473.
94 H. Suga, S. Seki, J. Non-Cryst. Solids *16* (1974) 171; Faraday Discussions, *69* (1980) 221.
95 C.A. Angell, J.M. Sare, E.J. Sare, J. Phys. Chem. *82* (1978) 2622.
96 C.A. Angell, W. Sichina, Ref. 128a, p. 53.
97 A.B. Bestul, Glastechn. Ber. *32K* (1959) VI/59; C.A. Angell, D.L. Smith, J. Phys. Chem. *86* (1982) 3845.
98 K.L. Ngai, Comm. Solid State Phys. *9* (1979) 127, *9* (1980) 141; R.W. Rendell, K.L. Ngai, in the same book as Ref. 145, p. 309; K.L. Ngai, R.W. Rendell, A.K. Rajagopal, S. Teitler, Ref. 128c, p. 150.
99 K. Adachi, H. Yoshida, F. Fukui, T. Kotaka, Macromolecules *23* (1990) 3138.

100 J.P. Jarry, L. Monnerie, Macromolecules *12* (1979) 927; P. Törmälä, J. Macromol. Sci. Rev. Macromol. Chem. *C17* (1979) 297; N. Kusumoto in R.F. Boyer, S.F. Kanath (Eds.), Molecular Motions in Polymers by ESR, Harwood Academic, Utrecht, 1980; see also P.D. Hyde, T.E. Evert, M.T. Cicerone, M.D. Edinger, Ref. 129, p. 42.
101 C. Schick, E. Donth, Phys. Scripta *43* (1991) 423.
102 F. Büche, J. Chem. Phys. *22* (1954) 603; Physical Properties of Polymers, Interscience, New York, 1962.
103 V.A. Kargin, V.A. Slonimskij, Ž. fiz. Chim. *23* (1949) 563 (in Russian).
104 P.E. Rouse jr., J. Chem. Phys. *21* (1953) 1272.
105 B.H. Zimm, J. Chem. Phys. *24* (1956) 269.
106 W. Kuhn, H. Kuhn, Helv. Chim. Acta *28* (1945) 1533; *29* (1946) 71; see the discussion of this point in Ref. 4, Chap. 6.
107 R. Cerf, J. Phys. Radium *19* (1958) 122.
108 D.J. Plazek, Polymer J. *12* (1980) 43; J. Polym. Sci. Polym. Phys. Ed. *20* (1982) 729; D.L. Plazek, D.J. Plazek, Macromolecules *16* (1983) 1469.
109 W. Pfandl, G. Link, F.R. Schwarzl, Rheologica Acta *23* (1984) 277; F.R. Schwarzl, in G. Astarita, G. Marucci (Eds.) Rheology, Vol. 1, Plenum, New York, 1980, p. 243; see also Ref. 156.
110 J.S. Higgins, Physica *136B* (1986) 201; J.S. Higgins, J.E. Roots, J. Chem. Soc. Faraday Trans 2, *81* (1985) 757; see also D. Richter, B. Ewen, J.B. Hayter, Phys. Rev. Lett. *45* (1980) 2121.
111 P.G. de Gennes, Physics *3* (1967) 181.
112 A.Z. Akcasu, M. Benmouna, C.C. Han, Polymer *21* (1980) 866.
113 S.F. Edwards, Proc. Roy. Soc. London *92* (1967) 9.
114 P.G. de Gennes, J. Chem. Phys. *55* (1971) 572.
115 M. Daoud, P.G. de Gennes, J. Polym. Sci. Polym. Phys. Ed. *17* (1979) 1971.
116 L. Leger, P.G. de Gennes, Ann. Rev. Phys. Chem. *33* (1982) 49; L. Leger, J.L. Viovy, Contemp. Phys. *29* (1988) 579.
117 J. Skolnick, A. Kolinski, Adv. Chem. Phys. *78* (1990) 223.
118 M. Doi, J. Polym. Sci. Polym. Lett. Ed. *19* (1981) 265; J. Polym. Sci. Polym. Phys. Ed. *21* (1983) 667; see also J. Noolandi, D.A. Bernard, Canad. J. Phys. *61* (1983) 1035.
119 D. Richter, B. Farago, L.J. Fetters, J.S. Huang, B. Ewen, C. Lartigue, Phys. Rev. Lett. *64* (1990) 1389; R. Butera, L.J. Fetters, J.S. Huang, D. Richter e.a., Phys. Rev. Lett. *66* (1991) 2088.
120 P.G. de Gennes, J. Phys. *42* (1981) 735.
121 K. Kremer, G.S. Grest, I. Carmesin, Phys. Rev. Lett. *61* (1988) 566; K. Kremer, G.S. Grest, J. Chem. Phys. *92* (1990) 5057.
122 E. Donth, Polym. Bull. *7* (1982) 417.
123 A.V. Tobolsky, Properties and Structure of Polymers, Wiley, New York, 1960.

124 M.L. Williams, R.F. Landel, J.D. Ferry, J. Am. Chem. Soc. *77* (1955) 3701; a historical account to master curves is: H. Markovitz, J. Polym. Sci. Polym. Symp. Ser. *50* (1975) 431.
125 H. Vogel, Phys. Z. *22* (1921) 645.
126 G.S. Fulcher, J. Am. Chem. Soc. *8* (1925) 339, 789.
127 G. Tammann, G. Hesse, Z. anorg. allg. Chem. *156* (1926) 245.
128a Ann. New York Acad. Sci. *279* (1976).
128b Ann. New York Acad. Sci. *371* (1981).
128c Ann. New York Acad. Sci. *484* (1986).
129 J. Non-Cryst. Solids *131–133* (1991).
129a J. Jäckle, Rep. Progr. Phys. *49* (1986) 171.
130 U. Bengtzelius, W. Götze, A. Sjölander, J. Phys. *C17* (1984) 5915.
131 E. Leutheusser, Phys. Rev. *A29* (1984) 2765.
132 T.R. Kirkpatrick, Phys. Rev. *A31* (1985) 939; W. Götze, Z. Phys. *B56* (1984) 139; *B60* (1985) 195.
133 G.H. Fredrickson, Ann. Rev. Phys. Chem. *39* (1988) 149; G.H. Fredrickson, H.C. Andersen, Phys. Rev. Lett. *53* (1984) 1244; J. Chem. Phys. *84* (1985) 5822.
134 T.R. Kirkpatrick, D. Thirumalai, P.G. Wolynes, Phys. Rev. *A40* (1989) 1045.
135 F.H. Stillinger, J. Chem. Phys. *89* (1988) 6461.
136 B. Frick, D. Richter, W. Petry, U. Buchenau, Z. Phys. B – Condens. Matter *70* (1988) 73; D. Richter, B. Frick, B. Farago, Phys. Rev. Lett. *61* (1988) 2465; F. Mezei, W. Knaak, B. Farago, Phys. Rev. Lett. *58* (1987) 571; W. Knaak, F. Mezei, B. Farago, Europhys. Lett. *7* (1988) 529.
137 e.g. G. Meier, B. Gerharz, D. Boese, E.W. Fischer, J. Chem. Phys. *94* (1991) 3050; W. Steffen, A. Patkowski, G. Meier, E.W. Fischer, J. Chem. Phys. *96* (1992) 4171; see also Ref. 191.
138 Y.-X. Yan, L.-T. Cheng, K.A. Nelson, J. Chem. Phys. *88* (1988) 6477; *91* (1989) 6052.
139 N.O. Birge, S.R. Nagel, Phys. Rev. Lett. *54* (1985) 2674; N.O. Birge, Phys Rev. *B34* (1986) 1631; P.K. Dixon, S.R. Nagel, Phys. Rev. Lett. *61* (1988) 341; Y.H. Jeong, Phys. Rev. *A36* (1987) 766.
140 F. Kremer, E.W. Fischer, A. Hofmann, A. Schönhals, to be submitted; D.W. Davidson, R.H. Cole, J. Chem. Phys. *19* (1951) 1484.
141 E. Donth, Glasübergang, Akademie-Verl., Berlin, 1981.
142 A. Schönhals, F. Kremer, E. Schlosser, Phys. Rev. Lett. *67* (1991) 999.
143 E.W. Fischer, priv. comm. 1990 (original version of the old bus model).
144 K. Binder, A.P. Young, Rev. Mod. Phys. *58* (1986) 801.
145 C.A. Angell, in K. Ngai, G.B. Wright (Eds.), Relaxation in Complex Systems, Nat. Techn. Inform. Service, US Dep. of Commerce, Springfield VA, 1985, p. 1; C.A. Angell, J. Non-Cryst. Solids *73* (1985) 1.
146 E. Donth, J. Non-Cryst. Solids *53* (1982) 325; see also Ref. 101.
147 E. Donth, Acta Polym. *30* (1979) 481.

148 R.G. Palmer, D. Stein, E. Abrahams, P.W. Anderson, Phys. Rev. Lett. *53* (1984) 958.
149 R.G. Palmer, Adv. Phys. *31* (1982) 669; R.G. Palmer, Lecture Notes in Physics *275* (1986) 275 (Heidelberg Colloq. Glassy Dynam.), Springer, Berlin, 1986.
150 U. Strom, P.C. Taylor, Phys. Rev. *B16* (1977) 5512.
150a H.-W. Hu, G.A. Carson, S. Granick, Phys. Rev. Lett. *66* (1991) 2758, J. Van Alsten, S. Granick, Phys. Rev. Lett. *61* (1988) 2570.
151 S.H. Glarum, J. Chem. Phys. *33* (1960) 639.
152 C.F. Böttcher, P. Bordewijk, Theory of Electric Polarization, Elsevier, New York, 1979.
153 E. Donth, J. Non-Cryst. Solids *131–133* (1991) 204.
154 J.H. Gibbs, E.A. DiMarzio, J. Chem. Phys. *28* (1958) 373; E.A. DiMarzio, J.H. Gibbs, P.D. Fleming, I.C. Sanchez, Macromolecules *9* (1976) 763, see also Ref. 244; H. Bässler, Phys. Rev. Lett. *58* (1987) 767; T.A. Vilgis, J. Phys. Condens. Matter *2* (1990) 3667.
155 T. Pakula, priv. comm. 1991; W. Pechhold, O. Grassl, W. von Soden, Colloid Polym. Sci. *268* (1990) 1089.
156 W. Pechhold, E. Sautter, W. von Soden, B. Stoll, H.P. Grossmann, Makromol. Chem. Suppl. *3* (1979) 247.
157 R.F. Boyer, J. Polym. Sci. *C14* (1966) 3, and other authors in this volume.
158 K. Adachi, Y. Imanishi, T. Kotaka, J. Chem. Soc. Faraday Trans. I *85* (1989) 1075; the undiluted cis PI system: Y. Imanishi, K. Adachi, T. Kotaka, J. Chem. Phys. *89* (1988) 7585.
159 D. Boese, F. Kremer, Macromolecules *23* (1990) 829.
160 G.S. Attard, J.J. Moura-Romas, G. Williams, J. Polym. Sci. Polym. Phys. Ed. *25* (1987) 1099. [A glass transition was also observed in suspensions of spherical colloidal particles with a diameter of 340 nm: P.N. Pusey, Phys. Rev. Lett. *59* (1987) 2083].
161 H. Kresse, S. Ernst, W. Wedler, D. Demus, F. Kremer, Ber. Bunsenges. *94* (1990) 1478; see also Ref. 253.
162 J.M. Stevels, in G.H. Frischat (Ed.), The Physics of Non-Crystalline Solids, Trans. Tech. Publ., 1977; J.M. Stevels, J. Non-Cryst. Solids *40* (1980) 69; G.M. Bartenev, D.S. Sanditov, Relaksacionnye Processy v Stekloobraznych Sistemach, Nauka, Novosibirsk, 1986 (in Russian).
163 G.P. Johari, M. Goldstein, J. Chem. Phys. *55* (1971) 4245.
164 J. Heijboer, in J.A. Prins (Ed.), Physics of Non-Crystalline Solids, North Holland, Amsterdam, 1965, p. 231.
165 T.F. Schatzki, J. Polym. Sci. *57* (1962) 496.
166 R.H. Boyd, S.M. Breitling, Macromolecules *7* (1974) 855.
167 B. Valeur, J.P. Jarry, F. Geny, L. Monnerie, J. Polym. Sci. Polym. Phys. Ed. *13* (1975) 667, 675; J.-L. Viovy, L. Monnerie, F. Merola, Macromolecules *18* (1985) 1130.

168 R.T. Bailey, A.M. North, R.A. Pethrick, Molecular Motion in High Polymers, Clarendon, Oxford, 1981.
169 J.M.G. Cowie, J. Macromol. Sci.-Phys. *B18* (1980) 569.
169a U. Pschorn, E. Rössler, H. Sillescu, S. Kaufmann, D. Schaefer, H.W. Spiess, Macromolecules *24* (1991) 398.
170 E.N. da C. Andrade, Proc. Roy. Soc. *A84* (1911) 1, *A90* (1914) 329.
171 N.G. McCrum, Polym. Comm. *25* (1984) 2.
172 A. Schönhals, E. Donth, phys. stat. sol. (b) *124* (1984) 515.
173 D. Stauffer, private comm. 1991; for percolation see e.g. J.W. Essam, Rep. Prog. Phys. *43* (1980) 833.
174 A.V. Tobolsky, J. Appl. Phys. *27* (1956) 673.
175 J.D. Ferry, in H.A. Stuart (Ed.), Die Physik der Hochpolymeren, Springer, Berlin, 1956.
176 D.J. Plazek in K. Ngai, G.B. Wright (Eds.) as in Ref. 145, p. 83.
177 F.R. Schwarzl, priv. comm. 1990.
178 J.P. Hansen, I.R. McDonald, Theory of Simple Liquids, 2nd Ed., Academic, London, 1986; see also K. Kawasaki, J.D. Gunton, in Ref. 17, p. 175.
179 W. Götze, in J.P. Hansen, D. Levesque, J. Zinn-Justin (Eds.), Liquids, Freezing and the Glass Transition, North Holland, Amsterdam, 1991.
180 J. Jäckle, J. Phys. Condens. Mat. *1* (1989) 267.
181 L. Sjögren, W. Götze, Ref. 129, p. 153; W. Götze, L. Sjögren, Ref. 129, p. 161.
182 G.F. Mazenko, Ref. 129, p. 120; S.P. Das, G.F. Mazenko, S. Ramaswamy, J.J. Toner, Phys. Rev. Lett. *54* (1985) 118.
183 A.C. Angell, J. Phys. Chem. Sol. *49* (1988) 863; Nucl. Phys. B. (Proc. Suppl.) *5A* (1988) 69; see also A.C. Angell, Ref. 129, discussion remark to the papers of Götze and Sjögren.
183a K. Adachi, G. Harrison, J. Lamb, A.M. North, R.A. Pethrick, Polymer *22* (1981) 1032.
184 D.C. Chameney, D.F. Sedgwick, J. Phys. *C5* (1972) 1903; see also Ref. 186.
185 Ref. 136; F. Mezei, W. Knaak, B. Farago, Phys. Rev. Lett. *58* (1987) 571; see also e.g. B. Frick, B. Farago, D. Richter, Phys. Rev. Lett. *64* (1990) 2921.
186 F. Fujara, W. Petry, Europhys. Lett. *4* (1987) 921; E. Bartsch, F. Fujara, M. Kiebel, H. Sillescu, W. Petry, Ber. Bunsenges. Phys. Chem. *93* (1990) 1252.
187 G. Meier, private comm. 1991.
188 G. Hohlweg, G. Strobl, submitted to Makromolecules.
189 P. Debye, A.M. Bueche, J. Appl. Phys. *20* (1949) 518.
190 M. Dettenmaier, E.W. Fischer, Kolloid Z.Z. Polymere *251* (1973) 922.
191 B. Gerharz, G. Meier, E.W. Fischer, J. Chem. Phys. *92* (1990) 7110.
192 V.K. Malinovsky, A.P. Sokolov, Solid State Comm. *57* (1986) 757.
193 E.W. Fischer, private comm. 1990.
194 I. Alig, G. Heinrich, E. Donth, Polymer *29* (1988) 1199; see also M.E. Bauer, W.H. Stockmayer, J. Chem. Phys. *43* (1965) 4319.

195 K. Schneider, E. Donth, Acta Polym. *37* (1986) 333.
196 G. Williams, Chem. Soc. Rev. *7* (1978) 89; R.H. Cole, Inst. Phys. Conf. Ser. *58* (1981) 1. For a microscopic picture of shear see e.g. T. Egami, K. Maeda, V. Vitek, Phil. Mag. *A41* (1980) 883; D. Srolovitz, K. Madea, V. Vitek, T. Egami, Phil. Mag. *A44* (1981) 847.
197 R.F. Boyer, 1st Ed. of Ref. 11, Suppl. 2 (1977) p. 745; R.F. Boyer, J. Macromol. Sci.-Phys. *B18* (1980) 461; R.F. Boyer, Macromolecules *14* (1981) 376, *15* (1982) 774; 1498.
198 T. Hatakyama, J. Macromol. Sci.-Phys. *B21* (1982) 299.
199 D.J. Plazek, J. Polym. Sci. Polym. Phys. Ed. *20* (1982) 1533; 1551; 1565; 1575.
200 I. Alig, F. Stieber, A.D. Bakhranov, Yu.S. Manucharov, V. Solovyev e.a. Polymer *30* (1989) 842, *31* (1990) 877.
201 H.R. Zeller, Phys. Rev. Lett. *48* (1982) 334.
202 C.R. Bartels, B. Crist, L.J. Fetters, W.W. Greassley, Macromolecules *19* (1986) 785.
203a E. Rössler, Phys. Rev. Lett. *65* (1990) 1595.
203b F. Fujara, B. Geil, H. Sillescu, G. Fleischer, Z. Phys. B, (1992), in press.
203c see the JCP paper of Ref. 121; T. Pakula, private comm. 1992.
204 A.Q. Tool, J. Am. Ceram. Soc. *29* (1946) 240.
205 A.Q. Tool, C.G. Eichlin, J. Am. Ceram. Soc. *14* (1931) 276; (see also Amer. Ceram. Soc. meeting, Atlanta City, 1924).
206 R.J. Roe, Lecture given at the Creta Conference 1990 (not in Ref. 129), see also R.J. Roe, J.J. Curro, Macromolecules *16* (1983) 428; J.J. Curro, R.J. Roe, J. Polymer Sci. Polym. Phys. Ed. *21* (1983) 1785.
207 I.L. Hopkins, J. Polym. Sci. *28* (1958) 631; L.W. Morland, E.H. Lee, Trans. Soc. Rheolog. *4* (1960) 233; see also Ref. 39.
208 O.S. Narayanaswamy, J. Amer. Ceram. Soc. *54* (1971) 491; R. Gardon, O.S. Narayanaswamy, J. Amer. Ceram. Soc. *53* (1970) 380; R. Gardon, J. Amer. Ceram. Soc. *64* (1981) 114.
209 M.A. De Bolt, A.J. Easteal, P.B. Macedo, C.T. Moynihan, J. Amer. Ceram. Soc. *59* (1976) 16.
210 O.V. Mazurin, J. Non-Cryst. Solids *25* (1977) 129; A.J. Kovacs, J.J. Aklonis, J.M. Hutchinson e.a., J. Polym. Sci. Polym. Phys. Ed. *17* (1979) 1097.
211 I.M. Hodge, Macromolecules *20* (1987) 2897; I.M. Hodge, A.B. Berens, Macromolecules *14* (1981) 1598.
212 G.W. Scherer, J. Amer. Ceram. Soc. *67* (1984) 504; G.W. Scherer, Relaxation in Glass and Composites, Wiley, New York, 1986.
213 I.M. Hodge, priv. comm. 1990.
214 J.M. O'Reilly, J.S. Sedita, Ref. 129, p. 1140.
214a E. Hempel, priv. comm. 1992. [Freezing-in of (density) fluctuation was, of course, early discussed, e.g. by N.L. Laberge, V.V. Visilescu, C.J. Montrose, P.B. Macedo, J. Am. Ceram. Soc. *56* (73) 506].

215 C.A. Angell, J.H.R. Clarke, L.V. Woodcock, Adv. Chem. Phys. *48* (1981) 397; C.A. Angell, L.M. Torrell, J. Chem. Phys. *78* (1983) 937.

215a M. Schulz, E. Donth, to be submitted.

216 M.H. Cohen, G.S. Grest, Phys. Rev. *B20* (1979) 1077; *B24* (1981) 4091; *B26* (1982) 2664; Adv. Chem. Phys. *48* (1981) 455.

216a L. Boehm, M.D. Ingram, C.A. Angell, J. Non-Cryst. Solids *44* (1981) 305.

217 M.V. Vol'kenštejn, Yu.A. Žaronov, Vysokomolek. Soed. *3* (1961) 1739; Yu.A. Žaronov, M.V. Vol'kenštejn, ibid. *4* (1962) 917; M.V. Vol'kenštejn, O.V. Pticyn, Ž. Techn. Fiz. *26* (1956) 2204 (in Russian).

218 B. Wunderlich, D.M. Bodily, M.H. Kaplan, J. Appl. Phys. *35* (1964) 95; S.M. Wolpert, A. Weitz, B. Wunderlich, J. Polym. Sci. *A-2,9* (1971).

219 K.-H. Illers, Makromol. Chem. *127* (1969) 1.

220 A.J. Kovacs, Fortschr. Hochpolym. Forsch. *3* (1963) 394.

221 L.C.E. Struik, Physical Aging in Amorphous Polymers and Other Materials, Elsevier, Amsterdam, 1978.

222 F.R. Schwarzl, G. Link, R. Greiner, F. Zahradnik, Progr. Coll. Polym. Sci. *17* (1985) 180.

223 B.E. Read, Polymer *22* (1981) 1580; B.E. Read, Ref. 129, p. 408.

224 G.P. Johari, J. Chem. Phys. *77* (1982) 4619; K. Pathmanathan, J.-Y. Cavaille, G.P. Johari, J. Polym. Sci. *B27* (1989) 1519.

225 L.C.E. Struik, Ref. 129, p. 395; Polymer *21* (1980) 962; Polymer *28* (1987) 1869.

226 M.M. Santore, R.S. Duran, G.B. McKenna, Polymer *32* (1991) 2377; G.B. McKenna, A.J. Kovacs, Polym. Engng. Sci. *24* (1984) 1138; see also: R.E. Robertson, R. Simha, J.G. Curro, Macromolecules *18* (1985) 2239; P. Destruel, B. Ai, H.-T. Giam, J. Appl. Phys. *55* (1984) 2726.

227 K. Schneider, Thesis, TH Merseburg 1984 (unpublished).

228 M.T. Clavaguera-Mora, Thermochim. Acta *148* (1989) 261; M.T. Clavaguera-Mora, M.D. Baró, S. Suriñach, J. Saurina, N. Clavaguera, Ref. 129, p. 479.

229 J.P. Sethna, Europhys. Lett. *6* (1988) 529; D.L. Stein, R.G. Palmer, Phys. Rev. *B38* (1988) 12035.

230 S.F. Edwards, P.W. Anderson, J. Phys. *F5* (1975) 965.

231 see e.g. Ref. 179.

232 K. Kishimoto, H. Suga, S. Seki, J. Non-Cryst. Solids *13* (1973/74) 357.

233 G. Rehage, Ber. Bunsengesellsch. physik. Chem. *81* (1977) 969; G. Rehage, W. Borchard, in R.N. Haward (Ed.), The Physics of Glassy Polymers, Appl. Sci. Publ., London, 1973.

234 M.W. Zemansky, Heat and Thermodynamics, 5th Ed., McGraw-Hill, New York, 1968.

235 J.E. McKinney, M. Goldstein, J. Res. Nat. Bur. Standards *78A* (1974) 331.

236 R.C. Zeller, R.O. Pohl, Phys. Rev. *B4* (1971) 2029.

237 T. Klitsner, A.K. Raychaudhuri, R.O. Pohl, J. Phys. *C6* (1981) 66; A.K. Raychaudhuri, R.O. Pohl, Solid State Comm. *37* (1980) 105; for a general review see:

W.A. Phillips, (Ed.), Amorphous Solids. Low Temperature Properties, Topics in Current Physics *24*, Springer, Berlin, 1981; S. Hunklinger, J. Phys. (France) *42* (1981) C5-595.
238 P.W. Anderson, B.I. Halperin, C.M. Varma, Phil. Mag. *25* (1972) 1.
239 W.A. Phillips, J. Low Temp. Phys. *7* (1972) 351.
240 U. Buchenau, Phil. Mag. *B65* (1992) 303.
241 G.P. Johari, Phil. Mag. *B41* (1980) 41.
242 M. Goldstein, J. Chem. Phys. *64* (1976) 4767; M. Goldstein in Ref. 128a.
243 W. Sommer, Kolloid Z. *167* (1959) 97.
244 M.C. Shen, A. Eisenberg, Rubber Chem. Technol. *43* (1970) 95, 156.
245 J. Brandrup, E.H. Immergut (Eds.), Polymer Handbook, 3rd Ed., Wiley, New York, 1989.
246 V.R. Privalko, Svojstva Polymerov v Bločnom Sostoyanii = Vol. 2 of ref. 38b.
247 S.M. Aharoni, Ref. 38c.
248 W. Vogel, Glaschemie, Dt. Verlag Grundstoffind., Leipzig, 1979.
249 G. Tammann, Der Glaszustand, L. Voss, Leipzig, 1933.
250 e.g. T. Egami, D. Srolovitz, J. Phys. F. Met. Phys. *12* (1982) 2141; T. Egami, J. Mat. Sci. *13* (1978) 2587.
251 G. Rehage, J. Frenzel, Brit. Polym. J. *14* (1982) 173.
252 H. Suga, S. Seki, Faraday Discuss. *69* (1980) 221 [see also G.P. Johari, J.W. Goodby, G.E. Johnson, Nature *297* (1982) 315].
253 W. Wedler, Thesis Univ. Halle, 1989.
254 C.A. Angell, La Recherche *5* (1982) 584; A. Hallbrucker, E. Mayer, G.P. Johari, Phil. Mag. *B60* (1989) 179; J. phys. Chem. *93* (1989) 4986.
255 K. Biljakovič, J.C. Lasjaunias, P. Monceau, Ref. 129, p. 1254; see also Phys. Rev. Lett. *62* (1989) 1512.
256 E. Donth, R. Conrad, Acta Polymer. *31* (1980) 47.
257 D.L. Questad, M. Oskooie-Tabrizi, J. Appl. Phys. *53* (1982) 6574; W. Heinrich, B. Stoll, Colloid Polym. Sci. *263* (1985) 873.
258 J.H.G. Cowie, Europ. Polym. J. *11* (1974) 297; P. Claudy, J. M. Létoffé, Y. Camberlain, J.P. Pascault, Polym. Bull. *9* (1983) 208.
259 B.E. Read, Polymer *3* (1962) 529; J.A. Faucher, J.V. Koleske, E.R. Santee, J.J. Stratta, C.W. Wilson, J. Appl. Phys. *37* (1966) 3962.
260 R. Becker, Faserforsch. Textiltechn. Z. Polymerforsch. *29* (1978) 361.
261 A. Eisenberg, H. Farb, L.G. Cool, J. Polymer Sci. *A2* (1966) 855.
262 Y. Ishida, Kolloid-Z. *168* (1960) 29; K. Adachi, Y. Ishida, J. Polym. Sci. Polym. Phys. Ed. *14* (1976) 2219.
263 E. Riande, H. Markowitz, D.J. Plazek, N. Raghupathi, J. Polym. Sci. Symp. *50* (1975) 405.
264 L. Wolinski, K. Witkowski, Z. Turzynski, Macromol. Chem. *180* (1979) 2399; *181* (1980) 1717.
265 D.J. Massa, J.L. Schrag, J.D. Ferry, Macromolecules *4* (1971) 210.

266 K. Osaki, J.L. Schrag, Polym. J. *2* (1971) 541; J.W.M. Noordermeer, O. Kramer, F.H.M. Nestler, J.L. Schrag, J.D. Ferry, Macromolecules *8* (1975) 539; see also J.L. Schrag e.a., in Ref. 129, p. 537.
267 L.A. Holmes, S. Kusamizu, K. Osaki, J.D. Ferry, J. Polym. Sci. *A-2* (1971) 2009.
268 J.S. Vrentas, J.L. Duda in Ref. 11 Vol. 5, p. 36.
269 J.C. Philips, J. Non-Cryst. Solids *41* (1980) 179.
270 M. Gordon, J.S. Taylor, J. Appl. Chem. *2* (1952) 493.
271 L.A. Wood, J. Polym. Sci. *23* (1958) 319; H.A. Schneider, Polymer *30* (1989) 771.
272 R.F. Boyer, Polym. Engng. Sci. *8* (1968) 161.
273 B. Fritzsche, R. Trettin, J. Sobottka, Plaste Kautschuk *20* (1973) 125.
274 J.B. Enns, R. Simha, J. Macromol. Sci.-Phys. *B13* (1977) 11; 25.
275 R.F. Boyer, J. Polym. Sci. *C50* (1975) 189; R.F. Boyer, Rubber Chem. Technol. *36* (1963) 1303.
276 K. Schmieder, K. Wolf, Kolloid Z.Z. Polym. *134* (1953) 149.
277 M. Takayanaga, Pure Appl. Chem. *15* (1967) 555.
278 R. Popli, M. Glotin, L. Mandelkern, R.S. Benson, J. Polym. Sci. Polym. Phys. Ed. *22* (1984) 407.
279 Y.P. Khanna, E.A. Turi, T.J. Taylor, V.V. Vickroy, R.F. Abbott, Macromolecules *18* (1985) 1302.
280 R. Lam, P.H. Geil, J. Macromol. Sci.-Phys. *B20* (1981) 37.
281 H.-J. Dalcolmo, private comm. 1990.
282 R.H. Boyd, Polymer *26* (1985) 323.
283 R. Popli, L. Mandelkern, Polym. Bull. *9* (1983) 260.
284 T. Villwock, Diplom Thesis Uni Mainz 1984, E.W. Fischer, private comm. 1990.
285 G. Kanig, Kolloid Z.Z. Polym. *190* (1963) 1.
286 A.J. Batchinski, Z. physik. Chem. *84* (1913) 644.
287 A.K. Doolittle, J. Appl. Phys. *22* (1951) 1471; A.K. Doolittle, D.B. Doolittle, J. Appl. Phys. *28* (1958) 901.
288 M.H. Cohen, D. Turnbull, J. Chem. Phys. *31* (1959) 1164; D. Turnbull, M.H. Cohen, J. Chem. Phys. *34* (1961) 120.
289 J.M. O'Reilly, J. Polym. Sci. *27* (1962) 429; M. Goldstein, J. Chem. Phys. *39* (1963) 3369.
290 P.K. Gupta, C.T. Moynihan, J. Chem. Phys. *65* (1976) 4136; see also J. Jäckle, J. Chem. Phys. *79* (1983) 4463.
291 A.K. Jonscher, Dielectric Relaxation in Solids, Chelsea, London, 1983; L.A. Dissado, R.M. Hill, Proc. Roy. Soc. London *390* (1983) 131; J. Mat. Sci. *16* (1981) 638.
292 An overview on glassy ionics can be obtained from Sec. 2.8 of Ref. 129, e.g. H. Jain, p. 961; M.D. Ingram, p. 955.
293 G.P. Johari, K. Pathmanathan, Phys. Chem. Glass. *29* (1988) 219.
294 M. Pollak, G.E. Pike, Phys. Rev. Lett. *28* (1972) 1449; U. Strom e.a., Phys. Rev. *B13* (1976) 3329.

295 see e.g. A. Pradel, M. Ribes in Ref. 129, p. 1063; S.W. Martin, H.K. Patel, F. Borsa, D. Torgeson, p. 1041, U. Strom, K.L. Ngai, O. Kanert, p. 1011.

295a J. Van Turnhout, Thermally Stimulated Discharge of Polymer Electrets, Elsevier, Amsterdam, 1975.

296 R.H. Partridge, J. Polym. Phys. *A3* (1965) 2817; V.G. Nikol'skij, V.A. Tochin, N.Ya. Buben, Fiz. Tverd. Tela. *5* (1963) 2248 (in Russian).

296a e.g. J. Wong, C.A. Angell, Glass: Structure by Spectroscopy, Dekker, New York, 1976; K.S. Evstropyev, E.A. Porai-Koshits, J. Non-Cryst. Solids *11* (1972) 170; P.H. Gaskell, J. Phys. C. Solid State Phys. *12* (1979) 4337; J.F. Sadoc, J. Non-Cryst. Solids *44* (1981) 1; D.R. Uhlmann, J. Non-Cryst. Solids *42* (1980) 119; *49* (1982) 439; G. Hägg, J. Chem. Phys. *3* (1935) 162; M.D. Ingram, Phys. Chem. Glass. *28* (1987) 215.

297 R. Koningsveld, L.A. Kleintjens, E. Nies, Croatica Chem. Acta *60* (1987) 53; see also I.C. Sanchez, Polymer *30* (1989) 471.

298 G. Scatchard, Chem. Rev. *28* (1931) 321.

299 H. Tompa, Polymer Solutions, Butterworth, London, 1956.

300 J. M. Prausnitz, R.N. Lichtenthaler, E.G. de Azevedo, Molecular Thermodynamics of Fluid-Phase Equilibria, 2nd Ed., Prentice Hall, Englewood Cliffs NY, 1986.

301 R. Koningsveld, see e.g. Ref. 297.

302 A.E. Nestorov, Yu.S. Lipatov, Termodinamika rastvorov i smesej polimerov, Naukova Dumka, Kiev, 1984 (in Russian).

303 P.J. Flory, R.A. Orwoll, A. Vrij, J. Amer. Chem. Soc. *86* (1964) 3515.

304 R. Koningsveld, L.A. Keintjens, M.H. Onclin, J. Macromol. Sci.-Phys. *B18* (1980) 363.

305 R. Simha, T. Somcynsky, Macromolecules *2* (1969) 342; R.K. Jain, R. Simha, Macromolecules *17* (1984) 2663.

306 J.V.L. Singer, K. Singer, Mol. Phys. *24* (1972) 357; I.R. McDonald, Mol. Phys. *24* (1972) 399.

307 J.S. Rowlinson, Liquids and Liquid Mixtures, 2nd Ed., Butterworth, London, 1969; see also Ref. 178.

308 A. Sariban, K. Binder, Macromolecules *21* (1988) 711; see also the first reference of Ref. 84.

309 L. Leibler, Ref. 74b.

310 A. Sariban, K. Binder, J. Chem. Phys. *86* (1987) 5859.

311 F.S. Bates, M. Muthukumar, G.D. Wignall, L.J. Fetters, J. Chem. Phys. *89* (1988) 535.

312 H.W. Kammer, T. Inoue, T. Ougizawa, Polymer *30* (1989) 888.

313 I. Prigogine, The Molecular Theory of Solutions, North Holland, Amsterdam, 1957.

313a D. Patterson, A. Robard, Macromolecules *11* (1978) 690.

314 A.J. Staverman (1937) see Ref. 297; W.H. Stockmayer, J. Chem. Phys. *18* (1950) 58.

315 E. Nies, R. Koningsveld, L.A. Kleintjens, Progr. Coll. Polym. Sci. *71* (1985) 2.

316 R.J. Roe, W.C. Zin, Macromolecules *17* (1984) 189.

317 R. Simha, H. Branson, J. Chem. Phys. *12* (1949) 253; W.H. Stockmayer e.a., J. Polym. Sci. *16* (1955) 517.

318 R. Koningsveld, W.H. Stockmayer, J.W. Kennedy, L.A. Kleintjens, Macromolecules *7* (1974) 73; L.A. Kleintjens, R. Koningsveld, W.H. Stockmayer, Brit. Polymer J. *8* (1976) 144.

319 P. Flory, J. Rehner, J. Chem. Phys. *11* (1943) 521.

320 H.M. James, E.J. Guth, J. Chem. Phys. *11* (1943) 455.

321 A.M. Hecht, F. Horkay, E. Geissler, M. Zriny, Polym. Comm. *31* (1990) 53; see also J.E. Mark, Rubber Chem. Technol. *55* (1982) 762.

322 J.W. Cahn, Trans. Metall. Soc. AIME *242* (1968) 166; J.W. Cahn, J.E. Hilliard, J. Chem. Phys. *28* (1958) 258; J.W. Cahn, J. Chem. Phys. *42* (1965) 93.

323 J.D. Gunton, M. San Miguel, P.S. Sahni, in C. Domb, J.L. Lebowitz (Eds.), Phase Transitions and Critical Phenomena, Vol. 8, Academic, New York, 1983, p. 267.

324 H.E. Stanley, Introduction to Phase Transitions and Critical Phenomena, University Press, Oxford, 1971.

325 P.G. de Gennes, J. Chem. Phys. *72* (1980) 4756.

326 P. Pincus, J. Chem. Phys. *75* (1981) 1996.

327 K. Binder, J. Chem. Phys. *79* (1983) 6387; Colloid Polym. Sci. *265* (1987) 273.

328 C.C. Han e.a., Polymer *29* (1988) 2002.

329 F.S. Bates, J.H. Rosendale, P. Stepanek, T.P. Lodge, P. Wiltzius, G.H. Frederickson, R.P. Hjelm jr., Phys. Rev. Lett. *65* (1990) 1893; J.F. Joanny, J. Phys. France *A11* (1978) L-117; P.G. de Gennes, J. Phys. France Lett. *38* (1978) L-441.

330 D. Schwahn, K. Mortenson, V. Yee-Madeira, Phys. Rev. Lett. *58* (1987) 1544; D. Schwahn, T. Springer, S. Janssen, E. Hädicke, J. Appl. Cryst. (in press 1991).

331 G. Meier, Lecture given at Merseburg in 1991.

332 e.g. Yu.B. Mel'nichenko, V.V. Klepko, V.V. Shilov, Polymer *29* (1988) 1010; Polymer Comm. *30* (1989) 315.

333 R.B. Griffiths, Phys. Rev. *158* (1967) 176; E. Donth, Z. Physik *207* (1967) 342.

334 H. Tanaka, T. Yokokawe, H. Abe, T. Hayashi, T. Nishi, Phys. Rev. Lett. *65* (1990) 3136.

335 K. Kawasaki, J.D. Gunton in C.A. Croxton (Ed.), Ref. 17.

336 J.W. Cahn, Trans. Metal. Soc. *242* (1968) 166.

337 K.B. Rundman, J.E. Hilliard, Acta Met. *15* (1967) 1025.

338 F.S. Bates, P. Wiltzius, J. Chem. Phys. *91* (1989) 3258.

339 T. Sato, C.C. Han, J. Chem. Phys. *88* (1988) 2057.

340 J.S. Langer, Ann. Phys. N.Y. *65* (1971) 53.

341 J.S. Langer, M. Baron, H.D. Miller, Phys. Rev. *A11* (1975) 1417.
342 J.D. Gunton, M. Droz, Lecture Notes in Physics *183*, Introduction to the Theory of Metastable and Unstable States, Springer, Berlin, 1983; see also A. Chakrabarti, R. Toral, J.D. Gunton, M. Muthukumar, Phys. Rev. Lett. *63* (1989) 2072.
343 H.E. Cook, Acta Met. *18* (1970) 297.
344 G.R. Strobl, Macromolecules *18* (1985) 558; M. Okada, C.C. Han, J. Chem. Phys. *85* (1986) 5317.
345 I.G. Voigt-Martin, K.H. Leister, R. Rosenau, R. Koningsveld, J. Polymer Sci. Polym. Phys. Ed. *24* (1986) 723.
346 e.g. T. Hashimoto, J. Kumaki, H. Kawai, Macromolecules *16* (1983) 641; see also *17* (1984) 2812, 2818; T. Hashimoto in R.M. Ottenbrite, L.A. Utracki, S. Inoue (Eds.), Current Topics in Polymer Science, Hanser, München, 1987; T. Hashimoto in S. Komura, H. Furukawa (Eds.), Dynamics of Ordering Processes in Condensed Matter, Plenum, New York, 1988.
347 H. Tanaka, T. Nishi, Phys. Rev. Lett. *59* (1987) 692.
348 T. Hashimoto, M. Itakura, N. Shimidzu, J. Chem. Phys. *85* (1986) 6773.
349 J.I. Frenkel, Kinetic Theory of Liquids, Oxford Univ. Press, London, 1946.
350 R. Becker, W. Döring, Ann. Phys. (Leipzig) *24* (1935) 719.
351 G. Fleischer, Polym. Bull *9* (1983) 152; Colloid Polym. Sci. *265* (1987) 89; M. Antonietti, J. Coutandin, H. Sillescu, Macromolecules *19* (1986) 793.
352 H. Watanabe, T. Kotaka, Macromolecules *20* (1987) 530, 535.
353 M. Lohfink, H. Sillescu, preprint (1992).
354 N.L. Thomas, A.H. Windle, Polymer *21* (1980) 613, see also ref. 268.
355 J. Piglowski, H.W. Kammer, J. Kressler, Polymer *30* (1989) 1705; Polym. Bull. *16* (1986) 493.
356 J. Kanetakis, G. Fytas, Macromolecules *22* (1989) 3452; F. Brochard, J. Jouffroy, P. Levinson, Macromolecules *16* (1983) 1638. For the theoretical arguments see F. Brochard, P.G. de Gennes, Physica *A118* (1983) 289.
357 K. Binder, Colloid Polym. Sci. *265* (1987) 273.
358 W. Hess, A.Z. Akcasu, J. Phys. France *49* (1988) 261.
359 E. Helfand, in K. Šolc (Ed.), Polymer Compatibility and Incompatibility. Principles and Practices (2nd Ed.), Harwood Acad. Publ., Chur 1986, p. 143; see also I.C. Sanchez, Polym. Engng. Sci. *24* (1984) 79; Ann. Rev. Mater. Sci. *13* (1983) 387; S. Krause, Macromolecules *11* (1978) 1288.
360 U.K. Chaturvedi, U. Steiner, O. Zak, G. Krausch, J. Klein, Phys. Rev. Lett. *63* (1989) 616.
361 H. Krömer, M. Hoffmann, G. Kämpf, Ber. Bunsenges. phys. Chem. *74* (1970) 859; 851; D.J. Meier, J. Polym. Sci. *C26* (1969) 81.
362 E. Helfand, Z.R. Wassermann, in I. Goodman (Ed.), Developments in Block Copolymers, Vol. 1, Elsevier, Appl. Sci. Publ., London, 1982, p. 99.
363 A.N. Semenov, Sov. Phys. – JETP (Engl. Transl.) *61* (1985) 733.

364 F.S. Bates, J.H. Rosendale, G.H. Fredrickson, C.J. Glinka, Phys. Rev. Lett. *61* (1988) 2229; F.S. Bates, G.H. Fredrickson, Ann. Rev. Phys. Chem. *41* (1990) 525.

365 K. Mori, H. Hasegawa, T. Hashimoto, Polym. J. *17* (1985) 799; B. Stühn, A.R. Rennie, Macromolecules *22* (1989) 2460; B. Stühn, J. Polym. Sci. B (1992); B. Stühn, R. Mutter, T. Albrecht, Europhys. Lett. (1992).

366 E. Donth, Polym. Comm. *31* (1990) 139.

367 M.G. Brereton, E.W. Fischer, Ch. Herkt-Maetzky, K. Mortenson, J. Chem. Phys. *87* (1987) 6144.

368 D.R. Uhlmann, J. Non-Cryst. Solids *38–39* (1980) 693.

369 A. Keller in B. Sedlaček (Ed.), Morphology of Polymers, de Gruyter, Berlin, 1986, p. 3.

370 J.D. Hoffman, J.I. Lauritsen, J. Res. Nat. Bur. Stand. *65A* (1961) 297.

371 J.D. Hoffman, G.T. Davies, J.I. Lauritsen, in N.B. Hannay (Ed.), Treatise on Solid State Chemistry, Vol. 3, Plenum, New York, 1976, p. 497.

372 Several authors in Faraday Disc. Chem. Soc. *68* (1979).

373 D. Turnbull, J.C. Fisher, J. Chem. Phys. *17* (1949) 71.

374 R. Becker, Ann. Phys. (Leipzig) *32* (1938) 128.

375 W.B. Hillig, Acta Metall. *14* (1966) 1868.

376 P.D. Calvert, D.R. Uhlman, J. Appl. Phys. *43* (1972) 944.

377 I.C. Sanchez, E.A. DiMarzio, J. Res. Nat. Bur. Stand. *A76* (1972) 213.

378 J.I. Lauritsen, J. Appl. Phys. *44* (1973) 4353.

379 J.G. Fatou, C. Marco, L. Mandelkern, Polymer *31* (1990) 1685.

380 J. Martinez-Salazar, P.J. Barham, A. Keller, J. Polym. Sci. Polym. Phys. Ed. *22* (1984) 1085. A regime III is described in J.D. Hoffman, Polymer *24* (1983) 3.

381 J. Rault, M. Sotton, C. Rabourdin, E. Robelin, J. Phys. (France) *41* (1980) 1459.

382 E.W. Fischer, Pure Appl. Chem. *50* (1978) 1319.

383 E.W. Fischer, H. Goddar, G.F. Schmidt, Die Makromol. Chem. *118* (1968) 144.

384 P.A. Spegt, Makromol. Chem. *139* (1970) 139; *140* (1970) 167; A.J. Kovacs, A. Gonthier, Coll. Polym. Sci. *250* (1972) 530; A.J. Kovacs, A. Gonthier, C. Straupe, J. Polym. Sci. Polym. Phys. Ed. *50* (1975) 283.

385 D.M. Sadler, Polym. Comm. *27* (1986) 140.

386 J.J. Point, M. Dosière, Polymer *30* (1989) 2292; J.D. Hoffman, R.L. Miller, Macromolecules *22* (1989) 3502.

387 G. Strobl, in R. Graham, A. Wunderlin, (Eds.), Laser and Synergetics, Springer, Berlin, 1987, p. 191.

387a G.R. Strobl, T. Engelke, H. Meier, G. Urban, Colloid Polym. Sci. *260* (1982) 394; J.D. Hoffman, R.L. Miller, Macromolecules *21* (1988) 3038; Y. Tanabe, G.R. Strobl, E.W. Fischer, Polymer *27* (1986) 1147; I.R. Harrison, Polymer *26* (1985) 3; R. Gehrke, C. Riekel, H.G. Zachmann, Polymer *30* (1989) 1581.

388 T. Nishi, T.T. Wang, Macromolecules *8* (1975) 909.

389 E.A. DiMarzio, C.M. Guttman, J.D. Hoffman, Macromolecules *13* (1980) 1194.

390 H. Baur, in H. Batzer (Ed.), Polymere Werkstoffe, Vol. 1, Thieme, Stuttgart, 1984, p. 137.
391 J. Rault, E. Robelin-Souffaché, J. Polym. Sci. *B27* (1989) 1349.
392 M. Hikosaka, Polymer *31* (1990) 458; *28* (1987) 1257.

Subject index